T0190504

Solid Mechanics and Its Applications

Volume 242

Aims and Scope of the Series

The fundamental questions arising in mechanics are: *Why?*, *How?*, and *How much?* The aim of this series is to provide lucid accounts written by authoritative researchers giving vision and insight in answering these questions on the subject of mechanics as it relates to solids.

The scope of the series covers the entire spectrum of solid mechanics. Thus it includes the foundation of mechanics; variational formulations; computational mechanics; statics, kinematics and dynamics of rigid and elastic bodies: vibrations of solids and structures; dynamical systems and chaos; the theories of elasticity, plasticity and viscoelasticity; composite materials; rods, beams, shells and membranes; structural control and stability; soils, rocks and geomechanics; fracture; tribology; experimental mechanics; biomechanics and machine design.

The median level of presentation is to the first year graduate student. Some texts are monographs defining the current state of the field; others are accessible to final year undergraduates; but essentially the emphasis is on readability and clarity.

More information about this series at http://www.springer.com/series/6557

Alan Rothwell

Optimization Methods in Structural Design

 Springer

Alan Rothwell
Formerly Delft University of Technology
Delft
The Netherlands

Additional material to this book can be downloaded from http://extras.springer.com.

ISSN 0925-0042   ISSN 2214-7764   (electronic)
Solid Mechanics and Its Applications
ISBN 978-3-319-85593-6   ISBN 978-3-319-55197-5   (eBook)
DOI 10.1007/978-3-319-55197-5

Printed on acid-free paper

This Springer imprint is published by Springer Nature
The registered company is Springer International Publishing AG
The registered company address is: Gewerbestrasse 11, 6330 Cham, Switzerland

To my wife, Janette, for her love, support and patience, and to our four children, Katherine, Sarah, Rachel and Paul.

Preface

The aim of this book is to present numerical optimization methods in structural design to students in engineering courses at final undergraduate level or in the first year of a postgraduate study. For others in industry or elsewhere who may be new to these highly practical techniques, the book can bridge the gap between familiar design practice and some of the advanced texts on optimization theory. While the specific application is to structural design, the principles involved can be applied far more widely. A 'how to do it' approach is followed throughout the book, with less emphasis at this stage on mathematical derivations. Extensive use is made of the 'Solver' optimization tool in Microsoft Excel[1], because of its ready availability. This provides an ideal means of illustrating the methods presented, how to set up an optimization problem and to demonstrate the usefulness of optimization techniques in general. With practice in the use of Solver, use of optimization modules in more extensive computer packages should present little difficulty.

The spreadsheet programs provided with this book are, in the earlier chapters, principally illustrations of optimization methods. In later chapters, these are of a more practical nature, in particular for reinforced shell structures and for the design of composite laminates. These topics are chosen to reflect the ever-increasing demand for lightweight structures in many branches of engineering. Weight reduction is not only to reduce operational costs, but also to offset the high cost of many modern, high-performance metallic materials and composites. Detailed instructions are given for use of the spreadsheets and on the use of Solver. Exercises, with solutions where appropriate, are provided with each chapter, many of them making some other use of Solver or further use of the spreadsheets. These are intended to give practice in setting up an optimization problem and generally to explore the characteristics of the optimization process. Many of the examples in the book, throughout the text and in the spreadsheets, will be seen to have a distinct aerospace flavour, this being simply a reflection of the author's main field of work over many years.

---

[1]Microsoft and Excel are registered trademarks of the Microsoft Corporation.

The early chapters of the book show the relationship between formal optimization and the traditional methods of design, it not being the intention to replace existing methods but rather to supplement them with an additional weapon in the armoury of the designer. Strength-to-weight ratios, limits of feasibility and the concept of structural efficiency are discussed. Classical optimization is then introduced, together with the Lagrange multiplier, fundamental to the discussion of numerical optimization methods in the following chapters. Numerical methods are introduced in sufficient detail to enable the reader to appreciate the processes taking place in some of the highly sophisticated 'black box' optimization routines in advanced computer packages. It is not the intention to describe these numerical methods in the detail necessary to enable the reader to program them efficiently, this being a task primarily for the programming specialist. The generalized reduced gradient method and the genetic algorithm, two of the methods available in Solver, are given due attention, the latter in a later chapter in the context of composite laminates. The remaining chapters of the book are devoted to applications—reinforced shell structures, with the design of a box beam and an aircraft fuselage section, as well as some extended discussion of the design of composite laminates. For these topics, relevant methods of analysis are covered in sufficient detail before proceeding to specific optimization problems and spreadsheet programs for their solution. Composite laminates are of particular interest because of the special problem introduced by the discrete nature of the individual plies of the laminate and because of the freedom to optimize the lay-up to match the application. A final chapter is given to optimization with finite element analysis, for which some special methods are necessary.

The level of knowledge required to follow the text is no more than in a usual engineering course. No specific demands are made, and the text should remain largely accessible to those from other disciplines, sufficient information being given 'to proceed from this point'. However, it is assumed that the reader already has a working knowledge of Microsoft Excel, with some Visual Basic, and also is familiar with matrix notation. With a less mathematical bias, he might in the first place go rather superficially over Chaps. 4 and 5 and with no experience of finite element methods might be tempted to miss Chap. 9. No attempt is made at completeness in the book, but rather to provide a sound understanding of basic principles and a good start for further study. For this, a list of further reading is included (reflecting perhaps more the author's personal choice). Specific reference to research papers is limited to where this is of particular relevance. For a more comprehensive reference list, the reader should turn to the several excellent, more advanced books on optimization theory included amongst the references at the end of each chapter.

This book is based on lectures given at Delft University of Technology in the Netherlands, while the author was professor of aircraft structures. He hopes that the reader will enjoy a study of optimization methods as much as he has and will be able to put them to good use in further study and engineering practice.

Delft, The Netherlands   Alan Rothwell

Contents

1 The Conventional Design Process 1
  1.1 Fully Stressed Design 3
     1.1.1 Structure Made of Different Materials 7
     1.1.2 Structure Under Alternative Loads 9
  1.2 Strength-to-weight Ratio 12
     1.2.1 Feasibility 15
  1.3 Comparison of Layouts 16
     1.3.1 Classification of Optimization Problems 19
  1.4 Spreadsheet Program 21
     1.4.1 'Seven-Bar Truss' 21
  1.5 Summary 24
  Exercises 25
  References 28

2 Optimality Criteria 29
  2.1 Circular Tube in Compression 30
     2.1.1 Efficiency Formula 33
     2.1.2 Material Limitation 38
  2.2 Criterion for Maximum Stiffness 40
  2.3 Spreadsheet Programs 44
     2.3.1 'Circular and Square Tubes' 44
     2.3.2 'Truss with Tubular Members' 48
  2.4 Summary 50
  Exercises 51
  References 53

3 The General Optimization Problem 55
  3.1 Box Beam Structure 56
     3.1.1 General Form of Design Space 58

3.2   The Lagrange Multiplier Method   61
        3.2.1   Interpretation of Lagrange Multipliers...............   67
3.3   Inequality Constrained Problems   70
        3.3.1   The Kuhn–Tucker Conditions......................   73
3.4   Spreadsheet Program   73
        3.4.1   Eccentrically Loaded Column......................   74
3.5   Summary   78
Exercises...   79
References..   81

4   Numerical Methods for Unconstrained Optimization   83
4.1   Unconstrained Optimization   84
        4.1.1   Steepest Descent Method   85
        4.1.2   Fletcher–Reeves Method.........................   90
        4.1.3   Quasi–Newton Methods   92
4.2   Line Search Methods   94
        4.2.1   Region Elimination and the Golden Section Method......   95
        4.2.2   Polynomial Interpolation.........................   97
4.3   Spreadsheet Program   100
        4.3.1   'Hooke and Jeeves Method'   100
4.4   Summary   103
Exercises...   105
References..   106

5   Numerical Methods for Constrained Optimization   107
5.1   Constraint-Following Methods   108
        5.1.1   Gradient Projection Method   109
        5.1.2   Generalized Reduced Gradient Method...............   120
        5.1.3   Other Methods for Constrained Optimization   126
        5.1.4   Substitution of Variables.........................   129
5.2   Penalty Function Methods   129
        5.2.1   Interior Penalty Function.........................   130
        5.2.2   Exterior Penalty Function   133
        5.2.3   Augmented Lagrangian Penalty Function   135
5.3   Spreadsheet Program   138
        5.3.1   'Penalty Function Method'   140
5.4   Summary   142
Exercises...   143
References..   145

6   Optimization of Beams   147
6.1   Beam Cross Section   148
        6.1.1   Thin-Walled Beams...........................   150
        6.1.2   Geometrically Similar Sections   153

    6.2   Optimum Spanwise Distribution 154
        6.2.1 Statically Determinate Beams 155
        6.2.2 Statically Indeterminate Beams 158
    6.3   Limit Design 160
        6.3.1 Yield Moment 161
        6.3.2 Limit Load 163
    6.4   Spreadsheet Programs 168
        6.4.1 'I-Section Beam' 169
        6.4.2 'Beam Under Lateral Load' 173
    6.5   Summary 178
    Exercises 179
    References 181

7 Reinforced Shell Structures 183
    7.1   Bending Stress 185
        7.1.1 Effect of Yielding 187
        7.1.2 Modelling of Discrete Stiffeners 189
    7.2   Shear Stress 192
        7.2.1 Torsional Stiffness 198
        7.2.2 von Mises Criterion 200
    7.3   Buckling Formulae 201
        7.3.1 Buckling in Compression 201
        7.3.2 Buckling in Shear 208
        7.3.3 Efficiency Formula for a Compression Panel 210
        7.3.4 Shear Web Efficiency 214
        7.3.5 Post-buckled Shear Webs 218
    7.4   Spreadsheet Programs 221
        7.4.1 'Stiffened Panel' 221
        7.4.2 'Rectangular Box Beam' 227
        7.4.3 'Circular Fuselage Section' 230
    7.5   Summary 235
    Exercises 236
    References 239

8 Composite Laminates 241
    8.1   Lamination Theory 243
        8.1.1 Transformed Stiffness Matrix 245
        8.1.2 Laminate Stiffness Coefficients 247
        8.1.3 Failure Criteria 252
        8.1.4 Change in Temperature 256
        8.1.5 Practical Restrictions on Lay-up 258
    8.2   Laminate Optimization 259
        8.2.1 Netting Analysis 260
        8.2.2 Iterative Redesign 264

      8.2.3   Numerical Optimization   267
      8.2.4   Genetic Algorithm.   270
   8.3   Spreadsheet Program   275
      8.3.1   'Composite Laminate'.   275
   8.4   Summary   278
   Exercises.   280
   References.   282

9 Optimization With Finite Element Analysis   283
   9.1   Sensitivity Analysis   284
   9.2   Reduction in Design Variables   290
   9.3   Spreadsheet Program   291
      9.3.1   'Design Variable Linking'.   291
   9.4   Summary   294
   Exercises.   295
   References.   296

Appendix   297

Recommended Further Reading   307

Solutions to Selected Exercises   309

Index   311

Principal Notation

A	Laminate stiffness matrix
A	Cross-sectional area
A_0	Enclosed area
A_s	Stiffener area
$\bar{A}_s$	Effective stiffener area
a	Side of a square-section beam
B	Width of a rectangular box structure
b	Width of a square tube
	Width of a beam
	Stiffener spacing
	Search interval
C	Shape efficiency for a beam
	Coefficient for the equivalent thickness of a panel
D	Dimension of a truss
d	Diameter of a circular tube
	Stiffener spacing on a shear web
E	Elastic modulus
E_s	Secant modulus
E_t	Tangent modulus
e	Eccentricity of applied load on a column
F	Force in a member of a truss
$F(\mathbf{x})$	Penalty function
$F(\mathbf{x}, \lambda)$	Lagrangian function
f	Force vector
$f(\mathbf{x})$	Objective function
G	Shear modulus
$g_j(\mathbf{x})$	Inequality constraint ($j = 1 \ldots m$)
H	Height of a truss
	Height of a rectangular box structure
h	Vector of dummy loads
h	Height of a beam

	Stiffener height
	Height of a shear web
$h_k(\mathbf{x})$	Equality constraint ($k = 1 \dots p$)
I	Second moment of area
$\mathbf{K}$	Stiffness matrix
K	Buckling coefficient
K_0	Theoretical buckling coefficient for a circular cylinder
K_F	Flexural buckling coefficient
k	Torsional stiffness
L	Span of a truss
	Length of a column
	Span of a beam
	Length of a panel
	Number of layers in a laminate
l	Length of a member of a truss
M	Bending moment
	Mass
M_e	Elastic bending moment
M_y	Yield moment
m	Number of inequality constraints
	Index in the Ramberg–Osgood formula
$\mathbf{N}$	Vector of in-plane forces per unit width
n	Number of design variables
	Coefficient depending on the layout of a truss
	Number of plies at a given angle θ
n_g	Coefficient for geometrically similar beams
n_h	Coefficient for beams with a height limitation
P	Applied load
P_E	Euler buckling load
P_L	Local buckling load
p	Number of equality constraints
	Loading intensity
$\mathbf{Q}$	Ply stiffness matrix
$\bar{\mathbf{Q}}$	Transformed stiffness matrix
Q	Shear force
q	Shear flow
R	Radius of a tube
	Restraint force
r	Penalty parameter
S	Ply strength in shear
$\mathbf{s}$	Vector of components of the search direction s_i
s	Stress in netting analysis
s_i	Components of the search direction ($i = 1 \dots n$)
$\mathbf{T}$	Transformation matrix

T	Twisting moment
t	Thickness
$\bar{t}$	Equivalent thickness
t_k	Layer thickness
t_{ply}	Ply thickness
t_{s}	Stiffener thickness
u	Displacement vector
V	Volume of material
W	Weight
W_k	Weight expressed in kilogram
X	Ply strength in the fibre direction
x	Vector of design variables x_i
x_i	Design variables ($i = 1 \ldots n$)
Y	Ply strength in the transverse direction

α	Parameter in the line search
	Coefficient of thermal expansion
β	Post-buckling ratio
γ	Shear strain
δ	Deflection
	Displacement in a collapse mode
ε	Strain
η	Efficiency
θ	Angle of a member of a truss
	Rate of twist
	Fibre angle (anticlockwise from the laminate x-axis)
λ	Vector of Lagrange multipliers λ_k
λ_k	Lagrange multipliers ($k = 1 \ldots p$)
μ	Parameter for the stiffeners on a shear web
	$\mu = 1 - \nu_{12}\nu_{21}$ for a composite laminate
ν	Poisson's ratio
ρ	Density
ρ_{w}	Specific weight
σ	Tensile or compressive stress
σ_0	Allowable stress
σ_2	0.2% proof stress
σ_{b}	Buckling stress in compression
σ_{c}	Allowable compressive stress
σ_{E}	Euler buckling stress
σ_{eq}	von Mises equivalent stress
σ_{F}	Flexural buckling stress
σ_{L}	Local buckling stress
σ_{R}	Reference stress in the Ramberg–Osgood formula

σ_t Allowable tensile stress
σ_y Yield stress
τ Shear stress
      Constant in the golden section method
τ' Equivalent shear stress
τ_b Shear buckling stress
φ Penalty term

Matrices and vector quantities are written in bold, e.g. **T** is the transformation matrix, and **x** is the vector of design variables x_i.

An 'asterisk' denotes an optimum value, e.g. $x_1{}^*$ is the optimum value of x_1.

∇f Gradient of a function $f(\mathbf{x})$

$\|\mathbf{y}\|$ Modulus of a vector $\mathbf{y}$ $\left[= \sqrt{y_1^2 + y_2^2 + \cdots}\right]$

$[\,]^{-1}$ Matrix inverse
$[\,]^{T}$ Matrix transpose

Other notation is introduced locally in the text where it occurs.

Notation for the lay-up of a composite laminate:

$$\left[\pm 45_2/(90/0)_2\right]_S = [\,+45/-45/+45/-45/90/0/90/0/$$
$$0/90/0/90/-45/+45/-45/+45\,]$$

(S = symmetric)
A 'bar' denotes a layer on the middle-plane:

$$[\pm 45/\bar{0}]_S = [45/-45/0/-45/45]$$

Margin of safety is defined as: $\left(\dfrac{\text{actual strength}}{\text{required strength}}\right) - 1 \quad (\geq 0)$

Chapter 1
The Conventional Design Process

Abstract The characteristics of the conventional design process, implying repeated analysis of a structure and resizing of its members until a satisfactory design is obtained, is illustrated by means of some simple truss structures. In this process, it is implicitly assumed that by satisfying as closely as possible all requirements placed on the design this will lead to the 'best' design. In terms of the maximum stress in the members, this is the well-known principle of the fully stressed design. Effective as this method often is, common situations are identified where this does not lead to an optimum, minimum weight design. Furthermore, the process may be very slowly convergent, in addition to which it offers no help when conditions other than a simple maximum stress apply or, for example, with the optimum shape of a structure. Minimum weight implies economy of material as well as operational savings directly related to reduction in weight. All this provides justification for the formal optimization methods in the remaining chapters of this book. While this chapter is concerned only with truss structures, conclusions reached can, in principle, be taken to apply more widely to the optimization of many other types of structure. A spreadsheet program for the numerical optimization of a simple seven-bar truss provides a first introduction to use of the Solver optimization tool in Microsoft Excel.

The principal aim of structural design is to produce a structure which can carry the loads on it from where they are applied to where the structure is supported and to do this in an efficient way. By 'efficient' is meant here with the least use of material, implying at the same time a structure of minimum weight. Economy of material is, of course, only one aspect of reducing cost, but in many cases reducing weight also makes a substantial contribution to reducing manufacturing, transport and operational costs. Specific requirements imposed on a design relate in the first place to the necessary strength of a structure under load, but in practice they are likely also to include limits on deformation and many other requirements such as those referring to fatigue or resistance to accidental damage. With reducing weight, a structure inevitably becomes thinner and more slender, leading eventually to buckling of the structure or of its individual components. Buckling is frequently one of the prime considerations in the design of a lightweight structure.

© Springer International Publishing AG 2017
A. Rothwell, *Optimization Methods in Structural Design*, Solid Mechanics and Its Applications 242, DOI 10.1007/978-3-319-55197-5_1

A structure designed for minimum weight is commonly designed in one of two ways: either as a shell structure, such as a box beam or some similar form of structure, or as one of many different types of truss structure consisting of a lattice of individual bars. In a shell structure, if it is relatively thin, reinforcement is generally required to support the shell and delay buckling, or in other words to enable a satisfactory stress level to be reached in the structure. In the box beam in Fig. 1.1, stiffeners are placed along the length of the beam to break up its width into smaller, more buckling-resistant panels and transverse stiffeners are placed on the side walls. Diaphragms are placed at intervals along the length to assist in transmitting load into the structure. In a truss structure, such as in Fig. 1.2, the material is concentrated into more compact, thicker members, resulting in better buckling resistance of the individual bars forming the truss, but with the added complexity of the joints between them.

The wing and fuselage of an aircraft are well-known examples of a box beam (albeit of entirely different shape but working in essentially the same way), while more examples abound in a wide variety of applications. Examples of truss structures can be found in many different branches of engineering, such as in bridges, cranes, roof structures, space vehicles and countless other structures. The optimization of a typical box beam will be considered in some detail in a later chapter. But before proceeding to formal optimization methods and related numerical procedures in the remaining chapters of this book, a simple truss structure is chosen in the present chapter to explore first the characteristics of what

Fig. 1.1 Box beam structure (main structure of an aircraft wing)

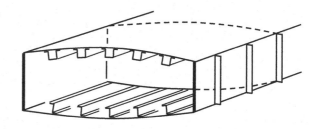

Fig. 1.2 Section of a typical truss structure

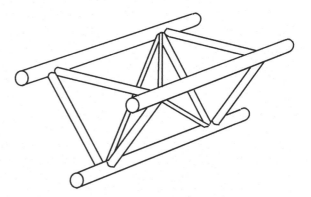

we shall refer to as the conventional design process. This will enable us better to understand the role of optimization in design and its relationship with traditional methods.

Conventional engineering design typically involves the repeated analysis and modification of an initial concept until what is considered to be a satisfactory design emerges. At each stage, the properties of the design are evaluated and compared with the specified requirements. Appropriate changes are made in an attempt to match the requirements as closely as possible. Precisely how this is done depends on the nature of the design, but invariably a change in some aspect of a design to meet a particular requirement influences the extent to which other requirements are met. The process has therefore to be repeated, perhaps many times, until further changes are sufficiently small. It is implicitly assumed that a design that just satisfies as many of the requirements as possible is in some sense the 'best' design. This is what is meant above by the conventional design process.

As already stated, a simple truss structure is used to illustrate the iterative design process described above. The assumption is made that resizing the members of a truss to just satisfy the required stress limits will result in a minimum weight design, at least when only strength requirements are considered. This means adjusting the cross-sectional area of all members until the maximum allowable stress is reached in each of them. This is the well-known principle of the fully stressed design. The validity of this principle—or situations under which a fully stressed design may not lead to a true optimum—is explored in the present chapter. At the same time, we can study the manner in which the fully stressed design procedure converges to a final design, whether or not this is a true optimum. The strength-to-weight ratios of some different layouts of truss structure are compared. From this, we can go on to the fundamental question of feasibility—what, for example, is the longest possible truss structure if supported at each end and required to be just able to carry only its own weight?

1.1   Fully Stressed Design

Consider a two-dimensional truss structure, such as the symmetric truss in Fig. 1.3, under a single applied load at mid-span. Note that the chosen truss is statically indeterminate and, as commonly assumed, its members are pin-jointed to their adjacent members at the nodes. All members are of the same material. Being statically indeterminate, the forces F_i in the members of the truss are not a simple function of the applied load but depend on the stiffness, in this case the cross-sectional areas A_i, of all the members. Forces F_i can be calculated by finite element analysis, or by a strain energy calculation (see, e.g., [1, 2] or [3]). A simple, iterative redesign procedure to minimize the weight of the truss is as follows. Starting from some chosen set of areas A_i and the corresponding forces F_i, calculate the stress σ_i in each member. Resize each member so that the stress in it, under the previously calculated force, is made equal to the allowable stress σ_0 of the material.

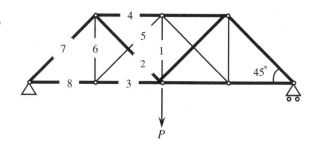

Fig. 1.3 Simply supported truss loaded at mid-span (also showing numbering of the members)

After all areas A_i have been increased or decreased as necessary, the forces F_i in the statically indeterminate truss are recalculated, and the process is repeated. This can be represented as follows:

$$F_i = f(A_i)$$

$$\sigma_i = \frac{F_i}{A_i}$$

$$A_i{}' = A_i \frac{\sigma_i}{\sigma_0}$$

where A_i' denotes the new value to replace the old one. For example, if the stress σ_i in one of the members is precisely twice the allowable stress σ_0, the new area A_i' must be made twice the previous A_i, and if σ_i is one-half of σ_0, then A_i' can be reduced to one-half of A_i. This process is continued until sufficiently converged, that is, until there is practically no further change in areas A_i and the stress in every member is equal to the allowable stress σ_0.

The outcome of this procedure for the chosen truss is shown in Fig. 1.4. The individual members are not identified in this figure, but it is seen that some members increase in area and others decrease, in some cases after an initial step in the other direction. The sequence of iterations is shown in more detail in Table 1.1. The area of the members has been non-dimensionalized by dividing by $P/2\sigma_0$, and

Fig. 1.4 Progressive changes in member areas A_i at each iteration for the truss in Fig. 1.3

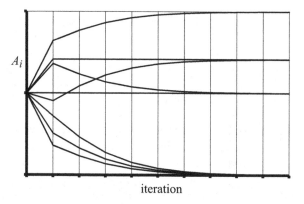

Table 1.1 Iteration history for the truss in Fig. 1.3

Iteration	Cross-sectional areas								Weight
	$\frac{A_1}{P/2\sigma_0}$	$\frac{A_2}{P/2\sigma_0}$	$\frac{A_3}{P/2\sigma_0}$	$\frac{A_4}{P/2\sigma_0}$	$\frac{A_5}{P/2\sigma_0}$	$\frac{A_6}{P/2\sigma_0}$	$\frac{A_7}{P/2\sigma_0}$	$\frac{A_8}{P/2\sigma_0}$	
0	1.000	1.000	1.000	1.000	1.000	1.000	1.000	1.000	1.000
1	0.718	0.906	1.359	1.641	0.508	0.359	1.414	1.000	0.997
2	0.462	1.087	1.231	1.769	0.327	0.231	1.414	1.000	0.968
3	0.274	1.220	1.137	1.863	0.194	0.137	1.414	1.000	0.946
4	0.153	1.306	1.076	1.924	0.108	0.076	1.414	1.000	0.933
5	0.081	1.357	1.041	1.959	0.057	0.041	1.414	1.000	0.924
6	0.042	1.384	1.021	1.979	0.030	0.021	1.414	1.000	0.920
7	0.021	1.399	1.011	1.989	0.015	0.011	1.414	1.000	0.918
8	0.011	1.407	1.005	1.995	0.008	0.005	1.414	1.000	0.916
9	0.005	1.410	1.003	1.997	0.004	0.003	1.414	1.000	0.916
10	0.003	1.412	1.001	1.999	0.002	0.001	1.414	1.000	0.915

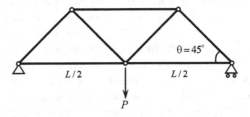

Fig. 1.5 Result of the iterative resizing procedure

initially, all members are chosen to have the same area. It is assumed throughout this text that applied loads already include the necessary factor of safety or that the allowable stress has been appropriately reduced. To eliminate the span of the truss and the density of the material, the weight of the truss has been normalized to a value of 1.0 at the start. The process has largely converged after 10 iterations, with new member areas and a reduction in weight. The extent of this reduction depends, of course, on the initial values chosen. Figure 1.5 shows the truss that results from this procedure (also indicated by the thicker lines in Fig. 1.3). Certain members have fallen away, leaving the statically determinate truss shown in the figure. Since the objective has been to reach the maximum allowable stress in each member—the principle of the fully stressed design—this is hardly surprising because for a statically determinate truss the forces in the members are independent of their cross-sectional areas. In other words, members can then be sized individually, according to the load in them, to ensure that each does reach the maximum stress. Only if minimum values are imposed on some or all of the cross-sectional areas of the members might other than a statically determinate structure be reached. In that case, the members that have been reduced to their specified minimum areas will not be fully stressed, and the final weight of the structure will, of course, be increased.

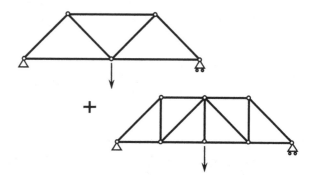

Fig. 1.6 'Partial structures' derived from the truss in Fig. 1.3

Further, it might be observed that if the iteration is terminated before fully converged, some members of the truss will have a stress greater than the allowable stress (i.e. those members increasing in area).

However, what is significant here is that the process has in this case managed to converge to what is indeed the *lightest* statically determinate structure that can be formed out of the original truss. Figure 1.6 shows two alternative such trusses. By superimposing the two figures, the original truss is obtained. It is easily verified that the left-hand structure (the one obtained by the iterative resizing process) is lighter than the other one, if both are fully stressed, implying a more efficient load path. The two structures in Fig. 1.6 are sometimes termed 'partial structures'. Any combination of the two partial structures will result in a structure with a weight intermediate between the two (the distribution of load between the two partial structures depending on matching their deflection at mid-span). The iterative procedure has therefore correctly selected the optimum, minimum weight structure. That is, of course, not to say that an altogether different layout of truss would not be more efficient. The advantage of starting with a statically indeterminate structure is, of course, that this selection can be made, and in other words, the best load path is chosen.

While this example demonstrates the effectiveness of the long established, iterative design process, it should not be assumed that it will always converge to the true optimum. Important situations exist where this may not occur, as will be seen in the examples in the following two sections of this chapter. For the present, it might be noted that the truss chosen here was of given layout, under a single load case, and of the same material throughout. Furthermore, convergence can in some cases be slow, requiring much repeated analysis (while this might, of course, be accelerated by making larger, arbitrarily chosen, changes in areas A_i). Also, it is clear that the process does not aim directly to minimize weight, but simply to achieve uniform stress throughout the structure, on the assumption that this does indeed correspond to minimum weight. Finally, it might have been inferred from above that a fully stressed design is *necessarily* statically determinate—in Sect. 1.3, we shall see a simple, statically indeterminate truss that is nevertheless fully stressed. However, such cases must be rare in more practical structures.

1.1.1 Structure Made of Different Materials

The truss structure studied in the previous section was made of a single material, with the same allowable stress throughout. To explore now in how far the iterative, fully stressed design approach does converge to the correct final result, we take the classic problem of the three-bar truss shown in Fig. 1.7. Consider two cases: (i) when all three bars are of the same material, and (ii) when the allowable stress of the outer bars is less than that of the inner bar. In both cases, all three bars have the same elastic modulus and density. By a simple strain energy calculation, formulae for the stresses in the bars of the symmetric truss can be obtained as follows. In the two outer bars:

$$\sigma_1 = \frac{P}{\left(\sqrt{2}A_1 + 2A_2\right)}, \tag{1.1}$$

and in the inner bar:

$$\sigma_2 = \frac{2P}{\left(\sqrt{2}A_1 + 2A_2\right)}. \tag{1.2}$$

The volume of the structure is

$$V = \left(\sqrt{2}A_1 + \frac{A_2}{2}\right)L.$$

Taking $P = 100$ kN and $L = 1000$ mm, the iterative resizing procedure of the previous section is followed in both cases, with the result given in Table 1.2.

It is seen that when all the bars are of the same material, the procedure converges correctly to the expected design, in which the outer bars are gradually removed and the whole load is taken by the inner bar (the obvious solution). However, in the second case when the outer bars have a *lower* allowable stress, the result is quite

Fig. 1.7 Three-bar truss made of different materials

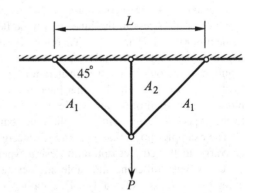

Table 1.2 Iterative resizing of the three-bar truss made of different materials in Fig. 1.7

$\sigma_{1_{max}} = \sigma_{2_{max}} = 300\,\text{N/mm}^2$			$\sigma_{1_{max}} = 150\,\text{N/mm}^2$ $\sigma_{2_{max}} = 450\,\text{N/mm}^2$		
A_1 (mm^2)	A_2 (mm^2)	V/1000 (mm^3)	A_1 (mm^2)	A_2 (mm^2)	V/1000 (mm^3)
100.0	100.0	191.4	100.0	100.0	191.4
97.6	195.3	235.7	195.3	130.2	341.2
61.6	246.3	210.2	242.6	107.8	397.1
35.4	283.3	191.7	289.5	85.8	452.2
19.1	306.3	180.2	332.2	65.6	502.6
10.0	319.2	173.7	368.5	48.5	545.4
5.1	326.1	170.3	397.4	34.9	579.4
2.6	329.7	168.5	419.3	24.5	605.3
1.3	331.5	167.6	435.4	17.0	624.2
0.6	332.4	167.1	446.8	11.6	637.6
0.3	332.9	166.9	454.7	7.9	647.0
0.2	333.1	166.8	460.1	5.3	653.4
0.1	333.2	166.7	463.8	3.6	657.7
0.0	333.3	166.7	466.3	2.4	660.7
0.0	333.3	166.7	468.0	1.6	662.7
0.0	333.3	166.7	469.1	1.1	664.0
0.0	333.3	166.7	469.9	0.7	664.9
0.0	333.3	166.7	470.4	0.5	665.5
0.0	333.3	166.7	470.7	0.3	665.9
0.0	333.3	166.7	471.0	0.2	666.1
0.0	333.3	166.7	471.4	0.0	666.7

different—the outer bars are now retained while the inner bar is removed. It is also seen that the volume V of the structure is increasing, not reducing, with each iteration, so in practice if would not be sensible to continue the iteration process after the first few steps. But by proceeding until little or no further change occurs, it becomes clear that an optimum truss has not been obtained, even though the resulting structure is statically determinate and fully stressed. The explanation for this behaviour in the second case is that the stress in the outer bars is necessarily one-half of the stress in the inner bar, regardless of the values of A_1 and A_2, as is evident in Eqs. (1.1) and (1.2). Therefore, if the allowable stress of the outer bars is less than one-half of the allowable stress of the inner bar, extra material will be required in the outer bars at each iteration to avoid failure, in turn attracting more load into those bars. On the other hand, if the allowable stress of the outer bars is more than one-half of the allowable stress of the inner bar, the procedure converges to the expected design with only the inner bar retained.

The example shows that an iterative redesign procedure cannot be *guaranteed* to converge to the correct optimum design when the structure is made of different materials, with different allowable stresses (a similar result for a 10-bar truss is given by Haftka and Gurdal [4]. This applies just as much when the materials are of different elastic modulus or different density, elastic modulus directly affecting the load distribution in the structure and density affecting the contribution of each member to the total weight. Although a very simple example has been chosen here, a similar conclusion has to apply to any structure made of different materials.

1.1.2 Structure Under Alternative Loads

It will be recalled that the truss structure in Sect. 1.1 carried a single applied load, at mid-span. However, any practical structure is likely to be subject to many different load cases, with loads applied at different locations and at different times during its use. To test the convergence of the iterative, fully stressed design approach for a structure under different load cases, we take the same three-bar truss as in the previous section, with all three bars of the same material, but now with obliquely applied loads P_1 or P_2, as shown in Fig. 1.8 (this problem is discussed in the classic review paper by Schmit [5]). Note that these loads are not applied at the same time. The resulting structure is required, therefore, to be able to carry either load on its own. To simplify the problem, we choose the two loads to be of equal magnitude:

$$P_1 = P_2 = P$$

so that the structure will be symmetric with area $A_3 = A_1$. We then have to consider the stress in only one outer bar and the inner bar. Formulae for these stresses are given below (these are again readily obtained by simple strain energy calculation). In the left-hand outer bar (member 1) under load P_1:

$$\sigma_1 = \left(\frac{\sqrt{2}A_1 + A_2}{\sqrt{2}A_1^2 + 2A_1A_2} \right) P, \tag{1.3}$$

and under load P_2:

Fig. 1.8 Three-bar truss under alternative loads P_1 or P_2

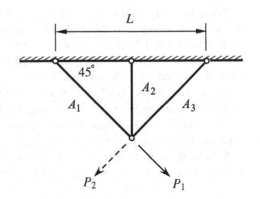

$$\sigma_1 = -\left(\frac{A_2}{\sqrt{2A_1^2 + 2A_1A_2}}\right)P. \tag{1.4}$$

In the inner bar (member 2) under either load P_1 or P_2:

$$\sigma_2 = \left(\frac{\sqrt{2}A_1}{\sqrt{2A_1^2 + 2A_1A_2}}\right)P. \tag{1.5}$$

In the above formulae, P has been substituted for P_1 or P_2 as appropriate. The volume of the structure is again:

$$V = \left(\sqrt{2}A_1 + \frac{A_2}{2}\right)L.$$

The same iterative resizing procedure as before is followed in Table 1.3. Note that stress σ_1 is compressive (i.e. negative) under load P_2. However, since the loads P_1 and P_2 have been chosen to be equal, and if the allowable compressive stress is taken to be equal to the allowable tensile stress σ_0, by comparing Eqs. (1.3) and (1.4) we see that the compressive stress in member 1 can never be critical. Equation (1.3) can then be used to resize member 1, and Eq. (1.5) for member 2. The sequence of iterations is shown in Table 1.3, starting with $A_1 = A_2 = 300$ mm^2, and taking $P = 100$ kN, $\sigma_0 = 300$ N/mm^2 and $L = 1000$ mm. The stress ratios at each iteration are given in the last two columns of the table. It is seen that after an initial decrease in volume at the first iteration (depending on the chosen initial values of A_1 and A_2) the volume continues to increase. Convergence becomes very slow as the iteration proceeds but, if followed towards the end, the design is seen to be reducing to a two-bar, statically determinate structure, the inner bar being eliminated and only the two outer bars carrying one or other of the two applied loads. The volume of the fully stressed truss (in column 3 of the table) is finally:

$$V = 471.4 \times 10^3 \text{ mm}^3.$$

While this may intuitively be considered a satisfactory result, the procedure has not in fact converged to the minimum volume. Since the stress ratio $\sigma_2/\sigma_0 < 1$ throughout Table 1.3, the inner bar inevitably reduces in area at each iteration until it disappears altogether. To avoid this, we choose a series of values of $R = A_2/A_1$ in Table 1.4 and calculate the required volume of the truss at each R. This is done now simply by scaling the structure to make the greater stress σ_1 equal to σ_0. We see a decrease in volume from $R = 0$ (no inner bar) to a minimum $V = 439.8 \times 10^3$ at $R = 0.5$, followed by increase in volume to $R = 1.0$. The above minimum volume is in fact very close to the true optimum.

Significant now is that the optimum design is *not* in this case a statically determinate structure, since all three bars are retained and that it is *not* fully stressed

Table 1.3 Iterative resizing of the three-bar truss under alternative loads in Fig. 1.8

A_1 (mm^2)	A_2 (mm^2)	$V/1000$ (mm^3)	$\frac{\sigma_1}{\sigma_0}$	$\frac{\sigma_2}{\sigma_0}$
500.0	500.0	957.1	0.4714	0.2761
235.7	138.1	402.4	1.0938	0.7735
257.8	106.8	418.0	1.0541	0.8153
271.8	87.1	427.9	1.0353	0.8441
281.4	73.5	434.7	1.0249	0.8651
288.4	63.6	439.6	1.0185	0.8812
293.7	56.0	443.4	1.0143	0.8938
297.9	50.1	446.4	1.0114	0.9040
301.3	45.3	448.8	1.0093	0.9124
304.1	41.3	450.8	1.0077	0.9194
306.5	38.0	452.4	1.0065	0.9255
308.5	35.1	453.8	1.0056	0.9306
310.2	32.7	455.1	1.0048	0.9351
311.7	30.6	456.1	1.0042	0.9391
313.0	28.7	457.0	1.0037	0.9426
314.2	27.1	457.9	1.0033	0.9457
315.2	25.6	458.6	1.0030	0.9485
316.2	24.3	459.3	1.0027	0.9510
317.0	23.1	459.9	1.0024	0.9533
317.8	22.0	460.4	1.0022	0.9554
333.3	0.0	471.4	1.0	1.0

Table 1.4 Volume of the three-bar truss in Fig. 1.8 at different values of ratio R

R	A_1 (mm^2)	A_2 (mm^2)	$V/1000$ (mm^3)
0	333.3	0.0	471.4
0.1	312.7	31.3	457.8
0.2	296.6	59.3	449.1
0.3	283.7	85.1	443.7
0.4	273.1	109.2	440.9
0.5	264.3	132.1	439.8
0.6	256.8	154.1	440.3
0.7	250.4	175.3	441.8
0.8	244.9	195.9	444.2
0.9	240.0	216.0	447.4
1.0	235.7	235.7	451.2

since the stress in the inner bar $\sigma_2 < \sigma_0$ under either P_1 or P_2. The explanation for this is that the tensile load in the inner bar makes a useful contribution to the load carrying capacity of the truss in *both* load cases, while it accounts for only a relatively small part of the total volume of the truss. However, a note of caution is necessary here. While it proved possible to find the minimum volume in this simple example with only two variables A_1 and A_2, the same procedure will rapidly prove impractical for larger, more realistic structures. A formal, numerical optimization will be found to offer a much more effective means of obtaining optimum values for the area of the bars and the corresponding minimum volume of the truss in this example.

We see that, in general, a minimum weight structure under more than a single load case need not be statically determinate or fully stressed and that an iterative, fully stressed design procedure does not necessarily converge to the correct minimum weight design. Taken together with the result of the previous section, we can conclude that the iterative resizing procedure can only be guaranteed to converge to a true optimum for a structure made of a single material and under a single load case. The formal proof that, under these conditions, an optimum truss structure is fully stressed was given by Michell [6]. Furthermore, we have up to now considered nothing other than a simple stress limitation. Again, we may assume that what applies here to a truss structure has to apply much more generally to other structures.

As we shall see in subsequent chapters of this book, numerical optimization offers a logical alternative to the iterative redesign process of the present chapter, by searching explicitly for a minimum weight design rather than simply satisfying the stress limits as in fully stressed design. Since we then no longer have to insist that all members are fully stressed, the particular difficulties experienced in this and the previous section are avoided. While optimization methods remain essentially iterative, convergence to the final solution will be greatly improved. The Solver optimization tool in Microsoft Excel, used throughout this book, is introduced in Sect. 1.4 of the present chapter.

1.2 Strength-to-weight Ratio

The strength-to-weight ratio of a structure is an essential parameter in many branches of structural design, not least in aerospace, and plays an important role in determining the feasibility of a design. We shall make use again of the truss structures in the previous sections and limit the discussion here to truss structures made of one material and under a single load case. For any chosen layout of truss, such as the one already shown in Fig. 1.3, individual member lengths l_i can be represented by

$$l_i = a_i L,$$

where L is the span of the truss (distance between supports, or some other representative dimension) and a_i are numerical coefficients. The forces F_i in the members can be represented by

$$F_i = b_i P,$$

where P is an applied load and b_i are again numerical coefficients. If there are more applied loads, each of these can be expressed in terms of a single chosen load P, but for simplicity, it will be assumed here that only a single load is applied. Under the conditions stated above, for minimum weight the truss will be statically determinate and fully stressed. With maximum allowable stress σ_0, the cross-sectional areas A_i of the members are then

$$A_i = \frac{F_i}{\sigma_0}$$

(being statically determinate no iterative calculation is now required).

The volume of material V in the truss can then be written as follows:

$$V = \sum A_i l_i = \sum \frac{b_i P}{\sigma_0} \cdot a_i L,$$

or:

$$V = n \cdot \frac{PL}{\sigma_0}, \tag{1.6}$$

where

$$n = \sum a_i b_i. \tag{1.7}$$

Expressed in non-dimensional form, we have

$$\frac{V}{L^3} = n \cdot \frac{1}{\sigma_0} \cdot \frac{P}{L^2},$$

in which we can identify the structural index P/L^2, in units of stress. We shall come across the structural index more often in subsequent chapters. If the specific weight (weight per unit volume) of the material is ρ_w, the weight W of the truss is

$$W = n \left(\frac{\rho_w}{\sigma_0} \right) PL. \tag{1.8}$$

Coefficient n in the above formula depends only on the layout of the truss, in other words the arrangement of its members. This can be used, therefore, to compare the efficiency of different layouts of truss—the smaller the n-value, the lower the weight. Equation (1.8) can also be used for the weight of a statically indeterminate truss, for example, if a minimum area of the bars is imposed, except that it will then no longer be fully stressed in all bars so the simple formula in Eq. (1.7) cannot be used to calculate n.

With Eq. (1.8), the strength-to-weight ratio becomes

$$\frac{P}{W} = \frac{1}{nL}\left(\frac{\sigma_0}{\rho_w}\right) \tag{1.9}$$

and, for given material properties and overall span, the strength-to-weight ratio again depends only on the layout of the truss. This can only be improved, therefore, by use of better materials (ratio σ_0/ρ_w), or by change in layout (n-value). For materials we are, of course, limited to those available, whereas layout demands a search in any particular problem for the optimum arrangement of the members of the truss. Other considerations emerge later, in Chap. 2, when we take into account buckling of those bars of a truss which are in compression.

It is perhaps interesting to note that the quantity $\left(\frac{\sigma_0}{\rho_w}\right)$ in Eq. (1.9) can be interpreted as follows. Consider a uniform bar of the material hanging simply under its own weight, as in Fig. 1.9. The stress at the upper end of the bar, regardless of its cross-sectional area, is

$$\sigma = L\rho_w,$$

and $\sigma = \sigma_0$ in the limit. The maximum length of the bar is therefore

$$L_{max} = \frac{\sigma_0}{\rho_w},$$

which is termed the 'material breaking length' (if σ_0 is regarded now as the tensile strength of the material, rather than simply an allowable stress). L_{max} is

Fig. 1.9 Bar hanging freely under its own weight

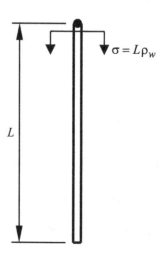

typically in the range 10 to 20 km for most metals, and in excess of 200 km for some fibres.

It will be realized that we are taking a highly idealized approach here to strength-to-weight ratio, and to the calculation of n-values. In reality, the weight of the necessary joints in the structure can be significant, and practical limitations on the design of the members of a truss (e.g. use of standard sections) will cause further increase in weight. Again, no consideration has been given as yet to the possibility of buckling of members in compression. The purpose here has not been to perform a detail design of some type of truss structure for an accurate strength-to-weight ratio, but simply to use the truss to illustrate the conventional design process and its limitations, and in due course to understand the relationship between this and formal optimization. Nevertheless, n-values as defined here will be used in the next section to introduce limits of feasibility and in Sect. 1.3 to explore the influence of the layout of the members of a truss on its minimum weight.

1.2.1 Feasibility

The strength-to-weight ratio of a structure can be used to investigate the important question of feasibility. Suppose a truss structure, of span L between two supports, is loaded entirely under its own weight. If it is just able to support its own weight without failure, this must represent the limit of feasibility, or the maximum span L_{max}, of the structure. Any additional load will cause it to fail. Likewise, if a similar structure is designed with a span a little less than the maximum span L_{max}, the additional load it can carry will be small compared with the weight of the structure itself, and the design will be uneconomic. From Eq. (1.8), for a truss structure with given n-value, taking the weight W of the structure itself to be the only load P on it at failure:

$$W = n\left(\frac{\rho_w}{\sigma_0}\right) PL_{max} = P,$$

or

$$L_{max} = \frac{1}{n}\left(\frac{\sigma_0}{\rho_w}\right). \tag{1.10}$$

This shows directly the influence of layout and choice of material on the feasibility of the design. However, it will be appreciated that this can only be a first estimate, because the load distribution assumed in the calculation of n is unlikely to match the actual distribution of the weight of the structure over its span.

1.3 Comparison of Layouts

While it was shown in Sect. 1.2 that the strength-to-weight ratio of a fully stressed truss of some given material and fixed span depends only on its layout, no layout variation has been considered up to now. Here, we shall take again the classic problem of a truss supporting a load at mid-span between two supports to explore the effect of variation of layout, in the first place by means of the simple truss structures in Fig. 1.10. Truss (a) is statically indeterminate and has 11 members. The other three statically determinate trusses are all derived from this one. Truss (b) has only 5 members, while truss (c) has 7 members and is unsymmetrical. Truss (d) has 6 members, but is actually a mechanism, in other words unstable. The n-value of all three statically determinate trusses (b), (c) and (d), when fully stressed, is plotted in Fig. 1.11 against the height-to-span ratio H/L. Perhaps surprisingly, in view of the differences in the three trusses noted above, it is found that all three follow exactly the same curve. The minimum $n = 2$ occurs at $H/L = 0.5$, with minimum volume:

$$V_{\min} = 2\frac{PL}{\sigma_0}.$$

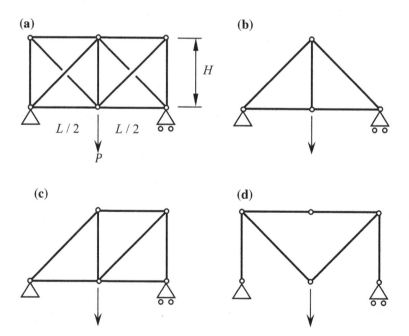

Fig. 1.10 Alternative layouts of truss for a load at mid-span between two supports

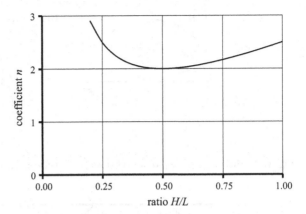

Fig. 1.11 Coefficient n for all four trusses in Fig. 1.10

Since all three statically determinate 'partial structures' (b), (c) and (d) have the same coefficient n, the statically indeterminate truss (a) must also have the same value. In other words, in this exceptional case, the three partial structures can be superimposed in any amount to make an equally efficient statically indeterminate truss.

From all this, it might be tempting to think that layout has little influence on strength-to-weight ratio, but in fact this is far from the truth. The truss that emerged in Sect. 1.1 (see Fig. 1.5) also has a coefficient $n = 2$, but if it is optimized by varying the angle θ of the sloping bars (and at the same time the height of the truss), the minimum volume is found to be

$$V_{\min} = \sqrt{3}\frac{PL}{\sigma_0} = 1.732\frac{PL}{\sigma_0} \tag{1.11}$$

at $\theta = 60$ °C. This result is confirmed in Sect. 1.4 using Solver. A still better truss is available in Fig. 1.12a, with

$$V = \frac{\pi}{2}\cdot\frac{PL}{\sigma_0} = 1.571\frac{PL}{\sigma_0}. \tag{1.12}$$

This is the classic 'Michell structure' for this problem, requiring in principle an infinite number of the radial 'spokes'. (Further treatment of Michell structures is beyond the scope of this book; for more information, see [6–8].) The above value of V is a theoretical minimum if no part of the structure is allowed 'under the bridge'. Otherwise, the structure in Fig. 1.12b gives an even smaller volume:

$$V = \left(\frac{1}{2} + \frac{\pi}{4}\right)\frac{PL}{\sigma_0} = 1.285\frac{PL}{\sigma_0}.$$

Useful as they are to illustrate some important principles, the trusses seen up to now can hardly be considered practical engineering structures, if only because for

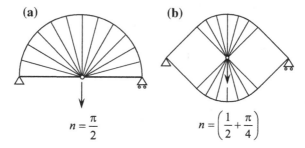

Fig. 1.12 Michell structures for a load at mid-span between two supports

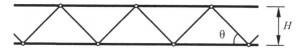

Fig. 1.13 Long truss structure with diagonal bracing

any significant span the height of the structure becomes very large and the members are likely to be too slender. A more practical design of truss, familiar from bridges and many other types of structure, is shown in Fig. 1.13. Treating the truss (provided it is sufficiently long) as equivalent to a continuous beam, we can first find an optimum angle θ for the bracing members, responsible for the overall shear force Q at any point. The force F in any bracing member is

$$F = \frac{Q}{\sin\theta},$$

so that the volume of the member (of length $H/\sin\theta$) when fully stressed is

$$V = \frac{Q}{\sigma_0 \sin\theta} \cdot \frac{H}{\sin\theta}.$$

Each bracing member relates to a length $H/\tan\theta$ along the span, so the volume of the bracing members per unit length of the truss is

$$V_Q = \frac{QH}{\sigma_0 \sin^2\theta} \bigg/ \frac{H}{\tan\theta} = 2\frac{Q}{\sigma_0} \cdot \frac{1}{\sin 2\theta}.$$

This has a minimum value:

$$V_Q = 2\frac{Q}{\sigma_0} \tag{1.13}$$

at $\theta = 45$ °C, the optimum angle of the bracing members if designed purely on the basis of a maximum allowable stress. This is, of course, a familiar result, and such structures are widely seen. The horizontal members of the truss are responsible for the bending moment M at any point along the span of the truss, and the volume of each per unit length at that point is

$$V_M = \frac{M}{H\sigma_0}. \tag{1.14}$$

Equations (1.13) and (1.14) can be used to estimate the minimum total volume of a truss of the type in Fig. 1.13, assumed to be carrying a uniformly distributed load between the two supports, with the corresponding bending moment distribution and shear force. Provided that the span-to-height ratio L/H is sufficiently large,

$$V_{\min} = \left(\frac{1}{2} + \frac{L}{6H}\right)\frac{PL}{\sigma_0}, \tag{1.15}$$

where P is the total load on the truss (the first term within the brackets referring to the bracing members and the second term to the horizontal members). Realistically this load must, of course, include the weight of the truss itself. If the truss in Fig. 1.13 has a ratio $L/H = 100$ and taking a typical material with $\sigma_0/\rho_w = 10 \times 10^3$ m, setting the load P on the truss equal to its weight $\rho_w V_{\min}$ the maximum span of such a truss would be:

$$L_{\max} = 583\,\text{m}.$$

In fact, this must be an overestimate of the maximum span, because Eq. (1.15) is for a uniformly distributed load on the beam whereas the actual weight of the beam will be concentrated towards the middle where the bending moment is greatest. Unless this is taken into account, an accurate bar-by-bar analysis of the truss, rather than the approximate method adopted here, would not be justified.

1.3.1   Classification of Optimization Problems

The various truss structures studied in this chapter offer a means of distinguishing between different classes of optimization problem. This distinction will be useful in the further chapters of this book. The different classes are illustrated in Fig. 1.14. At the highest level, we have optimization of *topology*, that is, the arrangement of the members of a structure and the connections between them. Two different topologies of truss structure are shown in Fig. 1.14a. With four nodes in the left-hand figure and five nodes in the other, no amount of movement of these nodes can transform the one into the other (even if the two inner bars of the right-hand truss are made vertical, two members and one node would have to be removed to correctly

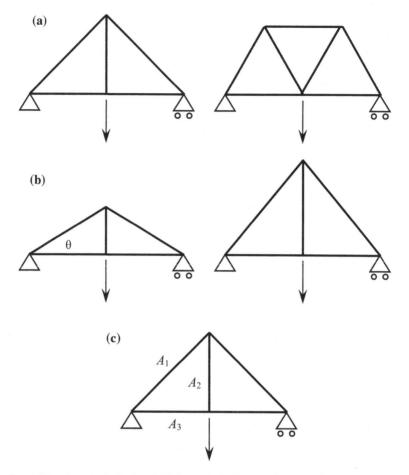

Fig. 1.14 a Topology optimization. **b** Shape optimization. **c** Sizing or optimization of members

reproduce the left-hand one). The various trusses in Fig. 1.10 also illustrate different topologies. While special methods exist for topology optimization, this no doubt remains one of the most challenging forms of optimization because of the discrete nature of the different structures, commonly demanding more of the ingenuity of the designer than any mathematical process.

At the next-level down, *shape* optimization, illustrated in Fig. 1.14b, refers in the case of a truss structure to the location of the nodes, for a truss of given topology. Simply by change in the angle θ, the one truss can readily be transformed into the other. Change in the H/L ratio for the trusses in Fig. 1.10 is another example of shape optimization. Frequently shape optimization can be handled simply by treating all or some of the overall dimensions as variables. The term 'layout' is commonly taken to refer to both topology and shape, to distinguish between these two classes of problem and the 'sizing' problem described below.

At the lowest level, *sizing*, or optimization of the dimensions of the individual members of a structure, is up to now by far the most common practical application of optimization methods. For the simple truss structure in Fig. 1.14c, this would imply optimization of the cross-sectional areas A_i of the members, for given topology and shape. The progressive change of member areas for the truss in Fig. 1.3 is, of course, an example of structural sizing.

While these classes of optimization are clear for truss structures, in other cases the distinction may become more blurred. For example, the diameter of a circular tube is treated simply a sizing variable, but change from a circular tube to a square one is clearly a change in shape. In the case of a stiffened panel—a thin sheet reinforced by a series of discrete stiffeners—change in stiffener spacing might be considered change in shape, but if this also results in a different number of stiffeners, this would be a change in topology. Even so, stiffener spacing is likely to be treated in practice simply as a sizing variable.

If the number of stiffeners in a panel is nevertheless treated as a variable, rather than the stiffener spacing, this introduces a further important distinction between different classes of optimization problem. The whole number of stiffeners has then to be treated as a *discrete* variable, whereas the stiffener spacing may well (at least in the first place) have been regarded as a *continuous* variable. A selection of tubes of different shapes and sizes, if they are individually numbered, may also be represented as a set of discrete variables, just as the number of plies in a composite laminate. Special methods exist for discrete variable optimization, as will be discussed in a later chapter.

1.4 Spreadsheet Program

The Solver tool in Microsoft Excel is a set of optimization routines that can be used to find the maximum or minimum of some function defined in a formula in an Excel spreadsheet. The quantities to be varied have to be defined in the spreadsheet, as well as any necessary limits on those variables and on other functions defined in the spreadsheet. Use of Solver is described in detail in the Appendix at the end of this book, together with instructions for loading the Solver add-in program. The mathematical procedures underlying different numerical optimization methods, including those available in Solver, are described in Chaps. 4 and 5.

1.4.1 'Seven-Bar Truss'

As introduction to the use of Solver, a spreadsheet program is presented in this section for the optimization of the seven-bar, pin-jointed truss structure in Fig. 1.15. The spreadsheet is shown in Fig. 1.16. The same spreadsheet serves as example of the use of Solver in the Appendix. The applied load P, span L, material allowable

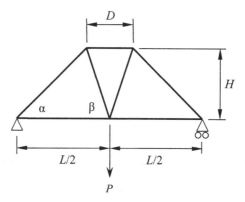

Fig. 1.15 Truss structure optimized in the spreadsheet 'Seven-bar Truss'

stress σ_0, density ρ and chosen initial values of dimensions D and H have to be inserted in the appropriate cells of the spreadsheet (i.e. by replacing their current values). Corresponding forces F_i in the members of the truss, the cross-sectional areas A_i and volumes V_i of the bars, coefficient n, the volume V of the truss and its strength–to-weight ratio P/W are automatically calculated.[1] Since the truss is statically determinate, it is assumed that all bars are fully stressed for calculation of the cross-sectional areas A_i. Formulae used in the spreadsheet for analysis of the truss are given in Table 1.5. They can also be seen in Excel formulae in the spreadsheet by clicking the appropriate cell. Parameters and variables to be entered in the spreadsheet are listed in Table 1.6.

Optimum values of variables D and H to minimize the volume V of the truss are found with Solver. The Solver Parameters dialog box is opened by clicking Solver on the Data tab. The cell references for the volume V and for the variables D and H have to be inserted in the Solver Parameters dialog box (in 'Objective' and 'Variable Cells', respectively). Limits $0 \leq D \leq L$ and $H \geq L/100$ are set on the variables D and H (in 'Constraints') to restrict these to realistic values. Choose the 'GRG Nonlinear' optimization method, and ensure that 'Min' (for minimum volume of the truss) has been selected. The Solver dialog box has already been set up for direct use of the spreadsheet. Click 'Solve' to optimize the truss.

Optimization gives $\alpha = \beta = 60\ °C$, with a coefficient n in agreement with Eq. (1.11). Further constraints can be added (or deleted) by means of the buttons on the right-hand side of the Solver dialog box. For example, a maximum value of H might be specified. If a constraint $D = 0$ is added, the two inner bars converge at the top of the truss, to give the truss in Fig. 1.10b. If constraints $D = L/2, H = L/4$ are specified, we have the truss in Fig. 1.5.

---

[1]Depending on the magnitude of the initial data and the calculated results, the width of some columns in the spreadsheet may have to be adjusted.

Seven-bar Truss

Parameters

$P = 10000$ N
$L = 1.000$ m
$\sigma_0 = 400$ MPa
$\rho = 2780$ kg/m³

Variables

$D = 0.200$ m
$H = 0.500$ m

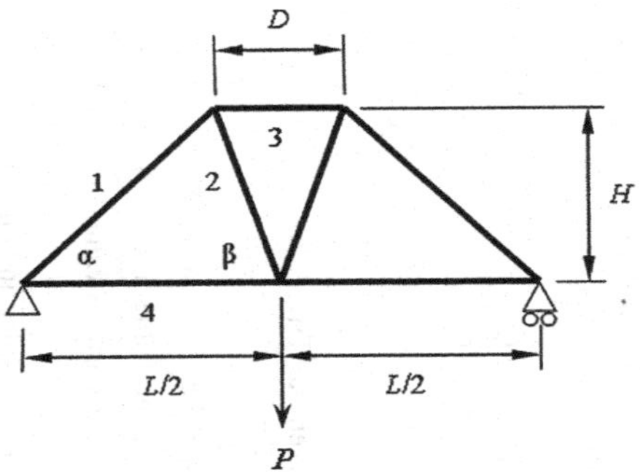

*Enter Parameters (in blue)
and Variables (in red).
To optimize the truss: click Solver
on the Data tab, then click Solve.*

Analysis

bar	l_i	F_i	A_i	V_i
1	0.6403	−6403	16.01	10250
2	0.5099	5099	12.75	6500
3	0.2000	−5000	12.50	2500
4	0.5000	4000	10.00	5000
	m	N	mm²	mm³

Optimization

$V =$	**46000**	mm³
$\alpha =$	51.34	deg
$\beta =$	78.69	deg
$n =$	1.840	
$P/W =$	7974	

Fig. 1.16 Spreadsheet 'Seven-bar Truss'

Table 1.5 Formulae used in the spreadsheet 'Seven-bar Truss'

$l_1 = \sqrt{H^2 + \frac{(L-D)^2}{4}}$	$F_1 = -\frac{P\,l_1}{2H}$
$l_2 = \sqrt{H^2 + \frac{D^2}{4}}$	$F_2 = \frac{P\,l_2}{2H}$
$l_3 = D$	$F_3 = -\frac{PL}{4H}$
$l_4 = \frac{L}{2}$	$F_4 = \frac{P(L-D)}{4H}$
$A_i = \frac{|F_i|}{\sigma_0}$ $V_i = A_i l_i$ $V = \sum V_i$	$n = \frac{V\sigma_0}{PL}$ $\frac{P}{W} = \frac{P}{V\rho_w}$

Table 1.6 Data entry for spreadsheet program 'Seven-bar Truss'

Parameters	
Applied load P (positive downwards)	Enter the value in cell C6
Span L	Enter the value in cell C7
Allowable stress σ_0 (both tension and compression)	Enter a positive value in cell C8
Density ρ	Enter the value in cell C9
Variables	
Dimensions D and H (see figure on spreadsheet)	Enter initial values in cells F6: F7 (H positive, nonzero)

1.5 Summary

Before proceeding to formal optimization methods in the further chapters of this book, a simple truss structure is used to explore the characteristics of the conventional, iterative design process. By this is meant the progressive modification of an initial design to just satisfy as many as possible of the design requirements, on the assumption that this will lead to a 'best' design. For a truss structure, subject only to stress limits, this means adjusting the cross-sectional area of the bars until the maximum allowable stress is reached in all of them—the principle of the fully stressed design. This implies that the outcome will normally be a statically determinate truss, if necessary by removal of unwanted members. It also implies that, in principle, all members of the structure will fail at the same time! This will no doubt be considered undesirable in practice, since adequate residual strength in a statically indeterminate structure is usually required after failure of one of its members. While this iterative process does frequently lead to an optimum, minimum weight design, it is demonstrated that for a truss made of different materials, or one subject to more than one load case, this need not always be so. These are, of course, commonly occurring conditions.

Based on a fully stressed design, an expression is obtained for the minimum weight of a truss in terms of a coefficient n, depending only on the layout of the truss. Some elementary layouts of truss are compared. The maximum possible span of a truss is reached when the load it can carry is just equal to its own weight, and this can be expressed in terms of this same coefficient n. The simple truss structures seen in this chapter also offer a convenient means of identifying three different levels of optimization: topology (layout of members), shape (location of nodes) and sizing (dimensions of members). For other types of structure, this classification may become somewhat less clear; nevertheless, it remains useful because different approaches are usually necessary. The distinction can also be made between continuous variable optimization and discrete variable optimization when considering, for example, the whole number of stiffeners in a panel, or the whole number of plies in a composite laminate.

Failure to converge to an optimum design of truss under some common conditions implies that for any other type of structure an iterative design procedure to satisfy stress limits at critical locations cannot be guaranteed to converge to a true optimum design, although in most cases it will lead at least to an improvement in the design. Furthermore, convergence of an iterative procedure may be slow. However, the purpose here is not to dismiss the conventional, iterative design process, which in its many forms will surely remain in widespread use, but rather to understand its limitations. At the same time, this does provide the justification for turning to the formal optimization methods in subsequent chapters. These can greatly enhance the traditional methods of design and are the foundation of an automated design procedure. We shall see that optimization offers a powerful alternative to the iterative design process, able to accommodate not only stress limitations but stiffness, buckling and all other requirements, not restricted by the nature of the often-conflicting design requirements, and searching directly for a true optimum design. There is no longer the need to know beforehand which requirements will prove to be the critical ones in the final design. The relationship between fully stressed design and formal optimization is explored further in a later chapter.

The 'Solver' optimization tool in Excel, used extensively throughout this book, is introduced in the final section of this chapter. A spreadsheet for the optimization of a seven-bar truss is used to demonstrate use of Solver.

Exercises

1.1 Verify the formula in Eq. (1.11) for the minimum volume of the truss structure in Fig. 1.5, and the optimum angle θ.
  Derive formulae for the forces in the members, the corresponding minimum cross-sectional areas of the members and the volume of the truss in terms of the angle θ. Try different values of θ to search for the minimum volume.

1.2 A truss structure has to carry two equal loads $P/2$ over a span L, the loads being placed at 1/3 and 2/3 of the span, respectively. The material of the truss has an allowable stress σ_0. Try different layouts to find a suitable layout of truss for this loading. Express the result in terms of the coefficient n in

Eq. (1.6), where P is the total load on the truss. Compare the value of n with that for the truss in Exercise 1.1 with a load P at mid-span.

Try a few simple layouts of statically determinate truss, with members at suitably chosen angles. Compare the volume of each layout.

1.3   Verify the formula in Eq. (1.15) for the minimum volume of the long truss structure shown in Fig. 1.13. If its height is limited to 1.0 m, calculate the maximum possible span when loaded only under its own weight. The truss is made of steel with an allowable stress of 1000 N/mm^2 and density 7850 kg/m^3.

In deriving Eq. (1.15), shear force and bending moment distributions are obtained by treating the truss as a continuous beam. Equations (1.13) and (1.14) can then be used to obtain formulae for the volume of the bracing members and horizontal members, respectively, by integration over the span of the truss. These together give the formula for minimum volume of the truss. Treating the weight of the truss as uniformly distributed over its span, use Eq. (1.15) with Eq. (1.10) to calculate the maximum possible span when the height is limited to 1.0 m.

1.4   Derive a formula for the minimum volume of a truss similar to the Michell truss in Fig. 1.12a, with only five radial 'spokes' at an angle of 45 °C to each other. Compare the result with Eqs. (1.11) and (1.12).

The forces in the bars can be solved by equilibrium at the nodes. Replace the circular arc by straight bars between the nodes. Note that by equilibrium the force in all five of these 'circumferential' bars is the same.

1.5   A steel cable is stretched between two towers each of height 100 m above the ground. It may be assumed that the cable forms a shallow parabolic curve, loaded only under its own weight. The tensile strength of the cable is 1000 N/mm^2 and its density is 7850 kg/m^3. What is the maximum possible distance between the towers if the cable is not permitted to touch the ground? *Derive a formula for the parabolic curve of the cable in terms of the height of the towers and the distance L between them, and from this the angle the cable makes at the towers. The vertical component of the force in the cable at each tower has to be equal to half the weight of the cable. Assume L to be much greater than 100 m, so that the actual length of the cable is approximately equal to L.*

1.6   A hollow steel tube of diameter 1.0 m is mounted vertically as a tower, fixed rigidly at its lower end and free at its upper end. It is loaded only by its own weight. Under these conditions, the tube can be treated as a column with an effective length $L_{\mathrm{eff}} = 1.122 L$ in Euler's formula:

$$P = \frac{\pi^2 EI}{L_{\mathrm{eff}}^2},$$

where L is the actual length of the tube, P is the load at its lower end and I is the second moment of area of the cross section of the tube. The elastic modulus E of the steel tube is 200 GN/m^2 and its density is 7850 kg/m^3. What is the maximum possible length of the tube if it is not to buckle under its own weight?

The load P at which buckling occurs is equal to the weight of the tube. For a thin circular tube $I = \pi R^3 t$, where R is the radius of the tube and t its thickness. Note that its maximum length is independent of the thickness of the tube.

1.7 Run Solver in the spreadsheet 'Seven-bar Truss' with the parameters and variables already entered in the spreadsheet to optimize the truss. Compare the coefficient n with that given by Eq. (1.11). Try a few different initial values of D and H to test convergence of the optimization.

Refer to Sect. 1.4 and the Appendix for information on the use of Solver.

1.8 In the spreadsheet 'Seven-bar Truss', add a constraint $D = 0$ to the Solver dialog box to create the truss in Fig. 1.10b, and run Solver to verify the result for this truss given in Eq. (1.11). Then, delete the constraint just added (to restore the original truss), add a new constraint to restrict the area of one (or more) of the bars to a suitable minimum value and run Solver again to observe the effect of this on the shape of the truss.

Refer to the Appendix for information on adding constraints in Solver. Ensure that the minimum area of the chosen bar is larger than its optimum value found in Exercise 1.7. Note that the spreadsheet maintains the symmetric shape of the truss.

1.9 Use Solver to find the minimum of the function:

$$f(x, y) = x + 2y$$

subject to constraints:

$$x + y - 4 \geq 0,$$

$$2x + y - 4 \geq 0,$$

$$-3x + y + 4 \geq 0 \text{ and}$$

$$x - 1 \geq 0.$$

Which constraints are active at the minimum?

Use the spreadsheet 'Seven-bar Truss' as a guide to setting up the spreadsheet and making the appropriate entries in the Solver dialog box. Examine the values of the four constraints after running Solver to see which are active, i.e. equal to zero.

1.10 Make a spreadsheet to optimize the three-bar truss under alternative loading in Sect. 1.1.2. Take load $P = 100$ kN, span $L = 1000$ mm and allowable stress (both in tension and compression) $\sigma_0 = 300$ N/mm^2. Compare the result with the minimum volume given in Sect. 1.1.2.

The stresses in the bars are given by Eqs. (1.3)–(1.5). Define constraints in the Solver dialog box to limit each of these to not greater than the allowable stress. Use Solver with variables A_1 and A_2 to minimize the volume of the truss.

References

1. Hoff NJ (1956) The analysis of structures. John Wiley & Sons, New York
2. Timoshenko SP, Young DH (1965) Theory of structures. McGraw-Hill International, New York
3. Megson THG (1999) Aircraft structures for engineering students. Arnold, London
4. Haftka RT, Gurdal Z (1992) Elements of structural optimization. Kluwer Academic Publishers, Dordrecht
5. Schmit LA (1981) Structural synthesis—its genesis and development. AIAA J 19(10):1249–1263
6. Michell AGM (1904) The limit of economy of material in frame structures. Phil Mag 8 (47):589–597
7. Cox HL (1965) The design of structures for least weight. Pergamon Press, Oxford
8. Hemp WS (1973) Optimum structures. Clarendon Press, Oxford

Chapter 2
Optimality Criteria

Abstract A first optimality criterion, that of the fully stressed design, was already introduced in the previous chapter. The buckling of a circular tube in compression is used to illustrate a second criterion, that of simultaneous buckling modes. In fact, when the tube forms part of a truss structure, this might be seen as a logical extension of the principle of the fully stressed design. This second optimality criterion leads directly to an efficiency formula, expressing the maximum stress that can be achieved in a thin tube or other component in terms of a suitable structural index and the elastic modulus of the material. The concept of the design space, widely used in subsequent chapters, is introduced with the circular tube. A third criterion is developed for the maximum stiffness of a structure, on the basis of a simple truss but taken in principle to apply more widely. It is shown that under certain conditions, a fully stressed design, with maximum strength-to-weight ratio, also has maximum stiffness. A spreadsheet program is presented for the optimization of circular and square tubes in compression, subject to dimensional restrictions and specified maximum allowable stress.

Optimality criteria are conditions that are assumed to be satisfied in an optimum design. When known to be valid, they can be used either directly to find an optimum or otherwise to reduce the size of an optimization problem. The first of these criteria, that of a fully stressed design, was introduced in the previous chapter, along with some necessary conditions to establish its validity. In short, a fully stressed design implies that the maximum allowable stress is reached in all parts of a structure, or in the case of a truss structure in each member. This was discussed without regard to the possibility of other modes of failure, such as buckling of some parts of the structure at a lower stress. A second optimality criterion relates specifically to the design of a structural component when buckling is the principal design condition. This is introduced here through the optimization of a circular tube loaded in compression, subject to both buckling and maximum stress limitations. This might be regarded as one of the members of a truss structure in the previous chapter. At the same time, the circular tube is used to introduce the concept of the 'design space', an invaluable aid in the visualization of a numerical optimization

© Springer International Publishing AG 2017   29
A. Rothwell, *Optimization Methods in Structural Design*, Solid Mechanics
and Its Applications 242, DOI 10.1007/978-3-319-55197-5_2

procedure and sometimes, as now for the circular tube, as a direct means of solving
an optimization problem. With this, an efficiency formula can be derived expressing
the maximum stress that can be achieved in terms of a structural index. A third
optimality criterion concerns the design of a structure for maximum stiffness, and
again conditions under which this is valid have to be established. This is introduced
later in the present chapter, in the context of a truss structure but taken to apply
more generally to other types of structure.

2.1 Circular Tube in Compression

In the previous chapter, the design of a truss structure was explored on the basis of a
specified maximum stress for all the members. However, for those members of the
truss which are in compression, if they are relatively slender the maximum stress
may be limited by buckling instead of by an allowable material stress. The maximum stress will depend then on the actual size and shape of cross section of each
member, as well as on its length. For a tubular member, such as the circular tube
considered here, buckling may be either in flexural buckling in a long-wave mode,
as illustrated in Fig. 2.1, or by local buckling in a short-wave mode.

The critical compressive load P_E for flexural buckling is given by the
well-known Euler's formula:

$$P_E = \frac{\pi^2 EI}{L^2},$$

where E is the elastic modulus of the material, I is the (minimum) second moment
of area of the cross section, and L is the length of the member (assumed to be
pinned at its ends). For a thin circular tube of radius R and thickness t, its
cross-sectional area is

$$A = 2\pi Rt$$

and second moment of area:

$$I = \pi R^3 t,$$

Fig. 2.1 Flexural buckling of a *bar* in compression

giving:

$$P_E = \frac{\pi^3 E}{L^2} R^3 t \tag{2.1}$$

and buckling stress:

$$\sigma_E = \frac{P_E}{A} = \frac{\pi^2 E R^2}{2L^2}.$$

The above formulae are sufficiently accurate if the thickness of the tube is small compared with its radius, and if R is taken to be the *mean* radius of the tube (measured to the mid-thickness). Use of Euler's formula above also implies that the tube is initially perfectly straight. If the tube is clamped at its ends, instead of pinned, an effective length $L/2$ should be used in the formula. Different effective lengths exist for other end conditions (see Young and Budynas [4], and many other texts).

It can be seen that the flexural buckling stress σ_E increases without limit with increasing radius R. If the design of the tube were based just on the stress σ_E, and leaving aside for the present any material strength limitation, this would imply that the strength-to-weight ratio of the tube also increases without limit. However, the restriction on R is through local, or short wavelength, buckling of the tube in which the cross section is deformed out of its initially circular shape into a pattern of small buckles, both around and along the length of the tube. The standard formula for this local mode of buckling is given as follows:

$$\sigma_L = KE\frac{t}{R},$$

where the buckling coefficient $K = 0.605$ for Poisson's ratio $v = 0.3$. (Formulae for Euler buckling and for the buckling of a thin, circular tube can be found in any standard textbook on buckling theory, including the classic text by Timoshenko and Gere [3], and more recent texts such as that by Megson [1].) It should be noted, however, that for very thin tubes the coefficient K is sensitive to imperfections in the circular form of the tube, as well as to the end conditions, and even for the thicker tubes assumed here may still have to be reduced. The above formula gives for the buckling load:

$$P_L = KE\frac{t}{R} \cdot A = 2\pi K E t^2, \tag{2.2}$$

independent of R.

We can now represent the design problem of the circular tube in a so-called design space, as shown in Fig. 2.2. The axes of the diagram are the radius R and thickness t of the tube—termed the 'design variables'. The design conditions, or 'constraints', imposed on the design are the Euler and local buckling loads, in

Fig. 2.2 Design space for a *circular* section tube in compression

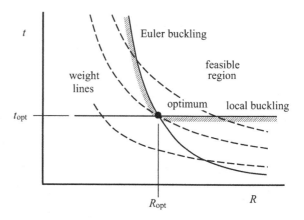

Eqs. (2.1) and (2.2), plotted as solid lines in the design space. Each of these lines represents critical combinations of R and t to precisely satisfy the particular constraint. The shading on the constraint lines marks the side on which the constraints are not satisfied, in other words the design is unsafe, or infeasible. The unshaded side is therefore the safe, or feasible, side. The two constraints together define the boundary of the feasible region. To locate the optimum, lines of constant cross-sectional area (so-called 'weight lines') are added to the design space, shown as broken lines. The cross-sectional area of the tube reduces towards the origin of the diagram, so it is clear that the optimum is located at the intersection of the two constraint lines, at which we have the smallest cross-sectional area and at which Euler and local buckling occur simultaneously.

This condition—that of simultaneous modes of buckling—is the second optimality criterion referred to earlier. In a sense, this might also be seen more generally as an extension of the principle of the fully stressed design—referring again to a truss structure, simultaneous failure of all members and now simultaneous failure in the different modes of buckling within a member. However, as before for a fully stressed design, the validity of this criterion cannot be guaranteed in every case, and has therefore first to be verified when applied to a different class of problem. For the circular tube, as well as for other shapes of cross section, use of this criterion leads to the definition of efficiency, and a direct solution for the optimum dimensions. However, when we include a material strength limitation we shall see that the condition of simultaneous buckling modes may no longer apply. Neither can it be guaranteed that *all* buckling modes will occur simultaneously. This is evidently so when the number of possible buckling modes exceeds the number of design variables available for optimization. Finally, it should be pointed out that while the discussion here has been about simultaneous buckling in the different modes, this is unlikely to occur in reality. Imperfections of various kinds reduce the buckling stress from its theoretical value, and in practice determine in which mode buckling will actually occur.

2.1.1 Efficiency Formula

In a practical problem, the design space in Fig. 2.2 might have been drawn for specific values of P, L and E. However, with the optimality criterion that we now have, explicit formulae can be derived for the optimum radius and thickness, and for the maximum stress that can be achieved in the tube. For simultaneous Euler and local buckling:

$$P = P_E = P_L,$$

and substituting from Eqs. (2.1) and (2.2)

$$P = \frac{\pi^3 E}{L^2} \cdot R^3 t = 2\pi K E t^2.$$

Here, we have two equations which can be solved for t and R to give:

$$t^* = \left(\frac{P}{2\pi K E}\right)^{1/2} \tag{2.3}$$

and

$$R^* = \left(\frac{2K}{\pi^5} \cdot \frac{PL^4}{E}\right)^{1/6}, \tag{2.4}$$

where the asterisk conventionally denotes optimum values of the given dimensions. The maximum stress is then as follows:

$$\sigma_{max} = \frac{P}{2\pi R^* t^*} = \left(\frac{\pi K}{4}\right)^{1/3} E^{2/3} \left(\frac{P}{L^2}\right)^{1/3},$$

which can be written as follows:

$$\sigma_{max} = \eta \, E^{2/3} \left(\frac{P}{L^2}\right)^{1/3}, \tag{2.5}$$

where the 'efficiency' η of the tube is

$$\eta = \left(\frac{\pi K}{4}\right)^{1/3}.$$

If the buckling coefficient K is taken to have its maximum theoretical value $K = 0.605$, we obtain an efficiency $\eta = 0.780$.

The above formulae define the maximum stress that can be achieved in an optimized circular tube in compression, together with the optimum radius and thickness, and show how these depend on the modulus E, load P and length L. The term P/L^2 in Eq. (2.5) is referred to as the 'structural index' and represents the non-material parameters of the problem. The different powers in the formula show the sensitivity of the maximum stress σ_{max} to the particular parameters. For example, while the individual buckling loads are directly proportional to the modulus E, the effect of change in E on the maximum stress in an optimized tube depends only on $E^{2/3}$. Similarly, the effect of any reduction in K, to compensate for imperfections in the tube, is felt only as $K^{1/3}$. Whereas the weight of a bar in tension is directly proportional to both the load on it and its length ($W \propto PL$), from the efficiency formula it is deduced that for a circular tube in compression the relation becomes

$$W \propto \left(\frac{P}{L^2}\right)^{-1/3} PL.$$

An effective length L can be used in all the formulae, if it is necessary to allow for different end conditions. Finally, it should be emphasized that the term efficiency, as used here, is simply an efficiency *coefficient*—the larger its value the greater the maximum stress that can be achieved. Its maximum value is *not* 1.0, and it should not be expressed as a percentage.

Similar efficiency formulae can be derived for other shapes of cross section, as shown in Table 2.1. For a solid section, there is, of course, no local buckling mode, so the design condition is simply $P = P_E$ to solve for the required radius or other cross-sectional dimension. For a square section tube, the second moment of area to calculate the Euler buckling load is $I = 2b^3t/3$, where b is the *mean* side of the square and t is the thickness of the tube. For local buckling, each side can be treated individually as a thin plate, for which the local buckling stress is:

$$\sigma_L = KE\left(\frac{t}{b}\right)^2,$$

Table 2.1 Efficiency formulae for circular and square sections

Circular section tube	$\sigma_{max} = \left(\frac{\pi}{4}\right)^{\frac{1}{3}} K^{\frac{1}{3}} E^{\frac{2}{3}} \left(\frac{P}{L^2}\right)^{\frac{1}{3}} = 0.780 E^{\frac{2}{3}} \left(\frac{P}{L^2}\right)^{\frac{1}{3}}$ with $K = 0.605$
Square section tube	$\sigma_{max} = \left(\frac{\pi^2}{24}\right)^{\frac{1}{3}} K^{\frac{1}{3}} E^{\frac{2}{3}} \left(\frac{P}{L^2}\right)^{\frac{1}{3}} = 0.907 E^{\frac{2}{3}} \left(\frac{P}{L^2}\right)^{\frac{1}{3}}$ with $K = 3.62$
Solid circular section	$\sigma_{max} = \left(\frac{\pi}{4}\right)^{\frac{1}{2}} E^{\frac{1}{2}} \left(\frac{P}{L^2}\right)^{\frac{1}{2}} = 0.886 E^{\frac{1}{2}} \left(\frac{P}{L^2}\right)^{\frac{1}{2}}$
Solid square section	$\sigma_{max} = \left(\frac{\pi^2}{12}\right)^{\frac{1}{2}} E^{\frac{1}{2}} \left(\frac{P}{L^2}\right)^{\frac{1}{2}} = 0.907 E^{\frac{1}{2}} \left(\frac{P}{L^2}\right)^{\frac{1}{2}}$

where buckling coefficient $K = 3.62$ for Poisson's ratio $v = 0.3$ (plate buckling is discussed further in Chap. 7). The different form of the local buckling formula above for a square tube results in different powers in the efficiency formula. This means that the efficiency coefficient of a circular tube cannot be compared directly with that of the square tube, or other flat-sided sections, instead of which the maximum stress has to be calculated from the appropriate efficiency formula in each case, at the required structural index, and the stresses compared.

Open sections, such as angle and I-sections, have significantly lower efficiency, as seen in Table 2.2 (with values taken from [2]. The I section in the table when optimized buckles simultaneously in all four modes, that is, flexural buckling about each axis and local buckling of the web and of the flanges. The angle section after optimization buckles simultaneously in local buckling and in flexural buckling about the axis shown. With only two variables, it is not possible for buckling to take place simultaneously in all three modes (local buckling and flexural buckling about both axes). The optimized X-section buckles simultaneously in flexural buckling about any axis and in torsional—local buckling. For many other open sections, such as a channel section, the shear centre does not coincide with the centre of gravity of the section, and these are subject to coupled flexural—torsional buckling with further reduction in efficiency.

In the foregoing text, by 'efficiency' is implied the *maximum* efficiency of a particular shape of section. Of course, for a tube or bar with given limits on dimensions (for example, if some minimum thickness is imposed), the maximum stress that can be reached will be less. This can be expressed as reduced or 'achieved' efficiency. For example, with Eq. (2.5) for a circular tube, if σ is the reduced maximum stress the achieved efficiency becomes:

$$\eta = \frac{\sigma}{E^{2/3}(P/L^2)^{1/3}}. \tag{2.6}$$

Finally, it is perhaps interesting to observe that, for a tube of any shape, if *all* its cross-sectional dimensions as well as its length are increased by, say, a factor of two, then both flexural and local buckling stresses are unchanged. An already optimized tube, with simultaneous buckling modes, therefore remains an optimum.

Table 2.2 Maximum efficiency η for some thin-walled open sections (values for angle and I sections taken from Rees [2])

$\sigma_{max} = \eta\, E^{\frac{2}{3}} \left(\frac{P}{L^2}\right)^{\frac{1}{3}}$		
I	$\eta = 0.705$	Buckles simultaneously about both axes
$\diagdown\diagup$	$\eta = 0.439$	Equal flange width and thickness, buckles about axis shown
$\times$	$\eta = 0.205$	Equal flange width and thickness, buckles simultaneously in flexural buckling and torsional/local buckling

The same applies to a solid bar, but now with only flexural buckling. Since in both cases, the cross-sectional area is increased by a factor of four, the maximum load is also increased by a factor of four (=2 squared), whereas the volume and therefore the weight is increased by a factor of 8 (=2 cubed). This is a demonstration of the 'square-cube law' in structural design.

Example 2.1 Find the optimum diameter and thickness of a circular tube, with simply supported length $L = 1000$ mm, to carry a compressive load $P = 10,000$ N. Take the elastic modulus $E = 72,000$ N/mm^2 for an aluminium alloy material, and local buckling coefficient $K = 0.605$.

From Eq. (2.3), the optimum thickness is

$$t^* = \left(\frac{P}{2\pi KE} \right)^{1/2} = 0.191 \text{ mm},$$

and from Eq. (2.4) the optimum (mean) radius is

$$R^* = \left(\frac{2K}{\pi^5} \cdot \frac{PL^4}{E} \right)^{1/6} = 28.6 \text{ mm}.$$

The corresponding outer diameter is 57.4 mm.

The thickness found above may in practice be considered too small for a tube of this diameter. Suppose we choose now a minimum thickness $t = 1.0$ mm. From Eq. (2.1), the required Euler buckling load is

$$P_E = \frac{\pi^3 E}{L^2} R^3 t = 10,000 \text{ N}.$$

Solving for R gives $R = 16.48$ mm, with corresponding outer diameter 34.0 mm.

With reduced radius and increased thickness, the local buckling condition is clearly more than satisfied.

The cross-sectional area of the tube is now

$$A = 2\pi Rt = 103.6 \text{ mm}^2,$$

and the compressive stress

$$\sigma = \frac{P}{A} = 96.6 \text{ N/mm}^2.$$

Fig. 2.3 Five-bar truss

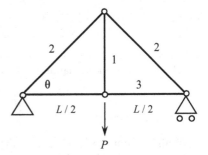

With the formula for 'achieved efficiency' in Eq. (2.6), we find:

$$\eta = \frac{\sigma}{E^{2/3}(P/L^2)^{1/3}} = 0.259.$$

We see that the effect of the chosen minimum thickness is to reduce substantially the efficiency of the tube, from its maximum value $\eta = 0.780$ to its present value $\eta = 0.259$. Note also that, while the compressive stress is comparatively low in this example, no account has yet been taken of a material stress limitation.   ■

Example 2.2 Find the optimum angle θ of the truss in Fig. 2.3, with members composed of circular tubes, taking into account the maximum compressive stress due to buckling of members in compression. Take load $P = 1000$ N, span $L = 1000$ mm, modulus $E = 72,000$ N/mm² and the allowable tensile stress $\sigma_t = 400$ N/mm².

An efficiency formula enables the maximum stress that can be achieved in a compression member to be calculated directly, without actually performing the design. The two sloping members (both numbered 2 in the figure) are in compression. The compressive force in each of these is

$$F_2 = \frac{P}{2 \sin \theta}$$

and their length is

$$l_2 = \frac{L}{2 \cos \theta},$$

giving a structural index

$$\frac{F_2}{l_2^2} = \frac{2P}{L^2} \cdot \frac{\cos^2 \theta}{\sin \theta}.$$

Since both members are circular tubes, with efficiency $\eta = 0.780$, the maximum stress in these members is found by substituting the above formula for the structural index into the efficiency formula, Eq. (2.5), to give:

$$\sigma_{max} = 0.780E^{2/3}\left(\frac{2P}{L^2}\cdot\frac{\cos^2\theta}{\sin\theta}\right)^{1/3}.$$

The forces in the remaining tension members are

$$F_1 = P, \qquad F_3 = \frac{P}{2\tan\theta}.$$

With maximum stress σ_{max} in the compression members and stress σ_t in the tension members, the volume of the truss becomes:

$$V = \left(\frac{1}{\sigma_t}\cdot\tan\theta + \frac{1}{\sigma_{max}}\cdot\frac{1}{\sin\theta\cos\theta} + \frac{1}{\sigma_t}\cdot\frac{1}{\tan\theta}\right)\frac{PL}{2}.$$

Varying θ to minimize the volume V gives an optimum angle of the truss in this example

$$\theta^* = 35.9°$$

and maximum compressive stress

$$\sigma_{max} = 176\,\text{N/mm}^2.$$

The optimum angle θ^* is less than the 45° found earlier ($H/L = 0.5$ in Fig. 1.11) when all bars were assumed to have the same allowable stress. The explanation for this is that reducing the angle θ reduces the length of the compression members but increases the load in them. The optimum angle θ^* is therefore a compromise between the two, with reduction in length having the greater effect. Note that both optimality criteria have now been used in this example. All members are fully stressed in the sense that their area is based either on the maximum tensile stress σ_t of the material in the tension members or on the maximum compressive stress σ_{max} in the compression members, the latter based on simultaneous flexural and local buckling of the tubular members. Even in this small problem, this illustrates how optimality criteria can be used to reduce the size of an optimization problem. All constraints—tensile strength and both buckling modes—have been eliminated, or better said directly satisfied, and only one design variable, angle θ, remains. ∎

2.1.2 Material Limitation

Up to now, no reference has been made to the allowable compressive stress of the material of a compression member, which may limit the stress predicted by the efficiency formula in Eq. (2.5). This is illustrated in Fig. 2.4, where the maximum compressive stress for four different shapes of cross section is plotted against

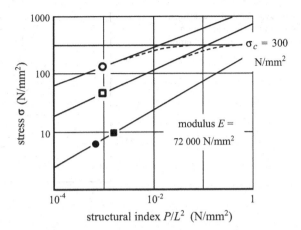

Fig. 2.4 Maximum compressive stress for *circular* and *square* section *tubes* and *bars* with material stress limitation σ_c

structural index, together with an allowable compressive stress $\sigma_c = 300$ N/mm^2 for the material. Note the logarithmic scale which, by the nature of the efficiency formula, gives straight lines in this plot but also tends to obscure the difference in performance of the different cross sections (lines for the solid circular and square sections are indistinguishable in the figure). The superiority of tubular sections over solid ones is nevertheless clear in this figure. In reality, there is a loss of modulus as the maximum allowable stress of the material is approached, due to progressive yielding, resulting in blending of the lines as suggested by the dotted lines in the figure.

A material limitation

$$P_M = \sigma_c A = \sigma_c \cdot 2\pi R t$$

has been added in Fig. 2.5 to the previous design space in Fig. 2.2. This new constraint clearly coincides with a line of constant cross-sectional area $A = P/\sigma_c$, meaning that all combinations of R and t between points 1 and 2 represent an optimum design. There is therefore no unique optimum, and unless point 1 or point 2 is chosen, neither buckling mode is critical. However, if a material with a larger allowable stress had been chosen, the curve representing the material limitation on the design space would be lowered and the problem would revert to the original one of simultaneous buckling modes. If, say, a minimum thickness limitation $t_{\min}$ or a maximum radius $R_{\max}$ is imposed (for manufacturing or other practical reasons), then these can also be added to the design space. It is clear that many different versions of the same diagram can exist, producing different design conditions in the now more highly constrained optimum.

The situation in Fig. 2.5 is representative of the majority of optimization problems, albeit here a very simple one. With only two design variables R and t, it is clearly not possible to satisfy all constraints simultaneously, whether they arise from buckling or material stress limitations, dimensional restrictions or perhaps other conditions that might be imposed. The principal task in optimization is

Fig. 2.5 Design space for a
circular section *tube* with
compressive stress limitation

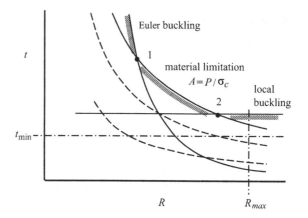

generally constraint *selection*, in other words to determine which constraints will prove to be active at the optimum and which will be more than satisfied. Of course, there can be no single answer to this—in the case of the circular tube considered here, this depends entirely on the actual values of the parameters P, L and σ_c, and on other limitations such as t_{min} and R_{max}.

2.2 Criterion for Maximum Stiffness

The two optimality criteria already discussed are concerned only with the *strength* of a structure, whether this be determined by material stress limits or by buckling. Other than up to now, the design of a structure for maximum stiffness cannot be achieved in a member-by-member, iterative resizing process as in a fully stressed design, since its stiffness depends on the properties of all parts of the structure. To establish a criterion for maximum stiffness, we again consider a simple truss structure, such as in Fig. 2.6, loaded by a single force P. If the deflection at the point of loading and in the same direction as the applied load is δ, then the stiffness of the structure is P/δ. It is this stiffness that we wish to maximize. By simply

Fig. 2.6 Deflection under a
single applied load

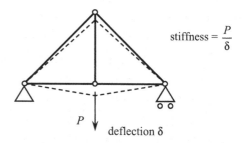

increasing the amount of material in the structure, its stiffness can be increased without limit, therefore we have to maximize the stiffness for some given total volume V of the structure.

The deflection δ can conveniently be found by the principle of conservation of energy—the work done by the applied load P is equal to the total elastic strain energy stored in the members of the truss:

$$\frac{1}{2}P\delta = \sum \frac{\sigma_i^2}{2E}V_i, \tag{2.7}$$

where σ_i is the stress in each member, V_i is the corresponding volume of each member and E is the elastic modulus of the material, for the present assumed to be the same for all members. Substituting for the stress

$$\sigma_i = \frac{F_i}{A_i} = \frac{F_i l_i}{V_i} \tag{2.8}$$

we obtain

$$\delta = \frac{1}{PE}\sum \frac{(F_i l_i)^2}{V_i}, \tag{2.9}$$

where F_i is the force in a member and A_i, l_i are its cross-sectional area and length, respectively. Note that forces F_i are constant if, as assumed here, the truss is statically determinate.

Putting $V = \sum V_i$ for a given volume of material, we can substitute

$$V_1 = V - V_2 - V_3 - \cdots$$

into Eq. (2.9) to give

$$\delta = \frac{1}{PE}\left[\frac{(F_1 l_1)^2}{(V - V_2 - V_3 - \cdots)} + \frac{(F_2 l_2)^2}{V_2} + \frac{(F_3 l_3)^2}{V_3} + \cdots\right]. \tag{2.10}$$

(Choice of V_1 for elimination in the formula above is entirely arbitrary.) Differentiating with respect to the remaining variables V_i for minimum δ:

$$\frac{\partial(\delta)}{\partial V_2} = \frac{1}{PE}\left[\frac{(F_1 l_1)^2}{(V - V_2 - V_3 - \cdots)^2} - \frac{(F_2 l_2)^2}{V_2^2}\right] = 0,$$

$$\frac{\partial(\delta)}{\partial V_3} = \frac{1}{PE}\left[\frac{(F_1 l_1)^2}{(V - V_2 - V_3 - \cdots)^2} - \frac{(F_3 l_3)^2}{V_3^2}\right] = 0,$$

and so on, from which

$$\frac{(F_1 l_1)^2}{V_1^2} = \frac{(F_2 l_2)^2}{V_2^2} = \frac{(F_3 l_3)^2}{V_3^2} = \cdots.$$

Note that elimination of V_1 above, then substituting back, is simply a device to ensure that differentiation takes account of a required total volume V. A more elegant way would be by use of Lagrange multipliers, to be introduced in the next chapter. Referring back to Eq. (2.8), the above condition becomes simply

$$\sigma_1^2 = \sigma_2^2 = \sigma_3^2 = \cdots = \sigma_0^2,$$

where σ_0 denotes here an arbitrarily chosen stress level. Uniform stress in all members is now the criterion for maximum stiffness, implying in principle a statically determinate structure, as was assumed at the start.

With the same stress in all members, Eq. (2.9) can be simplified to

$$\delta = \frac{V}{PE} \cdot \sigma_0^2. \tag{2.11}$$

For any fully stressed design of truss, with a maximum stress σ_0, its total volume V is given by Eq. (1.6):

$$V = n \cdot \frac{PL}{\sigma_0},$$

where as before, coefficient n depends only on the layout of the truss. For given P, V and span L, we can rewrite this as:

$$\sigma_0 = nPL/V,$$

and substituting for σ_0 into Eq. (2.11) gives

$$\delta = \frac{n^2 L^2}{VE} \cdot P.$$

For minimum deflection or, in other words, maximum stiffness P/δ for given volume V of material, it is clear that we require the smallest n value. We can conclude, therefore, that the optimum layout of truss (maximum strength-to-weight ratio) also has maximum stiffness, with regard to deflection at the point of loading, provided of course that the members are indeed uniformly stressed.

It should be noted that the criterion for maximum stiffness developed above applies strictly to a truss under a single applied load and to deflection at the point of loading in the direction of the applied load. If there is more than one load, such as an additional load applied at the top of the truss as in Fig. 2.7, the left-hand side of Eq. (2.7) has to be replaced by $\frac{1}{2}(P_1\delta_1 + P_2\delta_2)$, or in general by $\frac{1}{2}\sum P_i\delta_i$. The rest

Fig. 2.7 Deflection under
two applied loads

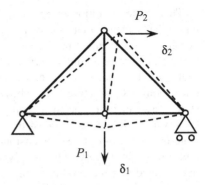

of the analysis is then unchanged, which implies that the condition of uniform stress in all members leads to a minimum of $\sum P_i \delta_i$. This is the minimum of the sum of the individual deflections weighted in proportion to the magnitude of the applied loads, but does not in practice provide any useful definition of maximum stiffness.

If the members of the truss are of different materials, each with elastic modulus E_i, Eq. (2.7) for a single load P becomes

$$\frac{1}{2}P\delta = \sum \frac{\sigma_i^2}{2E_i} V_i. \tag{2.12}$$

Following the same procedure as before, but by keeping the individual elastic moduli within each term of Eq. (2.10), we replace the condition of uniform stress by one of the uniform strain energy densities (strain energy per unit volume)

$$\frac{\sigma_1^2}{E_1} = \frac{\sigma_2^2}{E_2} = \frac{\sigma_3^2}{E_3} = \cdots.$$

If the members also have different densities ρ_i, and it is required to maximize the stiffness of the structure for a given mass rather than for a given volume, then Eq. (2.12) becomes

$$\frac{1}{2}P\delta = \sum \frac{\sigma_i^2}{2E_i\rho_i} M_i,$$

where M_i is the mass of an individual member. Following again the same procedure, but differentiating now with respect to mass rather than volume, the criterion for maximum stiffness becomes

$$\frac{\sigma_1^2}{\rho_1 E_1} = \frac{\sigma_2^2}{\rho_2 E_2} = \frac{\sigma_3^2}{\rho_3 E_3} = \cdots.$$

This is the condition of uniform specific strain energy (strain energy per unit mass). This means that, when made of different materials (different modulus or density), a uniformly stressed design is now no longer the optimum design for stiffness. Our third optimality criterion, for maximum stiffness for a given total mass of the structure, has finally become one of the uniform strain energies per unit mass throughout the structure. The above criterion, while developed here for truss structures, can be expected to apply more generally to other types of structure. However, it should be borne in mind that, even for a structure made of a single material, in a two- or three-dimensional state of stress, the condition must be that of uniform strain energy, not simply uniform stress. Furthermore, it has to be remembered that the condition applies strictly to stiffness measured at the point of loading under a single applied load. In practice, of course, material strength, buckling and other constraints on the design may override this criterion for maximum stiffness to a greater or lesser extent.

2.3 Spreadsheet Programs

The spreadsheets illustrate the inclusion of behavioural constraints such as buckling and material stress limits into the optimization process. In Sect. 2.3.1, we take the now familiar problem of a thin tube loaded in compression, for which we already have a theoretical efficiency formula based on simultaneous buckling modes. By including maximum stress and dimensional limits as well as buckling constraints, the spreadsheets offer a more general solution. In Sect. 2.3.2, the spreadsheet for a seven-bar truss in the previous chapter is extended by taking into account the buckling of the compression members.

2.3.1 'Circular and Square Tubes'

The spreadsheets use Solver to optimize the cross-sectional dimensions of a circular or square tube loaded in axial compression, with buckling and allowable stress constraints, also allowing practical limits to be set on the dimensions (outer diameter or side of the square and thickness). The two tubes are on separate sheets of the workbook, as shown in Figs. 2.8 and 2.9.

The applied compressive load P, effective simply supported length L (depending on the required end conditions), elastic modulus E and maximum allowable compressive stress σ_c of the material have to be inserted in the appropriate cells. Maximum and minimum values of the outer diameter d or side b and thickness t have to be specified (these cells may not be left blank). Constraints to be satisfied are Euler buckling and local buckling, using the formulae in Sect. 2.1, and the maximum allowable stress of the material. The local buckling coefficient for a circular tube is set to $K = 0.605$ (maximum theoretical value) on the spreadsheet,

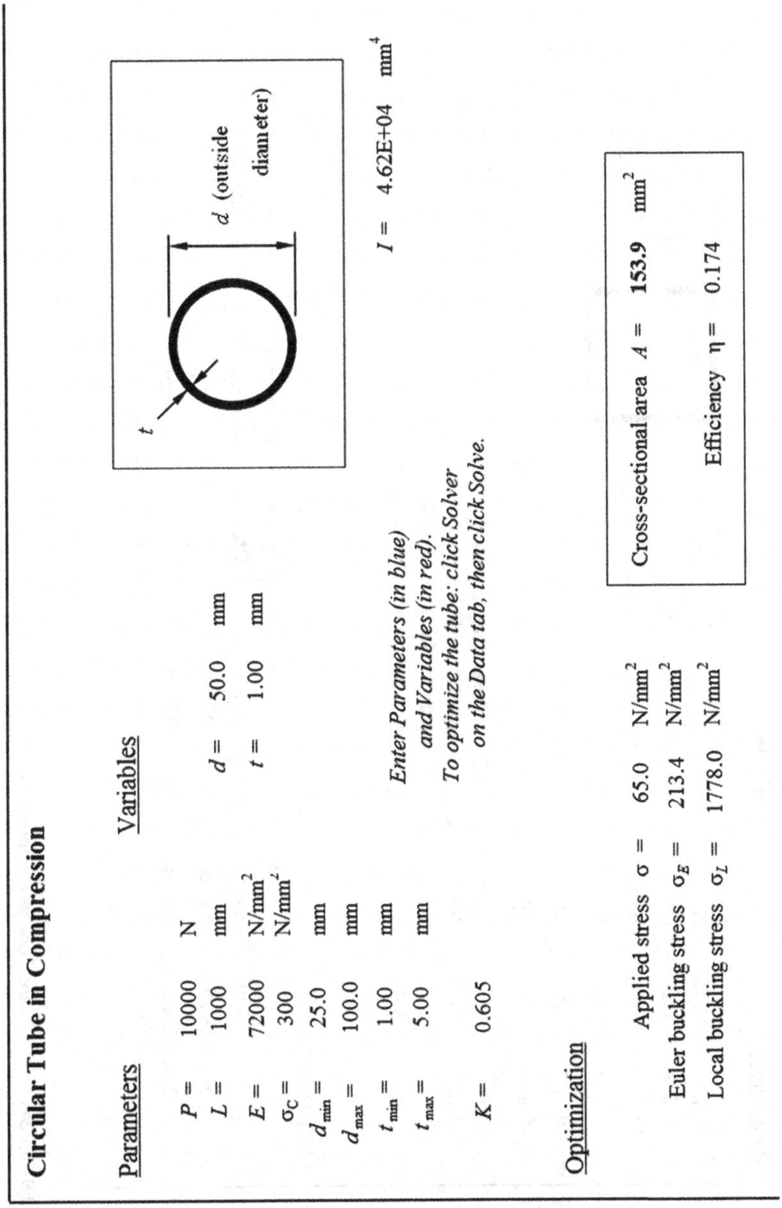
Circular Tube in Compression
Parameters
P = 10000 N
L = 1000 mm
E = 72000 N/mm²
σ_C = 300 N/mm²
d_min = 25.0 mm
d_max = 100.0 mm
t_min = 1.00 mm
t_max = 5.00 mm
K = 0.605
Variables
d = 50.0 mm
t = 1.00 mm
t
d (outside diameter)
Enter Parameters (in blue)
and Variables (in red).
To optimize the tube: click Solver
on the Data tab, then click Solve.
I = 4.62E+04 mm⁴
Optimization
Applied stress σ = 65.0 N/mm²
Euler buckling stress σ_E = 213.4 N/mm²
Local buckling stress σ_L = 1778.0 N/mm²
Cross-sectional area A = 153.9 mm²
Efficiency η = 0.174

Fig. 2.8 Spreadsheet 'Circular Tube in Compression'

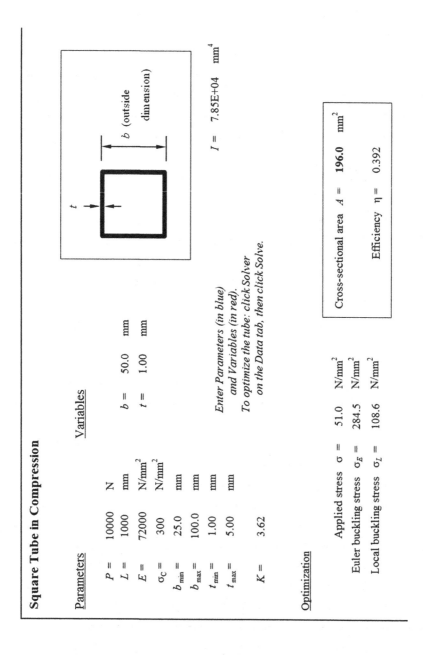

Fig. 2.9 Spreadsheet 'Square Tube in Compression'

but may be reduced to allow for imperfections if required. For a square tube, the local buckling coefficient is $K = 3.62$. The material is assumed to be perfectly elastic up to the maximum allowable stress. These constraints, together with limits on cross-sectional dimensions, are already set up in the Solver dialog box on the relevant sheet. Suitable initial values of the outer dimensions d or b and thickness t have to be entered in the appropriate cells. With these as variables, and subject to the above constraints, the cross-sectional area of the section is minimized by the GRG nonlinear method. To represent a solid circular or square bar, an extra constraint $t = d/2$ or $t = b/2$, respectively, may be added to the list of constraints in the Solver dialog box (in which case the local buckling stress has to be ignored). Parameters and design variables to be entered in the spreadsheets are listed in Table 2.3.

After optimization, the optimum cross-sectional dimensions replace the initial values of d or b and t, and corresponding values of stress σ, Euler buckling stress σ_E, local buckling stress σ_L and the achieved efficiency η (see Sect. 2.1.1) are calculated. Note that the efficiency is not calculated if either buckling stress exceeds the allowable compressive stress of the material, or if t/d or t/b exceeds 0.1. Comparison of the achieved efficiency with the known maximum efficiency for the circular or square tube shows the loss of efficiency due to the practical limits set on dimensions. The spreadsheets can be used to plot figures similar to Fig. 2.4 for any range of materials and structural index. If limits on dimensions do not intervene, and at a compressive load below that at which the material stress limit is reached, values of efficiency are obtained in agreement with those in Table 2.1.

An extended version of the present spreadsheet for a circular tube, with eccentrically applied compressive load and the effect of yielding of the material before the critical buckling load is reached, is presented in the next chapter.

Table 2.3 Data entry for spreadsheet programs 'Circular and Square Tubes in Compression'

Parameters	
Compressive load P	Enter the value in cell C6 as a *positive* number
Effective simply supported length L	Enter the value in cell C7
Elastic modulus E, allowable compressive stress σ_c	Enter values in cells C8:C9 as *positive* numbers
Min. and max. *outer* diameter d (or min. and max. *outer* width b)	Enter values in cells C10:C11 (cells may not be left blank)
Min. and max. thickness t	Enter values in cells C12:C13 (cells may not be left blank)
Local buckling coefficient K	Enter a reduced value in cell C15 for a circular tube if required
Variables	
Diameter d (or side b) and thickness t	Enter initial values in cells F7:F8

2.3.2   'Truss with Tubular Members'

The spreadsheet illustrates the use of Solver to optimize the seven-bar truss in Sect. 1.4, now with compression members made of circular tubes subject to both buckling and material stress limitations. Whereas in Example 2.2 an efficiency formula is used to predict the maximum stress in the compression members of a truss, in the spreadsheet Euler and local buckling and the maximum allowable stress are treated as separate constraints in each member. This allows upper and lower limits to be specified for the dimensions of any of the members, if so required. The spreadsheet is shown in Fig. 2.10.

The applied load P, span L, material allowable stresses σ_t and σ_c in tension and compression, elastic modulus E, specific weight ρ_w and local buckling coefficient K have to be entered in the spreadsheet. Constraints are the allowable stresses of the material, and Euler and local buckling of the compression members using the formulae in Sect. 2.1. These constraints are already set up in the Solver dialog box. Design variables are the dimensions D and H of the truss, the diameter d and thickness t of the compression members (1 and 3 in the diagram) and the cross-sectional area of the tension members (2 and 4). Ratios D/L and H/L are limited to not less than 0.01. With suitable initial values of the design variables, the volume of the truss is minimized by the GRG nonlinear method, subject to the above constraints. Parameters and design variables to be entered in the spreadsheet are listed in Table 2.4.

After optimization, initial values of the design variables are replaced by their optimized ones, together with the volume of the truss, its strength-to-weight ratio P/W, the stress σ in each member, the maximum forces in the bars in the different failure modes and the deflection δ of the truss at the point of loading. The spreadsheet shows the influence of buckling of the compression members on the optimum layout and strength-to-weight ratio of the truss. Since the truss is statically determinate, unless dimensional constraints imposed the tension members will reach the allowable tensile stress. Depending on the magnitude of the applied load, and again if no dimensional constraints are imposed, the compression members will reach either the allowable compressive stress or a reduced stress due to buckling. In the latter case, for maximum efficiency, buckling will occur simultaneously in Euler and local buckling. As in the spreadsheets in Sect. 2.3.1, no account is taken of reduction in modulus with yielding, the allowable compressive stress being treated as a simple cut-off for the buckling stress. It will be observed that the thickness of the tubes in compression frequently comes out impractically small. Only limits on the minimum thickness of the compression members are specified in the spread-sheet. Additional constraints for maximum or minimum dimensions can readily be added in the Solver dialog box, if required. The tubular members can be replaced by solid circular bars by adding constraints to specify the thickness equal to one-half of the diameter of the tubes.

Finally, a comment on the initial values of the design variables is worthwhile at this stage. As in all optimization problems, these should be chosen as far as possible

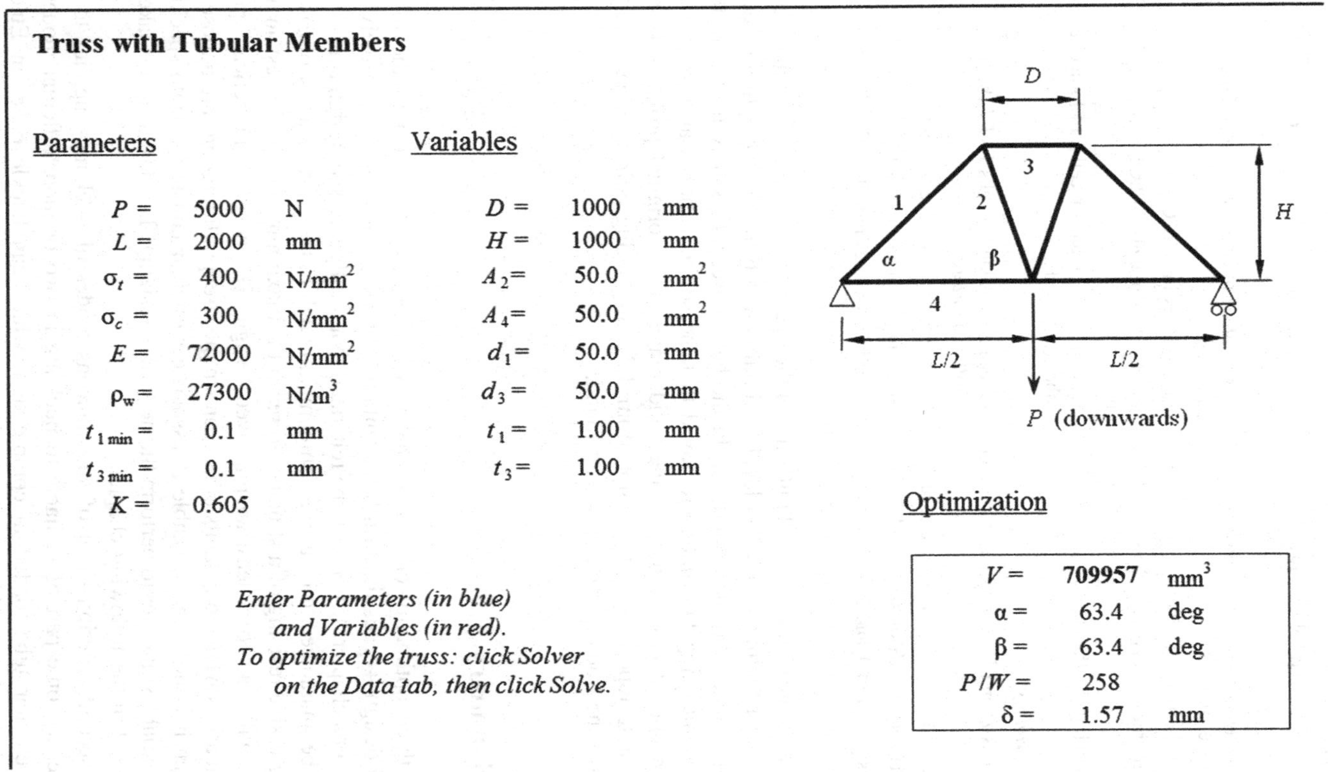
Truss with Tubular Members
Parameters
P = 5000 N
L = 2000 mm
σ_t = 400 N/mm²
σ_c = 300 N/mm²
E = 72000 N/mm²
ρ_w = 27300 N/m³
t_1 min = 0.1 mm
t_3 min = 0.1 mm
K = 0.605
Variables
D = 1000 mm
H = 1000 mm
A_2 = 50.0 mm²
A_4 = 50.0 mm²
d_1 = 50.0 mm
d_3 = 50.0 mm
t_1 = 1.00 mm
t_3 = 1.00 mm
Enter Parameters (in blue)
and Variables (in red).
To optimize the truss: click Solver
on the Data tab, then click Solve.
D
H
1
2
3
4
α
β
L/2
L/2
P (downwards)
Optimization
V = 709957 mm³
α = 63.4 deg
β = 63.4 deg
P/W = 258
δ = 1.57 mm

Fig. 2.10 Spreadsheet 'Truss with Tubular Members'

Table 2.4 Data entry for spreadsheet 'Truss with Tubular Members'

Parameters	
Applied load P (vertical downwards only)	Enter a positive value in cell C6
Span of truss L	Enter the value in cell C7
Allowable tensile stress σ_t, allowable compressive stress σ_c	Enter values in cells C8:C9 as *positive* numbers
Elastic modulus E, specific weight ρ_w	Enter values in cells C10:C11
Minimum thickness t_1, t_3 of bars 1 and 3	Enter values in cells C12:C13 (may not be left blank)
Local buckling coefficient K	Enter a reduced value in cell C14 for a circular tube if required
Design variables	
Dimensions D and H of truss (see figure on spreadsheet)	Enter initial values in cells F6:F7 (both positive, nonzero)
Cross-sectional area A_2, A_4 of bars 2 and 4	Enter initial values in cells F8:F9
Outer diameter d_1, d_3 of bars 1 and 3	Enter initial values in cells F10:F11
Thickness t_1, t_3 of 1 and 3	Enter initial values in cells F12:F13

within the range of the expected outcome of the optimization. Large differences can lead to Solver finding some other local minimum away from the true optimum, or it may fail to reach any solution at all. In the latter case, this is may be due to the starting point being far enough removed that the solution moves away from the optimum until it is no longer in the valid range of the original problem. Good practice is to repeat the optimization from different starting points, to gain confidence that the true optimum has been found.

2.4 Summary

Of the three optimality criteria we have now seen, the first—that of the fully stressed design—needs little further attention here, as it was treated extensively in the previous chapter. It will only be reiterated that a fully stressed design cannot in general be guaranteed to be a true optimum, although in the great majority of cases the progressive resizing of a structure to reach the maximum allowable stress in all its parts will lead to at least an improved design. The need for numerical optimization arises when constraints other than simple strength limitations are imposed, and when it is no longer possible to associate particular constraints with specific design variables in order to perform the necessary resizing. This last aspect is taken up in detail in the following chapter.

The second criterion—that of simultaneous modes of buckling—applies to a structure, or some part of it, liable to buckling in two or more different modes. A thin circular tube loaded in compression, which can buckle either in Euler buckling or in local buckling, is used to demonstrate that the optimum corresponds to buckling simultaneously in both of these modes. With this condition, efficiency formulae are derived by means of which the maximum stress that can be achieved is readily predicted. Again, it cannot be guaranteed that the condition of simultaneous buckling modes will be valid in all cases. Numerical optimization becomes necessary when a structure is subject to buckling constraints as well as material strength limitations, and to dimensional and other constraints.

The third criterion—that of uniform (specific) strain energy density for maximum stiffness—was demonstrated for a truss structure, but under highly restricted conditions. These are that the stiffness relates to deflection under a single load, measured at the point of application of that load. For a truss structure made of a single material, uniform strain energy density amounts to uniform stress throughout the structure, in other words a fully stressed design. When the conditions above are not met, for example if the stiffness is measured at some point other than the point of loading, or if buckling and other constraints intervene, numerical optimization is again necessary.

The three optimality criteria introduced in this and the previous chapter, deduced for simple truss structures, might be assumed to apply in principle to any type of structure. In fact, they are most likely what we would intuitively have assumed in the first place. However, while the use of optimality criteria may be quite appealing, it is seen that they are limited in application and their validity cannot be guaranteed. Numerical optimization methods avoid the limitations of optimality criteria and, above all, provide a consistent, general approach to the design of a structure. One of the main reasons for studying optimality criteria in some detail here is that, not surprisingly, many of the same characteristics reappear in designs obtained by numerical optimization. A good understanding of optimality criteria is therefore of great assistance in assessing the results of a numerical optimization and in understanding what has led to the solution obtained. The application of numerical optimization methods, together with the underlying numerical procedures, will be the principal task of the further chapters of this book.

Exercises

2.1   Verify the efficiency formula for a square section tube in Table 2.1. For a typical aluminium alloy, with an allowable compressive stress of 300 N/mm^2 and elastic modulus 72,000 N/mm^2, at what value of structural index is the maximum stress limited by the allowable stress of the material?
      To derive the efficiency formula, assume that the thickness of the tube is small compared with its width, i.e. use simplified formulae for A and I similar to those for the circular tube in Sect. 2.1.

2.2   Derive the efficiency of a thin tube of hexagonal cross section (6 equal sides) and simply supported length L under a compressive load P. To compare the hexagonal tube with circular and square tubes, plot the maximum stress of the three sections for a chosen material over a realistic range of structural index.

Verify that the second moment of area of a thin hexagonal section about any axis through its middle point is $5b^3t/2$, where t is the thickness of the tube and b is the mean width of each side. For the local buckling stress, use the formula: $\sigma_L = 3.62E\ (t/b)^2$.

2.3 Verify the relation $W \propto (P/L^2)^{-1/3}PL$ in Sect. 2.1.1 for the minimum weight of a circular tube in compression. Derive a similar relation for a square tube.

Use the efficiency formulae in Table 2.1.

2.4 Draw the design space for the three-bar truss made of different materials in Sect. 1.1.1.

Draw the design space with variables A_1 and A_2. Use the formulae in Eqs. (1.1) and (1.2) for the stress in the bars, with allowable stresses as in Table 1.2 and other data in Fig. 1.7. Plot the maximum stress constraints for both the single material and for the two different materials, Notice that in the second case, the stress in the outer bars will be the critical design condition unless these two bars are removed altogether.

2.5 Draw the design space for the three-bar truss under alternative loads in Sect. 1.1.2. Verify the minimum volume of the truss given in that section.

Draw the design space with variables A_1 and A_2. Use the formulae in Eqs. (1.3)–(1.5) for the stress in the bars. 'Goal Seek' can be used in Excel to solve for A_2 for a series of values of A_1 in Eq. (1.3). Take $P = 100$ kN, $\sigma_0 = 300$ N/mm^2 and $L = 1000$ mm, as in Sect. 1.1.2.

2.6 Use the spreadsheet 'Circular Tube in Compression' with different values of minimum thickness to explore the effect of this on the achieved efficiency of the tube.

Use the values of P, L and E already on the spreadsheet (or any other convenient values). Set $d_{min} = 0$, and d_{max} and σ_c large enough to ensure they have no effect. Make a plot of η and d over a range of minimum thickness.

2.7 Use the spreadsheet 'Circular Tube in Compression' to make a plot of maximum stress against structural index for circular tubes made of various different materials.

The allowable compressive stress and elastic modulus of different grades of aluminium alloy, steel, titanium and other materials are widely available. Choose P and L for values of structural index that give realistic stress levels for the materials.

2.8 Modify the spreadsheet 'Square Tube in Compression' for the optimization of a hexagonal tube in compression.

Use the formulae in Exercise 2.2. Compare the efficiency with the previously calculated value.

2.9 Use the spreadsheet 'Truss with Tubular Members' to find the optimum layout of the seven-bar truss under buckling constraints.

Use the material and other data already on the spreadsheet. Try different starting points for the optimization to verify that the same result is obtained. Note the deflection of the truss and the stress in each member of the optimized truss.

2.10 Verify the deflection δ given in the spreadsheet for the optimized truss in the previous exercise.

Deflection δ can be calculated from Eqs. (2.7) to (2.9).

2.11 Use the spreadsheet 'Truss with Tubular Members' to verify that the optimum design for stiffness, in the absence of constraints, has equal stress in all members and has the same shape as the optimum design for minimum weight.

Use the material and other data already on the spreadsheet (or any other convenient values). First remove all constraints currently in the Solver dialog box. Add a new constraint to set the volume V of the truss to any reasonable value. Deflection δ at the point of loading is given on the spreadsheet. Change the objective to deflection δ in the dialog box, and use Solver to optimize the truss for minimum δ.

2.12 Modify the spreadsheet 'Truss with Tubular Members' for a load applied vertically upwards at mid-span.

Members 2 and 4 are now in compression, members 1 and 3 in tension. Compare the strength-to-weight ratio P/W with that for a downwards applied load.

References

1. Megson THG (1999) Aircraft structures for engineering students. Arnold, London
2. Rees DWA (2009) Mechanics of optimal structural design: minimum weight structures. Wiley, New Jersey
3. Timoshenko SP Gere JM (1961) Theory of elastic stability. McGraw-Hill, New York (reprinted by Dover Publications, 2009)
4. Young WC, Budynas RG (2001) Roark's formulas for stress and strain. McGraw-Hill International, New York

Chapter 3
The General Optimization Problem

Abstract The problems studied in the previous two chapters have all been special cases of a more general optimization problem. In most practical problems, there is no close relationship between individual design variables and constraints, as was the case in the fully stressed design of a simple truss structure. A box beam is used to illustrate the complex relations between design variables and constraints in a more representative optimization problem. Since we generally cannot know in advance which constraints will prove to be active at the optimum, the task of numerical optimization is both to select the active constraints and to locate the optimum on those constraints. The general form of the optimization problem is defined, and a distinction drawn between an intersection optimum and a mathematical optimum. The classical method of Lagrange multipliers might be considered the mathematical basis of optimization, while applying only to equality constrained problems. For inequality constrained problems, Lagrange multipliers are used to identify those constraints that have been correctly selected as active at the optimum. The Kuhn–Tucker conditions are the necessary conditions for an inequality constrained optimum. A spreadsheet program is presented for the optimization of an eccentrically loaded column, taking into account the effect of yielding of the material at higher stresses and extending the scope of the spreadsheet in the previous chapter.

In previous chapters, we studied some particular examples of truss structures, both to explore the conventional, iterative resizing procedure and to test the validity of optimality criteria that can be applied to such structures. These examples are all special cases of a more general optimization problem. In practice, the relations between design variables and constraints are likely to be much more complex than in the simple structures up to now, and because it is mostly not known beforehand which constraints are going to be active in an optimized design, a different approach becomes necessary. A box beam (representative of the torsion box of an aircraft wing, bridge deck, or similar structure) is used to illustrate the general optimization problem. The standard notation of optimization is introduced in terms of the design variables and constraints of the box beam. This is treated for now in a purely

© Springer International Publishing AG 2017

A. Rothwell, *Optimization Methods in Structural Design*, Solid Mechanics and Its Applications 242, DOI 10.1007/978-3-319-55197-5_3

qualitative way. In a later chapter, a numerical solution for a box beam subject to stress, buckling and stiffness constraints will be developed.

In the great majority of cases, we have to employ numerical rather than analytical methods for the solution of practical optimization problems. However, a detailed discussion of different numerical methods will be deferred until the next two chapters. In mathematical terms, for a problem with only *equality* constraints (conditions that have to be identically satisfied, rather than those which may be satisfied in a 'not greater than' or 'not less than' sense), the classical method of Lagrange multipliers enables a maximum or minimum to be found simply by differentiation. This might be regarded as the fundamental theory of optimization. While it will be clear that few practical problems are actually amenable to analytical differentiation, we shall see later that Lagrange multipliers play an essential role in identifying the critical constraints in the numerical optimization of inequality constrained problems. The Lagrange multiplier method is introduced later in this chapter.

3.1 Box Beam Structure

The rectangular box beam shown in Fig. 3.1 is assumed to carry a vertical load causing a bending moment about the horizontal axis. The upper and lower panels are thin plates with discrete stiffeners to resist buckling. Transverse bulkheads, or ribs, are placed at intervals along the length of the beam to provide support for these panels, which carry a major part of the bending moment on the beam. The side panels, or shear webs, carry the shear force and together with the upper and lower panels provide torsional stiffness. Typical design variables are the plate thickness at different locations around the cross-sectional, stiffener areas, and the spacing of stiffeners and ribs. In greater detail, additional variables would be associated with, for example, the individual dimensions of the stiffeners (although stiffeners may well be treated separately in a more detailed design of the stiffened panels). Typical constraints refer to the stress in each part of the structure, buckling of the panels, torsional stiffness and constraints such as minimum thickness.

Looking first at the constraints, we know that the bending stress at any point in the plate and stiffeners depends on the plate thickness and stiffener areas around the entire cross section. The buckling stress of the upper and lower panels depends critically on the local stiffener spacing and the rib spacing, as well as on the plate thickness and stiffener area. The torsional stiffness of the box beam depends on the

Fig. 3.1 Rectangular box beam

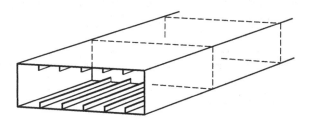

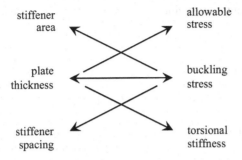

Fig. 3.2 Typical relations between design variables and constraints (buckling stress depends on all variables on *LHS*, and plate thickness enters into all constraints on *RHS*)

plate thickness not only in the given cross section but along its whole length. Looking now at the design variables, we see that plate thickness enters into all the constraints referred to above. Change in plate thickness to satisfy a material stress constraint results in changes in buckling stress and torsional stiffness as well. Change in stiffener area changes both the stress in the plate and stiffeners and the buckling stress. Change in stiffener spacing changes not only the buckling stress but also the number of stiffeners and therefore the stress in the plate and stiffeners. Figure 3.2 illustrates the complex relations between the design variables and constraints described above.

In general, there is no specific relationship between constraints and design variables to enable each constraint to be satisfied individually, as in an iterative resizing procedure, and as already pointed out, we generally do not know in advance which constraints will prove to be active in an optimized design. This is quite different to the design of a truss structure in Chap. 1, where the stress in each member depends simply on the cross-sectional area of that member or is treated in that way in an iterative resizing procedure for a statically indeterminate truss. In general, this lack of any special relationship between individual design variables and constraints, and in particular uncertainty over the critical constraints, means that an analytical approach to optimization is rarely possible. We have then to turn to numerical methods of optimization to search for an optimum design amongst a complex set of design variables and constraints. For this, we should define the optimization problem in its most general form:

Minimize (or maximize) an objective function:

$$f(\mathbf{x})$$

in terms of n design variables:

$$\mathbf{x} = x_1, x_2, \ldots x_n,$$

subject to m inequality constraints:

$$g_j(\mathbf{x}) \geq 0, \quad j = 1, 2, \ldots m$$

and p equality constraints:

$$h_k(\mathbf{x}) = 0, \quad k = 1, 2, \ldots p,$$

with side constraints:

$$x_{i_{\max}} \geq x_i \geq x_{i_{\min}}.$$

For the box beam, the objective function $f(\mathbf{x})$ is the weight W of the structure, expressed in terms of design variables $\mathbf{x}$. These are the plate thicknesses, stiffener areas and so on. Inequality constraints $g_j(\mathbf{x})$, again each a function of design variables $\mathbf{x}$, are those for which it is sufficient if they are satisfied, but not necessarily identically satisfied. These are, for example:

$$g_j(\mathbf{x}) = \sigma_t - \sigma \geq 0,$$

or

$$g_j(\mathbf{x}) = \sigma_b - \sigma \geq 0,$$

where σ_t is the allowable tensile stress of the material, σ_b is the buckling stress of a particular panel, and σ is the stress under the applied load at the appropriate point in the cross section. Note that inequality constraints are defined here as positive when feasible and zero when identically satisfied. Equality constraints $h_k(\mathbf{x})$ are less common in structural design but might, for example, refer to some required stiffness of a structure, rather than that it merely has sufficient stiffness. The so-called side constraints are simply limits imposed directly on the design variables, such as minimum plate thickness or maximum stiffener area.

3.1.1 General Form of Design Space

Realistic optimization problems are likely to have many design variables, requiring a multidimensional design space. While this presents no particular difficulty to the computer in a numerical optimization, clearly the design space cannot actually be drawn on paper. However, some of the main characteristics of a general, multidimensional design space can still be illustrated in a two-dimensional design space, as in Fig. 3.3. Some typical inequality constraints $g_j \geq 0$ are represented in the figure, lines $g_j = 0$ being the limits imposed by these constraints, with the shading indicating the infeasible side of a constraint and the feasible region above and to the

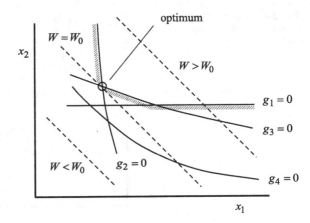

Fig. 3.3 Inequality constraints in a two-dimensional design space (*shading marks* the infeasible region)

right of the constraint lines. Line $g_1 = 0$ represents a side constraint (the minimum value of variable x_2), as the name implies directly limiting the extent of the design space. The other constraints are so-called behavioural constraints (such as stress or stiffness constraints). 'Weight lines' (lines of constant value of the objective function) are drawn as broken lines in this and the following figures. These are shown as linear, but of course this need not be so. It is implied in these figures that we are looking for the *minimum* of some function, such as minimum weight W. The line $W = W_0$ shows the optimum at the intersection of constraints g_2 and g_3. These are the active constraints at the optimum. Other constraints, such as g_1, are active in other parts of the design space, defining the full extent of the feasible region. Some constraints, such as g_4, may be inactive in the whole design space, that is fully overruled by other constraints. If any of the constraints is an equality constraint, then the optimum must, of course, lie on that constraint. All other constraints must still be satisfied, otherwise there would be no feasible solution to the optimization problem. The location of constraints relative to one another in the design space in any real problem, with more design variables, can scarcely be visualized in advance. This means that, as stated earlier, it is generally not known beforehand which constraints will be active at the optimum. The prime task of a numerical optimization procedure is then both to select the correct set of active constraints and to locate the optimum on those constraints.

The type of optimum illustrated in Fig. 3.3 is an 'intersection optimum' (at the intersection of constraints) with in principle the number of *active* constraints m_0 equal to the number of design variables n. We have a 'mathematical' optimum if $m_0 < n$, as illustrated in Fig. 3.4. More difficult to visualize is that in a multidimensional design space, the optimum may be a mathematical optimum along the intersection of a number of constraints. All constraints in Fig. 3.3 are 'convex' constraints, such that only a single, unique optimum exists. Figure 3.5 illustrates a non-convex constraint, with two (or more) local minima. It can also occur that there is no unique optimum, either if there are more local minima of equal value or if the optimum lies at an indeterminate point along a constraint. In contrast to the design

Fig. 3.4 Design space showing a mathematical optimum

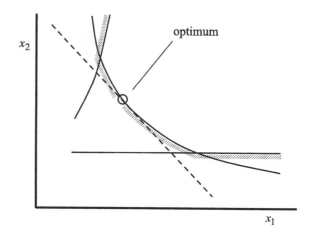

Fig. 3.5 Design space showing a non-convex constraint

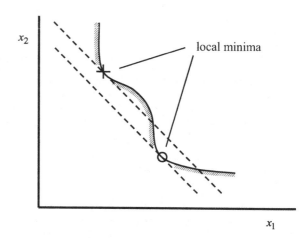

space shown in Fig. 3.3, the design space in Fig. 3.6a illustrates the special case of a fully stressed, statically determinate truss from Chap. 1. Since each constraint depends on only one variable, the constraints are simply the vertical or horizontal lines shown in the figure, with again $m_0 = n$. For a statically indeterminate truss, in which the stress in each member is usually only weakly dependent on the dimensions of the other members, the constraints are generally near-vertical or near-horizontal, as shown in Fig. 3.6b. It is this weak dependence that in many cases enables an iterative calculation for a truss and other structures to converge rapidly enough to the optimum.

It might be concluded from the discussion up to now that the optimality criteria developed earlier have become irrelevant to the more general design problem. However, as implied earlier, many of the same characteristics will be found back in the optimization of more complex structures, such as the box beam used as illustration here. Inevitably, the maximum stress limit will be reached in many parts of a

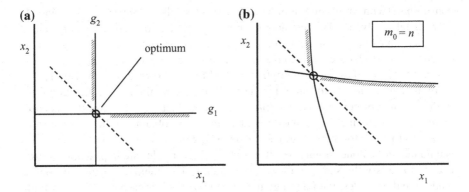

Fig. 3.6 Design spaces showing typical constraints for **a** a fully stressed, statically determinate truss and **b** a statically indeterminate truss

structure, as in a fully stressed design. Where there are buckling constraints, it can be anticipated that the buckling stress will be reached in different parts of the structure, often in simultaneous buckling modes. For maximum stiffness, the solution is likely to at least tend towards a uniform strain energy distribution throughout the structure. In fact, optimality criteria can still be employed in many optimization problems when it is known with certainty that particular conditions do apply, as a means of eliminating some of the design variables. We have already seen a simple example of this is in the spreadsheet 'Seven-bar Truss' in Chap. 1. Being statically determinate, the cross-sectional areas of the members are expressed directly in terms of the known forces in those members, thereby eliminating the member areas from the set of design variables. This leaves only those variables that define the shape of the truss in the optimization. In Chap. 2, for a truss structure with some members subject to buckling in compression, an efficiency formula based on simultaneous modes of buckling is used to eliminate the individual dimensions of members of the truss. For a box beam, we might, for example, select stiffener spacing and stiffener area to satisfy panel buckling constraints at each step of an optimization procedure. In this way, these variables are eliminated or, better stated, become dependent on the remaining variables.

3.2 The Lagrange Multiplier Method

While in practice we shall be mainly concerned with numerical optimization methods, it is worthwhile to understand the mathematical basis of optimization, if only because of its role in many numerical methods. The classical method of Lagrange multipliers finds the maximum or minimum of a function subject only to *equality* constraints, that is constraints all of which have to be identically satisfied. This implies that there are at least as many variables as there are independent constraints (with the same number of variables, the problem is in principle immediately solved). The method cannot be applied to inequality constrained problems unless it is known beforehand which constraints will be active at the optimum, in which case these constraints can of course be treated as equalities. The description of the Lagrange multiplier method here follows that of Walsh [1].

In the conventional procedure for the maximum or minimum of a function $f(\mathbf{x})$, in the absence of constraints, we differentiate the function with respect to each of the n variables in turn and set the derivatives to zero to obtain a sufficient number of equations to solve for all the variables. However, when constraints are imposed, this can no longer be done, simply because some of the variables are in principle already needed to satisfy the same number of constraints. The solution to this is to define p additional variables, the Lagrange multipliers λ_k, one for each equality constraint $h_k = 0$. These are combined with the original function $f(\mathbf{x})$ to form the Lagrangian function:

$$F(\mathbf{x}, \lambda) = f(\mathbf{x}) - \sum_{k=1}^{p} \lambda_k \, h_k(\mathbf{x}). \qquad (3.1)$$

The minus sign in the above function is introduced for convenience later (in many texts, this is a plus sign). Since all the h_k are zero when the constraints are satisfied, it is seen that $F(\mathbf{x}, \lambda) = f(\mathbf{x})$ at the constrained maximum or minimum, regardless of the values of λ_k. The properties of the Lagrangian function will now be explored to see how the originally constrained problem is transformed into an unconstrained one. This is assuming now that $f(\mathbf{x})$ and all the $h_k(\mathbf{x})$ are differentiable functions. To simplify the text, in what follows, it is assumed that we are seeking the *minimum* of the function $f(\mathbf{x})$.

Since only $(n - p)$ variables x_i can be independent, these are separated into two sets as follows: variables x_1 to x_p to satisfy the p equality constraints and the remaining variables x_{p+1} to x_n to minimize the function $f(\mathbf{x})$. The first set x_1 to x_p is treated as dependent variables, that is dependent on the second set x_{p+1} to x_n. The second set of variables is then independent variables. Any change in the second set of variables must be accompanied by corresponding changes in the first set, to continue to satisfy the constraints. In principle, the allocation of variables to the first and second set is entirely arbitrary. For the present purpose, it does not matter what these variables actually represent. Definition of the two sets of variables is given again in Table 3.1.

Table 3.1 Dependent and independent variables

$x_1 \ldots x_p$	Variables used to satisfy $h_k(\mathbf{x}) = 0$, $k = 1 \ldots p$ (dependent on $x_{p+1} \ldots x_n$)
$x_{p+1} \ldots x_n$	Independent variables used to minimize $f(\mathbf{x})$

The Lagrangian function in Eq. (3.1) can now be differentiated as follows:

$$\frac{\partial F}{\partial x_i} = \frac{\partial f}{\partial x_i} - \sum_{k=1}^{p} \lambda_k \frac{\partial h_k}{\partial x_i}. \tag{3.2}$$

In the range $1 \leq i \leq p$, we choose the p Lagrange multipliers λ_k (so far arbitrary) so that:

$$\frac{\partial F}{\partial x_i} = 0$$

at the constrained minimum (as indeed for an unconstrained minimum). Examining now Eq. (3.2) in the range $p+1 \leq i \leq n$,, we conclude that:

$$\frac{\partial F}{\partial x_i} = 0$$

in this second range as well, since $\partial f / \partial x_i = 0$ for a minimum of the original function $f(\mathbf{x})$ with respect to the (second) set of independent variables and $\partial h_k / \partial x_i = 0$ in this same range because the constraints $h_k = 0$ are deemed to be satisfied by the (first) set of dependent variables. Constraints h_k are, therefore, independent of the second set of variables. Equation (3.2) then reduces to:

$$\frac{\partial F}{\partial x_i} = 0$$

at a constrained minimum for both sets of x_i. Differentiating Eq. (3.1) now with respect to the Lagrange multipliers λ_k, we obtain:

$$\frac{\partial F}{\partial \lambda_k} = h_k = 0,$$

since $h_k = 0$ when the constraints are satisfied. The conditions for an equality constrained minimum are finally:

$$\frac{\partial F}{\partial x_i} = 0, \quad \frac{\partial F}{\partial \lambda_k} = 0 \tag{3.3}$$

for all values of i and k. We now have the required $(n+p)$ equations to solve for all the x_i and λ_k. Differentiation of function F with respect to both the x_i and the λ_k is perhaps an elegant way to write the conditions for a constrained minimum, but it will already be clear that the second of these is no more than stating simply $h_k = 0$, as required. While we refer above to minimization of the function, it is readily seen that conditions (3.3) apply equally to the maximum of a function.

Notice that Eq. (3.3) reduce to ordinary minimization of function $f(\mathbf{x})$ in the absence of constraints, since the second of Eq. (3.3) disappears. Just as for an unconstrained problem, Eq. (3.3) are necessary but not sufficient conditions for an equality constrained minimum. An extensive account of these conditions is given by Reklaitis et al. [2], Walsh [1] and many other texts. In fact, the Lagrangian function is not a minimum with respect to both x_i and λ_k but is actually a saddle point, that is a minimum with respect to x_i and a maximum with respect to λ_k (or vice versa if we have a maximum rather than a minimum of the original function). This means that we cannot simply minimize function $F(\mathbf{x}, \lambda)$ with respect to both $\mathbf{x}$ and λ to find the minimum of the original constrained function. Equation (3.3) represents only a stationary value of the Lagrangian function. In principle, we require the 'Hessian' matrix of second derivatives of $F(\mathbf{x})$ to determine whether the result is a minimum, maximum or other stationary point. If the Hessian matrix is positive definite, we have at least a local minimum, and if it is negative definite, we have a local maximum. In any but very small problems, to obtain the Hessian matrix demands a substantial amount of calculation. In practice, whether the outcome is a minimum or a maximum is mostly clear from the nature of the problem.

Example 3.1 Minimize the function:

$$f(\mathbf{x}) = x_1^2 + x_2^2 + x_3^2$$

subject to equality constraints:

$$h_1(\mathbf{x}) = x_1 + x_2 - a = 0,$$
$$h_2(\mathbf{x}) = x_3 - b = 0.$$

The problem is illustrated in Fig. 3.7, from which it is clear that the point sought is the nearest point to the origin on the line of intersection of the two constraints (shown with heavy lines). The Lagrangian function is:

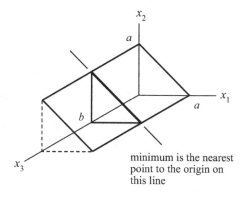

Fig. 3.7 Three-dimensional design space for the problem in Example 3.1 (constraints are shown with *heavy lines*)

minimum is the nearest point to the origin on this line

$$F(\mathbf{x}, \lambda) = x_1^2 + x_2^2 + x_3^2 - \lambda_1(x_1 + x_2 - a) - \lambda_2(x_3 - b).$$

Differentiating this function with respect to x_i and λ_k, we obtain:

$$\frac{\partial F}{\partial x_1} = 2x_1 - \lambda_1 = 0, \quad \frac{\partial F}{\partial \lambda_1} = -(x_1 + x_2 - a) = 0,$$

$$\frac{\partial F}{\partial x_2} = 2x_2 - \lambda_1 = 0, \quad \frac{\partial F}{\partial \lambda_2} = -(x_3 - b) = 0,$$

$$\frac{\partial F}{\partial x_3} = 2x_3 - \lambda_2 = 0.$$

Solving these five equations gives:

$$x_1 = \tfrac{a}{2}, \quad x_2 = \tfrac{a}{2}, \quad x_3 = b,$$
$$\lambda_1 = a, \quad \lambda_2 = 2b,$$
$$f_{\min} = \tfrac{a^2}{2} + b^2.$$

at the minimum. The above values of x_1, x_2, x_3 are easily verified from Fig. 3.7. The meaning of the Lagrange multipliers that are obtained is investigated in the following section.

Of course, in this simple example, a solution could also be obtained by substituting $x_2 = a - x_1$ and $x_3 = b$ in the original function and differentiating with respect to x_1 for the minimum of the now unconstrained function. In fact, this amounts to selecting two dependent variables x_2 and x_3 to satisfy the two constraints, with the single remaining independent variable x_1 used to minimize the function, as in Table 3.1. ∎

Example 3.2 Find the optimum thickness distribution (t_1, t_2, t_3, t_4) for maximum torsional stiffness of the rectangular torsion box in Fig. 3.8.

The rate of twist of a thin-walled, rectangular box (angle of twist per unit length) is given by:

$$\theta = \frac{T}{4B^2H^2G}\left(\frac{B}{t_1} + \frac{B}{t_2} + \frac{H}{t_3} + \frac{H}{t_4}\right).$$

where T is the applied twisting moment, and G is the shear modulus of the material (this formula can be deduced from formulae in Chap. 7, or see [3] and many other

Fig. 3.8 Rectangular torsion box in Example 3.2

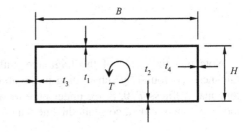

texts). For maximum torsional stiffness, we have to minimize the rate of twist. The volume of material (per unit length) is:

$$V = B(t_1 + t_2) + H(t_3 + t_4).$$

For some given total volume of material V_0, we have an equality constraint:

$$V - V_0 = 0 .$$

We write the Lagrangian function as:

$$F = C\left(\frac{B}{t_1} + \frac{B}{t_2} + \frac{H}{t_3} + \frac{H}{t_4}\right) - \lambda[B(t_1 + t_2) + H(t_3 + t_4) - V_0],$$

where, for simplicity, we define:

$$C = \frac{T}{4B^2 H^2 G}.$$

Differentiating the Lagrangian function:

$$\frac{\partial F}{\partial t_1} = -\frac{CB}{t_1^2} - \lambda B = 0, \quad \frac{\partial F}{\partial t_2} = -\frac{CB}{t_2^2} - \lambda B = 0,$$

$$\frac{\partial F}{\partial t_3} = -\frac{CH}{t_3^2} - \lambda H = 0, \quad \frac{\partial F}{\partial t_4} = -\frac{CH}{t_4^2} - \lambda H = 0,$$

$$\frac{\partial F}{\partial \lambda} = B(t_1 + t_2) + H(t_3 + t_4) - V_0 = 0 .$$

From the first four equations, we obtain directly:

$$\frac{C}{t_1^2} = \frac{C}{t_2^2} = \frac{C}{t_3^2} = \frac{C}{t_4^2},$$

or

$$t_1 = t_2 = t_3 = t_4 = t, \quad \text{say}$$

and

$$\lambda = -\frac{C}{t^2}.$$

Note that the value of the Lagrange multiplier is in this case negative. The thickness t is found by substituting in the above equation for θ, for some required torsional stiffness T/θ. Since under pure torsion the shear flow is uniform around the whole torsion box, constant thickness amounts to equal shear stress in all panels or, in other words, a uniform strain energy density. This is the same result as found for a truss structure in Chap. 2. ■

In a similar way to the above example, the problem in Sect. 2.2 could otherwise be solved with the aid of the Lagrange multiplier. The deflection of the truss in Sect. 2.2 is:

$$\delta = \frac{1}{PE} \sum \frac{(F_i l_i)^2}{V_i},$$

and its volume is:

$$V = \sum V_i.$$

To minimize δ for given volume V_0 of material, we write the Lagrangian function:

$$F = \frac{1}{PE} \sum \frac{(F_i l_i)^2}{V_i} - \lambda \left(\sum V_i - V_0 \right)$$

and differentiate with respect to each V_i in turn. The result is, of course, exactly the same except that now a value is also obtained for the Lagrange multiplier.

3.2.1 Interpretation of Lagrange Multipliers

In the examples of the previous section, we found not only the constrained minimum of the function and corresponding values of the variables, but also values of the Lagrange multipliers. We need now to understand the meaning of the Lagrange multipliers, and to do this, we consider the effect of a small increment ϵ_r in the required value of one of the constraints h_r. Since the Lagrangian function in Eq. (3.1) is a function of both the variables x_i and the Lagrange multipliers λ_k, by the usual rule of differentiation (treating ϵ_r as variable), we can write:

$$\frac{\partial F}{\partial \epsilon_r} = \sum_{i=1}^{n} \frac{\partial F}{\partial x_i} \cdot \frac{\partial x_i}{\partial \epsilon_r} + \sum_{k=1}^{m} \frac{\partial F}{\partial \lambda_k} \cdot \frac{\partial \lambda_k}{\partial \epsilon_r}. \tag{3.4}$$

However, we know that in the above equation both:

$$\frac{\partial F}{\partial x_i} = 0 \quad \text{and} \quad \frac{\partial F}{\partial \lambda_k} = 0$$

for all values of i and k at a constrained maximum or minimum, in which case we conclude that Eq. (3.4) reduces to:

$$\frac{\partial F}{\partial \epsilon_r} = 0.$$

Alternatively, we can differentiate the Lagrangian function in Eq. (3.1) explicitly with respect to ϵ_r:

$$\frac{\partial F}{\partial \epsilon_r} = \frac{\partial f}{\partial \epsilon_r} - \sum_{k=1}^{m} \left(\lambda_k \frac{\partial h_k}{\partial \epsilon_r} + h_k \frac{\partial \lambda_k}{\partial \epsilon_r} \right). \tag{3.5}$$

All constraints have to remain zero at a constrained maximum or minimum except the chosen constraint $h_r = \epsilon_r$, so in the above equation:

$$\frac{\partial h_k}{\partial \epsilon_r} = 1 \quad \text{if} \quad k = r, \quad \text{otherwise} \quad \frac{\partial h_k}{\partial \epsilon_r} = 0.$$

The summation in Eq. (3.5) now disappears, and since $h_r \rightarrow 0$ as $\epsilon_r \rightarrow 0$ in the limit, Eq. (3.5) reduces to:

$$\frac{\partial F}{\partial \epsilon_r} = \frac{\partial f}{\partial \epsilon_r} - \lambda_r.$$

It was established above that $\frac{\partial F}{\partial \epsilon_r} = 0$, so we obtain finally:

$$\lambda_r = \frac{\partial f}{\partial \epsilon_r} \tag{3.6}$$

at a constrained maximum or minimum of the function f. In more general terms, the value of the Lagrange multiplier λ_k defines the *sensitivity* of the optimum value of a function to the value of the corresponding constraint. Here, this is the rate of change of the maximum or minimum of function f with change ϵ_r in each of the constraint values. Positive λ_k indicates that the maximum or minimum value of f increases with increase in each constraint value h_k, and negative λ_k indicates that it decreases.

Example 3.3 Use the values of the Lagrange multipliers obtained in Examples 3.1 and 3.2 to confirm the meaning of the Lagrange multiplier given by Eq. (3.6).

Because of the simple algebraic solution:

$$f_{\min} = \frac{a^2}{2} + b^2$$

obtained in Example 3.1, the meaning of the Lagrange multiplier is readily confirmed as follows. Making small increments ϵ_1 and ϵ_2 in each of the constraints, these can be rewritten:

$$x_1 + x_2 = a + \epsilon_1,$$
$$x_3 = b + \epsilon_2,$$

and f_{min} is correspondingly increased to:

$$f_{min} = \frac{(a + \epsilon_1)^2}{2} + (b + \epsilon_2)^2.$$

The rate of change of f_{min} with ϵ_1 and ϵ_2 can now be obtained directly by differentiation:

$$\frac{\partial f_{min}}{\partial \epsilon_1} = a + \epsilon_1, \qquad \frac{\partial f_{min}}{\partial \epsilon_2} = 2(b + \epsilon_2).$$

In the limit, as $\epsilon_1 \to 0$, $\epsilon_2 \to 0$:

$$\frac{\partial f_{min}}{\partial \epsilon_1} = a, \qquad \frac{\partial f_{min}}{\partial \epsilon_2} = 2b.$$

With the meaning of the Lagrange multiplier defined as in Eq. (3.6), we have from above:

$$\lambda_1 = \frac{\partial f_{min}}{\partial \epsilon_1} = a, \qquad \lambda_2 = \frac{\partial f_{min}}{\partial \epsilon_2} = 2b.$$

This agrees with the values of the Lagrange multipliers already found in Example 3.1.

Similarly, in Example 3.2, we found uniform thickness t to be the optimum for torsional stiffness. Substituting in the previous formulae for V and θ, we obtain:

$$V = 2(B + H)t = V_0$$

at the constrained minimum, or:

$$t = \frac{V_0}{2(B + H)}$$

and:

$$\theta_{min} = \frac{2C(B + H)}{t} = \frac{4C(B + H)^2}{V_0}.$$

Again, with the meaning of λ as in Eq. (3.6), and treating V_0 as variable:

$$\lambda = \frac{d\theta_{min}}{dV_0} = -\frac{4C(B+H)^2}{V_0^2} = -\frac{C}{t^2}.$$

This agrees with the value of λ found previously in Example 3.2. Note that increase in volume results in *reduction* in the angle of twist, accounting for the negative λ.   ∎

As already stated, the values of the Lagrange multipliers obtained in Examples 3.1 and 3.2 are readily verified here simply by differentiation of the minimum of the function, because in both cases we have an analytical solution for the optimum. Generally, we do not have this, and values of the Lagrange multipliers, if required, have to be obtained in the course of numerical optimization.

3.3 Inequality Constrained Problems

If, in an inequality constrained optimization, it can be known with certainty which constraints will be active at the optimum, then of course these can be treated simply as equality constraints, and the inactive constraints ignored. The method of Lagrange multipliers in the previous section then still applies. However, it will not in general be known which are the active constraints, and the optimization task is then to select the correct set of active constraints and to seek the optimum on those constraints. In this process, we can still use the Lagrange multipliers to distinguish between those constraints that are correctly selected as active and those that are not.

Suppose first that we are seeking a *minimum* of function $f(\mathbf{x})$, subject to inequality constraints $g_j(\mathbf{x}) \geq 0$. If a constraint is correctly selected as active, then $g_j(\mathbf{x}) = 0$ on the constraint boundary, and any increase in the constraint value $g_j(\mathbf{x}) > 0$ to move away from the constraint and into the feasible region must lead to an *increase* in the objective function $f(\mathbf{x})$. Noting that the Lagrange multiplier $\lambda_j = \partial f / \partial g_j$, the above condition corresponds to a *positive* value of the Lagrange multiplier. However, if the Lagrange multiplier is negative, this means that moving away from the constraint boundary and into the feasible region leads to a decrease in $f(\mathbf{x})$, and in other words, the constraint has been incorrectly selected. The constraint is now preventing further reduction in $f(\mathbf{x})$ towards the true optimum and should be discarded as an active constraint. An inactive constraint $g_j(\mathbf{x}) > 0$ has no influence on the objective function $f(\mathbf{x})$, so we deduce that $\lambda_j = 0$ for inactive constraints.

The situation described above is illustrated in the design space in Fig. 3.9. Suppose we first select constraints $g_1 \geq 0$ and $g_2 \geq 0$ as active constraints to find a minimum of function $f(\mathbf{x})$. Moving away from constraint $g_1 = 0$ at point A and along constraint $g_2 = 0$ is seen to lead to a *reduction* in $f(\mathbf{x})$. Constraint g_1 is not, therefore, correctly selected as an active constraint. This will be indicated by the values of the Lagrange multipliers for the two constraints at point A: λ_1 will be

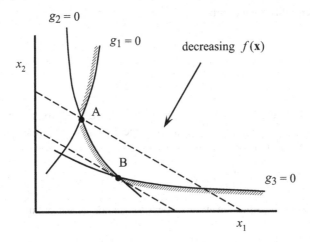

$g_2 = 0$

$g_1 = 0$

decreasing $f(\mathbf{x})$

x_2

A

B

$g_3 = 0$

x_1

Fig. 3.9 Search for the minimum of a function subject to inequality constraints

Table 3.2 Sign of the	Minimum of function		Maximum of function	
Lagrange multiplier for active constraints in inequality	Form of constraint		Form of constraint	
constrained problems ($\lambda_j = 0$	$g_j \geq 0$	$g_j \leq 0$	$g_j \geq 0$	$g_j \leq 0$
for inactive constraints)	λ_j positive	λ_j negative	λ_j negative	λ_j positive

negative for constraint g_1 and λ_2 positive for constraint g_2. Rejecting constraint g_1 and selecting instead constraints g_2 and g_3 as active constraints, we see that any movement from point B along either of the constraints leads to an increase in $f(\mathbf{x})$. The Lagrange multipliers for both these constraints will be *positive*, indicating that constraints g_2 and g_3 are correctly selected as active and that point B is the optimum point in the design space. Otherwise, an optimum might have been found between points A and B, in which case only constraint g_2 will have a positive Lagrange multiplier.

The reverse of this rule applies to finding a *maximum* of $f(\mathbf{x})$, in which case a constraint is correctly selected as active if the Lagrange multiplier for that constraint is *negative*. These rules are summarized in Table 3.2, also for the case of constraints defined in the alternative form $g_j(\mathbf{x}) \leq 0$.

Example 3.4 Minimize the function:

$$f(\mathbf{x}) = x_1^2 + x_2^2 + x_3^2$$

subject to inequality constraints:

$$g_1(\mathbf{x}) = x_1 x_2 - 1 \geq 0,$$
$$g_2(\mathbf{x}) = 2x_2 - 1 \geq 0,$$
$$g_3(\mathbf{x}) = x_3 - 1 \geq 0.$$

To make use of the Lagrange multiplier method, we have first to establish which constraints are active at the optimum. Assume first that all three constraints are active. The Lagrangian function is:

$$F = x_1^2 + x_2^2 + x_3^2 - \lambda_1(x_1 x_2 - 1) - \lambda_2(2x_1 - 1) - \lambda_3(x_3 - 1).$$

Differentiating the function with respect to x_i and λ_k, we obtain:

$$\frac{\partial F}{\partial x_1} = 2x_1 - \lambda_1 x_2 = 0, \qquad \frac{\partial F}{\partial \lambda_1} = -(x_1 x_2 - 1) = 0,$$
$$\frac{\partial F}{\partial x_2} = 2x_2 - \lambda_1 x_1 - 2\lambda_2 = 0, \qquad \frac{\partial F}{\partial \lambda_2} = -(2x_2 - 1) = 0,$$
$$\frac{\partial F}{\partial x_3} = 2x_3 - \lambda_3 = 0, \qquad \frac{\partial F}{\partial \lambda_3} = -(x_3 - 1) = 0.$$

Solving these six equations gives:

$$x_1 = 2, \quad \lambda_1 = 8,$$
$$x_2 = \tfrac{1}{2}, \quad \lambda_2 = -\tfrac{15}{2},$$
$$x_3 = 1, \quad \lambda_3 = 2.$$

It can be immediately verified that the three variables do satisfy the three constraints, when treated as equalities and that the problem is therefore fully constrained. However, since λ_2 is negative, we can conclude that the second constraint has in fact been incorrectly selected as an active constraint. Removing this constraint from the Lagrangian function, the previous equations reduce to:

$$\frac{\partial F}{\partial x_1} = 2x_1 - \lambda_1 x_2 = 0, \quad \frac{\partial F}{\partial \lambda_1} = -(x_1 x_2 - 1) = 0,$$
$$\frac{\partial F}{\partial x_2} = 2x_2 - \lambda_1 x_1 = 0,$$
$$\frac{\partial F}{\partial x_3} = 2x_3 - \lambda_3 = 0, \quad \frac{\partial F}{\partial \lambda_3} = -(x_3 - 1) = 0.$$

Solving these five equations gives:

$$x_1 = \pm 1, \quad \lambda_1 = 2,$$
$$x_2 = \pm 1,$$
$$x_3 = 1, \quad \lambda_3 = 2.$$

where x_1 and x_2 must have the same sign to satisfy the first constraint. Note that both λ_1 and λ_3 are now positive, as required. The second constraint $g_2(\mathbf{x}) \geq 0$ is more than satisfied provided that $x_2 = +1$, implying $x_1 = +1$ as well. We can conclude that we now have a correct solution, with $\lambda_2 = 0$ for the inactive second constraint and $f_{min} = 3$. Nevertheless, it will immediately be realized that a method such as this rapidly becomes impractical if there are a large number of constraints. The example is given here simply to illustrate the role of Lagrange multipliers in identifying active and inactive constraints. ∎

3.3.1 The Kuhn–Tucker Conditions

The Kuhn–Tucker conditions [4] are the necessary conditions for an optimum $\mathbf{x}^*$ of a function $f(\mathbf{x})$ subject to inequality constraints $g_j(\mathbf{x}) \geq 0$ and equality constraints $h_k(\mathbf{x}) = 0$. (In fact, the same conditions were later found to have been developed earlier by Karush [5].) The Kuhn–Tucker conditions express in concise form conditions for an equality and inequality constrained optimum introduced in the previous sections. The three conditions for a constrained minimum at a point $\mathbf{x}^*$ are as follows:

(1) $\mathbf{x}^*$ is feasible,
(2) $\lambda_j g_j(\mathbf{x}^*) = 0$ for all j, and
(3) $\frac{\partial f(\mathbf{x}^*)}{\partial x_i} - \sum_{j=1}^{m} \lambda_j \frac{\partial g_j(\mathbf{x}^*)}{\partial x_i} - \sum_{k=1}^{p} \lambda_{k+m} \frac{\partial h_k(\mathbf{x}^*)}{\partial x_i} = 0$ for all x_i,

with $\lambda_j \geq 0$ and λ_{k+m} unrestricted in sign.

The first condition states simply that the optimum must satisfy all constraints. The second condition states that if inequality constraints are not precisely satisfied (i.e. nonzero), then the corresponding Lagrange multipliers must be zero. The third condition is the same as Eq. (3.2) except for the additional term to include inequality constraints. Inactive constraints are automatically removed by the condition $\lambda_j = 0$ for those constraints.

The Kuhn–Tucker conditions are a formal statement of the necessary but not sufficient conditions for a local optimum, provided that the objective function $f(\mathbf{x})$ is continuously differentiable at the optimum and subject to some further 'regularity conditions' on the nature of the constraints. It will be realized that the Kuhn–Tucker conditions do not as such lead to a solution of the optimization problem, since the task remains to determine which inequality constraints are active at the optimum and, in a numerical solution, to search for the optimum on those constraints. However, they can if required be used to verify an optimum once it has been found.

3.4 Spreadsheet Program

The spreadsheet extends the scope of the spreadsheet for a circular tube in axial compression in the previous chapter by allowing eccentricity of the applied load and by taking into account loss of stiffness due to yielding of the material at higher stresses. An eccentrically applied load causes progressive bending deformation of the column with increase of load. Failure then occurs either by local buckling of the tube during this bending deformation, or by excessive deformation coupled with yielding of the material under a combination of bending and compressive stress, before the classical Euler buckling load is reached.

3.4.1 Eccentrically Loaded Column

The spreadsheet uses Solver to optimize the eccentrically loaded column in Fig. 3.10, taking into account the effect of yielding of the material at higher stresses. The column is a circular tube with given (effective) simply supported length and compressive load. The diameter and thickness of the tube are optimized for minimum weight. The spreadsheet is shown in Fig. 3.11.

Under an eccentrically applied compressive load (i.e. along an axis offset from the neutral axis of the column), bending of the column causes bending stresses in addition to the compressive stress due directly to the applied load. The maximum stress is given by the well-known secant formula (see Timoshenko and Gere [6]):

$$\sigma_{max} = \frac{P}{A}\left(1 + \frac{ec}{r^2}\sec\frac{L}{2r}\sqrt{\frac{P}{AE}}\right),$$

where P is the compressive applied load, A is the cross-sectional area of the column, e is the eccentricity of the applied load, c is the distance from the neutral axis to the outer edge of the column (in this case, the outer radius of the tube), L is its effective simply supported length, r is the radius of gyration of the cross section, and E is the modulus of elasticity of the material. Substituting $r^2 = I/A$, the last part of this formula may be rewritten as:

$$\frac{L}{2r}\sqrt{\frac{P}{AE}} = \frac{L}{2}\sqrt{\frac{P}{EI}},$$

where I is the second moment of area of the cross section. As this term approaches $\pi/2$, the stress σ_{max} increases without limit. This corresponds to the theoretical Euler buckling load of the column in Sect. 2.1. The maximum stress σ_{max} in the

Fig. 3.10 Eccentrically loaded column

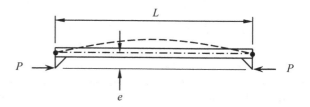

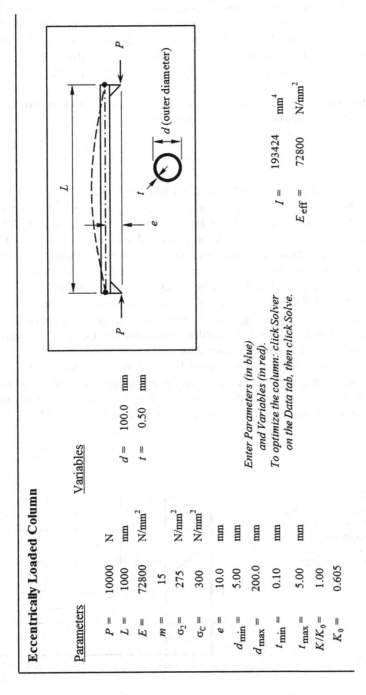

Fig. 3.11 Spreadsheet 'Eccentrically Loaded Column'

cross section is limited to a specified allowable stress σ_c of the material, which in effect limits the deformation of the column. Eccentricity e can also be used to estimate the effect of some initial bow, or lack of straightness, of a column if the maximum deviation from the axis of the column is added to any eccentricity of the applied load. The Euler buckling stress σ_E is given in the spreadsheet for reference only since, unless eccentricity $e = 0$, the maximum stress σ_{max} must always be less than σ_E.

The tubular section of the column is also subject to local, or short wavelength, buckling. The local buckling stress given in Sect. 2.1:

$$\sigma_L = KE\frac{t}{R} \, ,$$

with thickness t and mean radius R is used for this, but is taken now to refer to the *maximum* stress σ_{max} at some point around the circumference. Buckling of a cylindrical shell (here a circular tube) is known to be highly sensitive to small local imperfections. A reduction factor K/K_0 is introduced to represent this, where $K_0 = 0.605$ is the theoretical buckling coefficient. The extent of this reduction depends, of course, on the degree of imperfection. Figure 3.12 plots values of K/K_0 against δ/t, based on accumulated test results for a large number of practical cylindrical shells in ESDU Data Item 83034 [7], where δ is the maximum deviation from a cylinder with no imperfection. While Fig. 3.12 is principally for large, thin shells, rather than for the relatively thick tubes intended here, reduction factors K/K_0 may be used as a guide in the absence of other data.

At higher stresses, yielding of the material may occur. This is taken into account in the formula for σ_{max} by replacing E by the tangent modulus E_t. For a wide range of materials with a gradual yielding behaviour, the stress–strain relation can be expressed by the Ramberg–Osgood formula (Ramberg and Osgood [8]):

Fig. 3.12 Reduction in the local buckling coefficient of a circular tube due to initial imperfections (based on data in ESDU Data Item 83034, with permission from IHS ESDU)

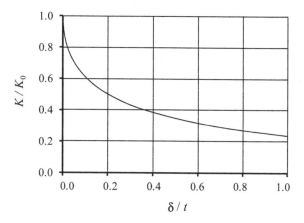

$$\epsilon = \frac{\sigma}{E} + 0.002 \left(\frac{\sigma}{\sigma_2}\right)^m,$$

where σ_2 is the 0.2% proof stress of the material and index m defines the sharpness of the yield curve. The tangent modulus, widely used for Euler buckling, is then:

$$E_t = \frac{d\sigma}{d\epsilon} = \frac{E}{1 + \frac{0.002Em}{\sigma_2} \left(\frac{\sigma}{\sigma_2}\right)^{m-1}}.$$

The tangent modulus is evaluated here at the *average* stress P/A in the column. The Ramberg–Osgood formula and use of a reduced modulus are discussed further in Chap. 7. Generalized stress–strain curves, based on the Ramberg–Osgood formula, are given in ESDU Data Item 76016 [9].

The tangent modulus is also conservatively substituted for E in the local buckling formula, but this is based now on the *maximum* stress σ_{max} in the column (an alternative formula for the effective modulus E_{eff} for local buckling is readily substituted in the spreadsheet if so required). In many practical problems, the thickness of the tube will be limited by some specified minimum thickness, and both the reduction factor for imperfections and the choice of effective modulus for local buckling become less significant.

The compressive load P, effective length L, material data referred to above, maximum and minimum values of outer diameter d and thickness t, and K/K_0 have to be entered in the spreadsheet, as well as suitable initial values of design variables d and t. Solver can then be run with the GRG Nonlinear method to optimize the diameter and thickness of the column. Note that d_{min}, d_{max}, t_{min} and t_{max} may not be left blank. Parameters and design variables to be entered are listed in Table 3.3.

With no limits on diameter or thickness, optimization will lead to a maximum stress σ_{max} equal to the local buckling stress σ_L, neither of which may exceed the allowable stress of the material. The achieved efficiency η, given in Sect. 2.1.1, is:

$$\eta = \frac{\sigma}{E^{2/3}(P/L^2)^{1/3}},$$

where σ is the average stress in the column. Note that efficiency η is based on the *initial* elastic modulus E. It is reduced in value by the stress σ which takes account of reduction in modulus with yielding as well as the effect of eccentric applied load and any dimensional restrictions. If the stresses σ_2 and σ_c are set sufficiently high, $K/K_0 = 1$, $e = 0$, and again if limits on dimensions do not intervene, the spreadsheet produces identical results to the earlier spreadsheet 'Circular Tube in Compression' in Sect. 2.3.1, with a maximum efficiency $\eta = 0.780$.

The spreadsheet can be used to produce plots of σ against P/L^2 for different materials, as in Fig. 2.4, but now with the effect of yielding at higher stresses, and

Table 3.3 Data entry for spreadsheet program 'Eccentrically Loaded Column'

Parameters	
Compressive load P	Enter value in cell C6 as a *positive* number
(Effective) simply—supported length L	Enter value in cell C7
Elastic modulus E, Ramberg-Osgood index m	Enter values in cell C8: C9
0.2% proof stress σ_2, allowable compressive stress σ_c	Enter values in cells C10: C11 as *positive* numbers
Eccentricity e	Enter value in cell C12 (enter zero if no eccentricity)
Min. and max. diameter and thickness $d_{\min}, d_{\max}, t_{\min}, t_{\max}$	Enter values in cells C13: C16 (cells may not be left blank)
Reduction factor for local imperfection K/K_0	Enter value in cell C17, $K/K_0 \leq 1$ (may not be left blank)
Theoretical local buckling coefficient K_0	Enter value in cell C18 ($K_0 = 0.605$, or other value depending on the end conditions)
Variables	
Outer diameter d, thickness t	Enter initial values in cells F7: F8 within the specified range

to investigate the effect of eccentricity of load on the minimum weight of a column to carry a given load. Attention is again drawn to the warning about the choice of initial values of design variables in the final paragraph of Sect. 2.3.2. Initial values that differ widely from the final optimum values can lead to Solver being unable to reach a solution.

3.5 Summary

A rectangular box beam is used to illustrate the general optimization problem and the complex relationship that is likely to exist between design variables and constraints in a realistic design. Constraints may depend on many design variables, while the same variables may enter into many different constraints. This implies that constraints cannot be satisfied individually in an iterative resizing procedure, as done previously for a truss structure. While it is in principle possible to find a solution that satisfies all constraints, in this more general situation, there is little reason to suppose that the resulting design will be an optimum. Furthermore, in a practical optimization problem, it is seldom known beforehand which of the inequality constraints are going to be active at the optimum and which are not. Active constraints are those that are identically satisfied; inactive constraints are those that are satisfied in a 'greater than' or 'less than' sense. Active and inactive constraints, as well as the distinction between an intersection optimum and a mathematical one, are illustrated in different design spaces. Equality constraints

must, of course, always be identically satisfied. A numerical optimization procedure
has both to select the correct set of active constraints and to search for the optimum
on those constraints. Numerical optimization methods are discussed in detail in
Chaps. 4 and 5, and a numerical optimization for the box beam described earlier is
developed in Chap. 7.

The Lagrange multiplier method provides a mathematical basis for optimization
in equality constrained problems, provided that both the constraints and the
objective function are differentiable. Values of the Lagrange multipliers indicate the
sensitivity of the optimum to individual constraint values, that is the change in the
optimum value of the objective function for small changes in that constraint. In
inequality constrained problems, Lagrange multipliers for inactive constraints take
zero values which, in effect, remove the inactive constraints from an inequality
constrained problem for as long as they remain inactive. In mathematical terms, the
Kuhn–Tucker conditions are the formal statement of the necessary conditions for an
optimum subject to both equality and inequality constraints.

Exercises

3.1   Use the Lagrange multiplier method to find the minimum of the function:

$$f(\mathbf{x}) = x_1^2 + x_2^2 + x_3^2$$

subject to the constraint:

$$h(\mathbf{x}) = x_1 + 2x_2 + 3x_3 - 7 = 0.$$

Follow the method of Example 3.1.

3.2   Find the minimum of the function:

$$f(\mathbf{x}) = 2x_2 + x_1$$

subject to inequality constraints:

$$g_1(\mathbf{x}) = 2x_2 - x_1 \geq 0,$$
$$g_2(\mathbf{x}) = x_2 - 2x_1 + 4 \geq 0,$$
$$g_3(\mathbf{x}) = x_2 + x_1 - 3 \geq 0,$$
$$g_4(\mathbf{x}) = x_1 - 1 \geq 0.$$

*Only two of the above constraints are active at the minimum. Draw the design
space to find the active constraints. With the active constraints known, treat
these as equalities and solve the problem analytically by the Lagrange*

multiplier method. Notice that the Lagrange multipliers are positive, confirming correctly chosen active constraints.

3.3   A thin circular tube of radius R and length L has flat, closed ends. Use the Lagrange multiplier method to find an expression for the minimum total surface area A of the material of the tube and the two ends, if it has a required internal volume V_0.

Derive expressions for the total surface area and for the internal volume to set up the Lagrangian function.

3.4   A rectangular container is to be made of materials that cost €20/m² for the bottom, €30/m² for the sides and €10/m² for the top. The volume of the container has to be 4 m³. Use the Lagrange multiplier method to calculate the minimum cost of the container and its corresponding dimensions.

Use the value of the Lagrange multiplier found above to estimate the minimum cost if the volume of the container is increased to 5 m³.

The Lagrange multiplier gives the rate of change of the objective function (cost) with the value of the constraint (volume).

3.5   Verify the principle of simultaneous buckling modes for a circular tube loaded in compression by deriving expressions for the Lagrange multipliers for this problem.

Use the Lagrange multiplier method to optimize the circular tube, with only flexural and local buckling constraints (use the formulae in Sect. 2.1). Derive expressions for the optimum radius and thickness and for the Lagrange multipliers. Observe that the Lagrange multipliers are positive for all values of P, L and E.

3.6   Set up a spreadsheet for the problem in Exercise 3.2, and use Solver to find the minimum of the function.

Compare the values of the Lagrange multipliers with those found in the exercise. These are found by selecting the Sensitivity Report in the Solver result box after optimization (see the Appendix). Notice that the Lagrange multipliers are zero for the inactive constraints.

3.7   Repeat Exercise 3.4, using Solver to find the minimum cost of the 4 m³ container and again after increase in volume to 5 m³.

Use the formulae derived in Exercise 3.4 to set up a spreadsheet for this problem. The cost has to be minimized, with variables the dimensions of the container and its volume as the single constraint. Compare the value of the Lagrange multiplier in the Sensitivity Report with the value found in Exercise 3.4. By repeating the optimization for a volume of 5 m³, the increase in cost can be compared with the estimated increase based on the Lagrange multiplier in Exercise 3.4.

3.8   Use the spreadsheet 'Eccentrically Loaded Column' to show the effect of eccentric load on the minimum cross-sectional area of a column of effective length 1000 mm, under a compressive load of 10,000 N.

Take a range of eccentricity from 0 to 50 mm, with $K/K_0 = 1.0$ for no local imperfection. Use the material data already present in the spreadsheet. Plot a *graph of cross-sectional area, diameter and thickness after optimization against eccentricity. Observe the different values of stress at each eccentricity.*

3.9 Use the spreadsheet 'Eccentrically Loaded Column' to show the effect of yielding on the minimum cross-sectional area of a column of effective length 1000 mm, over a range of compressive load from 5000 to 50,000 N.

Take $e = 0$ for a perfectly straight column and $K/K_0 = 1.0$ for no local imperfection. Use the material data already present in the spreadsheet. Note the cross-sectional area and the average stress at each load, and observe the reduction in modulus with increasing load. Plot a graph of stress after optimization against structural index P/L^2 on a log-log basis. Add a line to this graph for the same column with no yielding (efficiency $\eta = 0.780$).

3.10 Modify the spreadsheet 'Eccentrically Loaded Column' to optimize a square-section tube under eccentric load.

Use the formulae in Sect. 2.1.1 for the Euler and local buckling stresses of a square-section tube.

References

1. Walsh GR (1975) Methods of optimization. John Wiley & Sons, London
2. Reklaitis GV, Ravindran A, Ragsdell KM (1983) Engineering optimization. John Wiley and Sons, New York
3. Megson THG (1999) Aircraft structures for engineering students. Arnold, London
4. Kuhn HW, Tucker AW (1951) Nonlinear programming. Proceedings of 2nd Berkeley Symposium. University of California Press, pp 481–492
5. Karush W (1939) Minima of functions of several variables with inequalities as side constraints. M.Sc. dissertation, Department of Mathematics, University of Chicago
6. Timoshenko SP, Gere JM (1961) Theory of elastic stability. McGraw-Hill, New York (reprinted by Dover Publications, 2009)
7. ESDU 83034 (1988) Elastic local buckling stresses of thin-walled unstiffened circular cylinders under combined axial compression and internal pressure. Engineering Sciences Data, Structures series, section 25. IHS, London
8. Ramberg W, Osgood WR (1943) Description of stress-strain curves by three parameters. NACA Tech. Note 902
9. ESDU 76016 (1991) Generation of smooth continuous stress–strain curves for metallic materials. Engineering Sciences Data, Structures series, section 2. IHS, London

Chapter 4
Numerical Methods for Unconstrained Optimization

Abstract Unconstrained optimization is the search for the maximum or minimum of a function with no restriction on the values of the variables. At the same time, it forms the basis for methods of constrained optimization in the next chapter. Zero-order methods use only function values, progress made in the previous step pointing the way to the next step. The Hooke and Jeeves method is one such method, suitable for small problems with little programming effort. First-order methods employ the gradient of the function, usually obtained by finite difference, to derive a search direction. This is followed by a line search along this direction for the current maximum or minimum, performed either by progressive reduction of the region in which the maximum or minimum is to be found or by polynomial interpolation. In its simplest form, this is the steepest descent method. However, by the use of gradient data from the previous iteration, an improved search direction can be found, with faster convergence. This is the Fletcher–Reeves method. A more general formulation is based on a quadratic approximation to the objective function, referred to as a second-order method or quasi-Newton method. This involves progressively building up an approximation to the inverse of the Hessian matrix of second derivatives to deduce a search direction. A spreadsheet program for the Hooke and Jeeves method is also used in the next chapter for the penalty function method for constrained optimization.

In previous chapters, we made use of the Solver tool in various examples of unconstrained and constrained optimization. This was without further consideration of the methods underlying numerical optimization routines such as those in Solver. Numerical optimization provides the means of solving an optimization problem when it is too complex for analytical solution and in particular for large problems when the analysis itself is numerical or derived from some other form of numerical data. Numerical methods are, therefore, at the heart of most practical optimization problems. Methods for unconstrained optimization, that is minimization of a function in the absence of constraints, are introduced in the present chapter and in the following chapter methods for constrained optimization. While few practical problems are likely to be fully unconstrained, unconstrained optimization forms the

© Springer International Publishing AG 2017
A. Rothwell, *Optimization Methods in Structural Design*, Solid Mechanics and Its Applications 242, DOI 10.1007/978-3-319-55197-5_4

basis of the constrained optimization methods in the next chapter. In both chapters, the aim is to present a broad outline of numerical optimization methods, in sufficient detail to enable the reader to make good use of existing optimization codes and to set up an optimization problem in an efficient way. It is not the intention to present these in the detail necessary to implement such methods in the computer, since this is generally a substantial programming task reserved for the expert in the field.

Numerical optimization is essentially an iterative process, the most common methods consisting of the following two steps. The first is to identify a 'search direction', that is a direction in which to move in the design space to approach as closely as possible the required minimum. The second is to perform a 'line search' along this direction to locate the minimum point on the line. Following a successful line search, a new search direction is calculated, and the process is repeated until a sufficiently accurate minimum has been found. Various methods for determining a search direction will be introduced in Sect. 4.1 and in Sect. 4.2 methods for the line search. However, before this, we shall consider a simple 'direct search' routine, requiring no more than repeated evaluations of the function to be minimized. Finally, it might be pointed out that while we refer above only to the minimum of a function, not the maximum, optimization methods can of course be applied equally to either problem with no change in the methods used, if only by changing the sign of the objective function. For simplicity, we shall assume throughout this and the following chapter that it is the minimum of a function that we require.

4.1 Unconstrained Optimization

In principle, it would be possible to search for the minimum of an unconstrained function in a trial-and-error fashion, simply by evaluating enough points in the design space until a result judged to be sufficiently close to the minimum emerges. In a more structured approach, we might proceed as follows. Starting from some chosen initial point, we take a step in each design variable in turn to explore neighbouring points. At any point at which the function value is reduced, we take the next step from that point. This is continued until no further reduction is found. To refine the search, we then reduce the step size and resume the exploration of neighbouring points, now closer to the required minimum. Further reductions in step size are made until a sufficiently accurate minimum has been found. While this whole process is likely to involve a large number of individual steps, it can be accelerated in the following way. Provided that, after any complete round of the design variables in the 'exploratory' moves above, there has been some reduction in the function we can make a further 'pattern' move in the direction in which we have moved from the point at the beginning of that round to the current point. This is on the basis that progress has already been made in this direction, and it is reasonable to suppose that a further move in the same direction, by an amount equal to the progress made in that round, will speed up the search. This is the well-known method proposed by Hooke and Jeeves [1], at one time used extensively for many practical optimization problems (a useful description of the method is also given by Walsh [2]). The spreadsheet program in Sect. 4.3.1 demonstrates the use of the Hooke and Jeeves method for a problem to be defined by the user.

However, the purpose of any numerical optimization routine must be to locate the minimum of a function in the least number of steps, that is to follow a path through the design space leading as directly as possible to the desired optimum. Clearly to proceed by the step-by-step procedure described above, even with the pattern moves, requires the evaluation of the objective function many times. This can rapidly become very inefficient for larger problems, or when the function itself is expensive to compute. It will be observed that only function values are calculated in the above method (no derivatives of the function), and for this reason, it is referred to as a 'zero-order' method. More efficient zero-order methods are available, notably that of Powell [3], based on so-called conjugate directions, but the number of function evaluations remains high for large problems. For this reason, no further attention is given to zero-order methods in this chapter, other than for the previously mentioned program for the Hooke and Jeeves method in Sect. 4.3.1. However, the main reason for the brief discussion of the Hooke and Jeeves method here is to point out that exploration around the current point coupled with a pattern move is actually an attempt to determine a search direction with which to approach the minimum as quickly as possible. To find the most effective search direction is the prime goal of most conventional optimization methods. By first establishing a search direction, we proceed towards to the required minimum by allowing all variables to change by appropriate amounts at the same time, rather than one at a time, and this is essentially where the advantage lies. For the best search direction, we have to evaluate derivatives of the function, generally by finite difference. Since we use first derivatives, such methods are referred to as 'first-order' or gradient methods. The steepest descent method is the simplest form of gradient-based method, and its application to unconstrained optimization is discussed in the following section, before going on to discuss more elaborate methods in subsequent sections.

4.1.1 Steepest Descent Method

In the steepest descent method, also referred to as Cauchy's method, derivatives of the objective function are evaluated at some point in the design space and used to determine the best search direction from that point. By this is meant the direction in which the function decreases the most rapidly—in other words, the direction of steepest descent. A line search is then conducted along the search direction to locate the minimum point along this line. This point is not, of course, the actual minimum of the function, because the derivatives which have been used apply only at the point from which the line search is begun and generally change as we proceed along the search direction. Therefore, at the minimum point just found, we evaluate the derivatives again, find a new search direction, and conduct another line search. This rather intuitive procedure is continued until a sufficiently small change in the value of the function is found. This is then the required minimum. It has to be assumed that the function is 'smooth', that is continuous with continuous derivatives in the region of interest. The procedure is illustrated in Fig. 4.1, with contour lines of an unconstrained function shown as broken lines. In the figure, starting at some point 1, we perform a line search along the steepest descent direction to locate a minimum at point 2. At this point, a new search direction is calculated, followed by another line search to locate the next minimum at point 3 and so on.

The best search direction implies the greatest reduction δf in a function $f(\mathbf{x})$, $\mathbf{x} = x_1, x_2, \ldots, x_n$, at a nominal 'unit distance' from a given point. For this, we require the derivatives $\frac{\partial f}{\partial x_1}, \frac{\partial f}{\partial x_2}, \ldots, \frac{\partial f}{\partial x_n}$, as already stated generally evaluated by finite difference. For δf, we can write:

$$\delta f = \frac{\partial f}{\partial x_1} \delta x_1 + \frac{\partial f}{\partial x_2} \delta x_2 + \cdots + \frac{\partial f}{\partial x_n} \delta x_n.$$

If we define the search direction by its components:

$$s_1 = \delta x_1, s_2 = \delta x_2, \ldots, s_n = \delta x_n,$$

we have:

$$\delta f = f_1 s_1 + f_2 s_2 + \cdots + f_n s_n = \sum_{i=1}^{n} f_i s_i, \qquad (4.1)$$

where $f_1, f_2, \ldots, f_n$ are the derivatives of $f(\mathbf{x})$ with respect to $x_1, x_2, \ldots, x_n$. To minimize δf, that is for the greatest reduction in $f(\mathbf{x})$ subject to:

$$\sum_{i=1}^{n} s_i^2 = s_1^2 + s_2^2 + \cdots + s_n^2 = 1 \qquad (4.2)$$

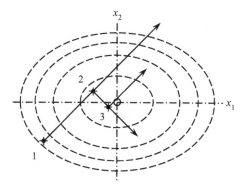

Fig. 4.1 Steepest descent method (contours of the function shown as *broken lines*)

to limit movement in any direction to unit distance from the given point, we can write the Lagrangian function:

$$F = \sum_{i=1}^{n} f_i s_i - \lambda \left[\sum_{i=1}^{n} s_i^2 - 1 \right]. \tag{4.3}$$

Differentiating with respect to each s_i and setting the result equal to zero, we obtain:

$$\frac{\partial F}{\partial s_i} = f_i - 2\lambda s_i = 0,$$

giving:

$$s_i = \frac{f_i}{2\lambda} = \frac{1}{2\lambda} \cdot \frac{\partial f}{\partial x_i}.$$

Recalling Eq. (3.6) in Chap. 3, the Lagrange multiplier λ represents the rate of increase in function f with increase in value of the constraint, in this case the distance from the given point. For movement in the direction of decreasing f, the Lagrange multiplier λ must be negative, therefore, but its actual numerical value is irrelevant because we are only interested in the relative values of $s_1, s_2, \ldots, s_n$. The components s_i of the required search direction are then directly related to the negative of the derivatives of f:

$$s_i = -\frac{\partial f}{\partial x_i}.$$

A line search is conducted along the search direction until the minimum along this line is found, as already indicated in Fig. 4.1. At that point, a new search direction is calculated, and another line search is performed. Numerical methods for the line search are given in Sect. 4.2.

The set of derivatives $\left\{ \frac{\partial f}{\partial x_1}, \frac{\partial f}{\partial x_2}, \ldots, \frac{\partial f}{\partial x_n} \right\}$ is termed the gradient ∇f of function $f(\mathbf{x})$. In terms of ∇f and the search direction $\mathbf{s} = \{s_1, s_2, \ldots, s_n\}$, Eqs. (4.1) and (4.2) can be written more compactly in matrix form as:

$$\begin{aligned} \text{minimise} &: \delta f = \mathbf{s}^t \nabla f, \\ \text{subject to} &: \mathbf{s}^t \mathbf{s} = 1. \end{aligned}$$

The Lagrangian function in Eq. (4.3) becomes:

$$F = \mathbf{s}^T \nabla f - \lambda (\mathbf{s}^T \mathbf{s} - 1),$$

(where superscript T denotes the transpose of **s**, i.e. a row vector). The resulting search direction:

$$\mathbf{s} = -\nabla f$$

is, of course, no different, but the notation introduced here is necessary in the following sections.

Example 4.1 Use the steepest descent method to find the minimum of the function:

$$f(\mathbf{x}) = x_1^2 + 2x_2^2 + 1.$$

While the purpose of this chapter is, of course, to study numerical optimization methods, a simple function has been chosen for this example so that the first few steps of the steepest descent method can be demonstrated analytically. The minimum of the above function is readily seen to be $f = 1$ at $x_1 = x_2 = 0$. However, to locate the minimum by the steepest descent method, we arbitrarily choose an initial point $x_1 = -2$, $x_2 = -1$. The derivatives of the function are as follows:

$$\frac{\partial f}{\partial x_1} = 2x_1, \quad \frac{\partial f}{\partial x_2} = 4x_2.$$

At the initial point, the components of the search direction are therefore:

$$s_1 = -\frac{\partial f}{\partial x_1} = 4, \quad s_2 = -\frac{\partial f}{\partial x_2} = 4.$$

The line search can also be performed analytically in this example. This is along the line:

$$x_2 = x_1 + 1$$

passing through the initial point $x_1 = -2$, $x_2 = -1$ and having a slope:

$$\frac{dx_2}{dx_1} = \frac{s_2}{s_1} = 1,$$

as required. On this line, substituting for x_2 from above:

$$f = x_1^2 + 2(x_1 + 1)^2 + 1,$$

with a minimum at:

$$x_1 = -\frac{2}{3}, \quad x_2 = \frac{1}{3}.$$

At this point, we now evaluate the components of a new search direction:

$$s_1 = -\frac{\partial f}{\partial x_1} = \frac{4}{3}, \quad s_2 = -\frac{\partial f}{\partial x_2} = -\frac{4}{3}.$$

As above, we define a line for the search direction:

$$x_2 = -x_1 - \frac{1}{3},$$

which, on substitution for x_2, gives:

$$f = x_1^2 + 2\left(-x_1 - \frac{1}{3}\right)^2 + 1$$

with a minimum at:

$$x_1 = -\frac{2}{9}, \quad x_2 = -\frac{1}{9}.$$

The next search direction now has to be evaluated at this point, and the procedure continued.

It is seen that we have already made substantial progress towards the minimum of the function. The procedure is continued further in Table 4.1 (the results of the present example were also used to draw Fig. 4.1). After seven iterations, we have approached the minimum to within 0.1 per cent of the initially chosen values. It will be appreciated that the analytical solution in this example is only for the purpose of illustrating the method. As already said, when implemented in the computer derivatives of the function to establish a search direction would normally be obtained by finite difference, and the line search would also be performed numerically, as described later in this chapter.   ∎

Iteration	x_1	x_2	$f(\mathbf{x})$	s_1	s_2
0	−2.0000	−1.0000	7.0000	−4.0000	−4.0000
1	−0.6667	0.3333	1.6667	−1.3333	1.3333
2	−0.2222	−0.1111	1.0741	−0.4444	−0.4444
3	−0.0741	0.0370	1.0082	−0.1481	0.1481
4	−0.0247	−0.0123	1.0009	−0.0494	−0.0494
5	−0.0082	0.0041	1.0001	−0.0165	0.0165
6	−0.0027	−0.0014	1.0000	−0.0055	−0.0055
7	−0.0009	0.0005	1.0000	−0.0018	0.0018

Table 4.1 Sequence of moves in the steepest descent method in Example 4.1

4.1.2   Fletcher–Reeves Method

While steepest descent is the simplest form of gradient-based method, it is by no means the most effective. It is shown in Table 4.1 that successive steps in the steepest descent method become increasingly small. In an extreme situation, this effect is referred to as 'zigzagging' to the minimum. Although in the previous example there was no noticeable difficulty, in a larger problem or with a less favourable function to be minimized convergence can become very poor. The method of Fletcher and Reeves [4] seeks to improve this by basing each successive search direction on both the current point and the previous one, in effect making use of the whole sequence of iterations as the process approaches the required minimum. The formula for the search direction at the next iteration, in the form in which it is usually presented, is:

$$\mathbf{s}^{(k+1)} = -\nabla f(\mathbf{x})^{(k+1)} + \frac{\left\|\nabla f(\mathbf{x})^{(k+1)}\right\|^2}{\left\|\nabla f(\mathbf{x})^{(k)}\right\|^2}\mathbf{s}^{(k)},$$

where superscripts $(k+1)$ and (k) refer to the next iteration and the last one, respectively, and $\|\nabla f(\mathbf{x})\|$ is the norm of $\nabla f(\mathbf{x})$:

$$\|\nabla f(\mathbf{x})\| = \sqrt{f_1^2 + f_2^2 + \cdots + f_n^2},$$

in which $f_1, f_2, \ldots, f_n$ are again the derivatives of $f(\mathbf{x})$ with respect to $x_1, x_2, \ldots, x_n$. The first iteration, since there has been no previous one, still has to be by steepest descent. The Fletcher–Reeves method makes a substantial improvement over the basic steepest descent method, requiring far fewer function evaluations, as illustrated in Example 4.2. It is left to the reader to consult the original reference or other text for the derivation of this formula.

Example 4.2 Repeat Example 4.1 using the Fletcher–Reeves method for the search direction.

As in the previous example, because a suitably simple function has been chosen, we can again illustrate the Fletcher–Reeves method analytically. The first iteration, by steepest descent, is identical to Example 4.1. The two terms of the gradient $\nabla f(\mathbf{x})$ at the initial point $x_1 = -2$, $x_2 = -1$ have already been calculated:

$$f_1^{(1)} = \frac{\partial f}{\partial x_1} = -4, \quad f_2^{(1)} = \frac{\partial f}{\partial x_2} = -4,$$

and at the first new point $x_1 = -\frac{2}{3}$, $x_2 = \frac{1}{3}$, we have:

$$f_1^{(2)} = -\frac{4}{3}, \quad f_2^{(2)} = \frac{4}{3},$$

Therefore, for the second iteration:

$$\frac{\left\|\nabla f(\mathbf{x})^{(2)}\right\|^2}{\left\|\nabla f(\mathbf{x})^{(1)}\right\|^2} = \frac{(f_1^{(2)})^2 + (f_2^{(2)})^2}{(f_1^{(1)})^2 + (f_2^{(1)})^2} = \frac{(-4/3)^2 + (4/3)^2}{(-4)^2 + (-4)^2} = \frac{1}{9}.$$

With the search direction $s_1^{(1)} = s_2^{(1)} = 4$ from the initial steepest descent step and the gradient terms $f_1^{(2)}$ and $f_2^{(2)}$ from above, the next search direction according to the Fletcher–Reeves method is:

$$s_1^{(2)} = -\left(-\frac{4}{3}\right) + \frac{1}{9} \times 4 = \frac{16}{9},$$

$$s_2^{(2)} = -\frac{4}{3} + \frac{1}{9} \times 4 = -\frac{8}{9}.$$

As before, we define a line for the new search direction:

$$x_2 = -\frac{x_1}{2}.$$

This line passes through the required point $x_1 = -\frac{2}{3}$, $x_2 = \frac{1}{3}$ and has slope:

$$\frac{dx_2}{dx_1} = \frac{s_2^{(2)}}{s_1^{(2)}} = -\frac{1}{2}.$$

Substituting for x_2 in the function:

$$f(\mathbf{x}) = x_1^2 + 2x_2^2 + 1$$

gives:

$$f = \frac{3}{2}x_1^2 + 1$$

with a minimum at:

$$x_1 = 0, \quad x_2 = 0.$$

At first sight, we have the rather remarkable result that the exact minimum has been reached in only two iterations. However, this should not be surprising because the Fletcher–Reeves method is based on the assumption that the function to be minimized is quadratic, which of course is the case in this example. Such favourable convergence should not be expected with a more general function. Nevertheless, even in a simple example such as this, we see the considerable advantage of the Fletcher–Reeves method over the basic steepest descent method. ■

4.1.3   Quasi–Newton Methods

The Fletcher–Reeves method (one of a class of methods referred to as 'conjugate gradient methods') is based on a quadratic approximation to the objective function. A more general formulation, again based on a quadratic approximation, is usually referred to as a 'quasi-Newton method'. To explain this term, in the classical Newton method, the 'Hessian' matrix $\mathbf{H}$ of second derivatives of the objective function is evaluated at each iteration. The change in gradient due to a move from a point $\mathbf{x}$ to point $\mathbf{x}'$ is $\mathbf{H}(\mathbf{x})(\mathbf{x}' - \mathbf{x})$. Therefore, to reduce the gradient $\nabla f(\mathbf{x})$ at point $\mathbf{x}$ to zero at a minimum point $\mathbf{x}'$, we set:

$$\mathbf{H}(\mathbf{x})(\mathbf{x}' - \mathbf{x}) = -\nabla f(\mathbf{x}),$$

or premultiplying by $\mathbf{H}(\mathbf{x})^{-1}$:

$$\mathbf{x}' = \mathbf{x} - \mathbf{H}(\mathbf{x})^{-1}\nabla f(\mathbf{x}).$$

Unless the original function is in fact quadratic, this has to be repeated until converged. However, if this procedure is adopted in any but a small problem, calculation of all second derivatives rapidly becomes impractical, in particular in a numerical optimization where these are calculated by finite difference. Furthermore, in some situations, the process may diverge rather than converge. In a quasi-Newton method, *estimates* of the Hessian matrix are progressively improved with the results of each previous iteration as the process continues. These are then used, not for a direct calculation of the optimum as in Newton's method, but for a new search direction expressed as:

$$\mathbf{s} = -\mathbf{B}\nabla f(\mathbf{x}),$$

where matrix $\mathbf{B}$ is an approximation to the inverse of the Hessian matrix, updated at each iteration.

Widely used formulae for updating $\mathbf{B}$ are the Davidon–Fletcher–Powell (DFP) formula [5, 6] and the Broyden–Fletcher–Goldfarb–Shanno (BFGS) formula [7–10]. These are stated without derivation below. In the usual notation, the DFP formula is:

$$\mathbf{B}^{(k+1)} = \mathbf{B}^{(k)} - \frac{\mathbf{B}^{(k)}\mathbf{y}\,\mathbf{y}^{\mathrm{T}}\mathbf{B}^{(k)}}{\mathbf{y}^{\mathrm{T}}\mathbf{B}^{(k)}\mathbf{y}} + \frac{\mathbf{p}\,\mathbf{p}^{\mathrm{T}}}{\mathbf{p}^{\mathrm{T}}\mathbf{y}}, \tag{4.4}$$

where

$$\mathbf{y} = \nabla f(\mathbf{x})^{(k+1)} - \nabla f(\mathbf{x})^{(k)}$$

are the differences in gradient from which improved estimates of matrix $\mathbf{B}$ of second derivatives are progressively built up, and:

$$\mathbf{p} = \mathbf{x}^{(k+1)} - \mathbf{x}^{(k)}.$$

Superscripts $(k+1)$ and (k) refer again to the next iteration and the last one, respectively. The BFGS formula is:

$$\mathbf{B}^{(k+1)} = \mathbf{B}^{(k)} + \left(1 + \frac{\mathbf{y}^{\mathrm{T}}\mathbf{B}^{(k)}\mathbf{y}}{\mathbf{p}^{\mathrm{T}}\mathbf{y}}\right)\frac{\mathbf{p}\,\mathbf{p}^{\mathrm{T}}}{\mathbf{p}^{\mathrm{T}}\mathbf{y}} - \frac{\mathbf{p}\,\mathbf{y}^{\mathrm{T}}\mathbf{B}^{(k)}}{\mathbf{p}^{\mathrm{T}}\mathbf{y}} - \frac{\mathbf{B}^{(k)}\mathbf{y}\,\mathbf{p}^{\mathrm{T}}}{\mathbf{p}^{\mathrm{T}}\mathbf{y}}. \qquad (4.5)$$

The initial matrix $\mathbf{B}$ is the identity matrix $\mathbf{I}$, so the first step in both cases is again steepest descent. In both methods, a new search direction is determined by means of one or other of the above formulae, followed directly by a line search in this direction. Both methods give a further significant improvement in convergence, at the expense of extra complexity, while a disadvantage of both compared with the Fletcher–Reeves method is the need to store matrix $\mathbf{B}$ in a large problem.

Example 4.3 Calculate a new search direction after the first iteration in Example 4.1 using the DFP update formula in Eq. (4.4).

The first iteration in Example 4.1 was by steepest descent. In the notation of Eq. (4.4):

$$\mathbf{x}^{(1)} = \begin{Bmatrix} -2 \\ -1 \end{Bmatrix}, \quad \mathbf{x}^{(2)} = \begin{Bmatrix} -2/3 \\ 1/3 \end{Bmatrix}$$

giving:

$$\mathbf{p} = \mathbf{x}^{(2)} - \mathbf{x}^{(1)} = \begin{Bmatrix} 4/3 \\ 4/3 \end{Bmatrix}.$$

Also:

$$\nabla f(\mathbf{x})^{(1)} = \begin{Bmatrix} -4 \\ -4 \end{Bmatrix}, \quad \nabla f(\mathbf{x})^{(2)} = \begin{Bmatrix} -4/3 \\ 4/3 \end{Bmatrix}$$

giving:

$$\mathbf{y} = \nabla f(\mathbf{x})^{(2)} - \nabla f(\mathbf{x})^{(1)} = \begin{Bmatrix} 8/3 \\ 16/3 \end{Bmatrix}.$$

The initial $\mathbf{B}$ matrix is the unit matrix $\mathbf{B} = \mathbf{I}$. With $\mathbf{p}$ and $\mathbf{y}$ from above, we calculate the following terms in the DFP update formula (expressed now in decimals):

$$\mathbf{B}^{(1)}\mathbf{y}\,\mathbf{y}^{\mathrm{T}}\mathbf{B}^{(1)} = \begin{bmatrix} 1 & 0 \\ 0 & 1 \end{bmatrix} \begin{Bmatrix} 2.667 \\ 5.333 \end{Bmatrix} \begin{Bmatrix} 2.667 \\ 5.333 \end{Bmatrix}^{\mathrm{T}} \begin{bmatrix} 1 & 0 \\ 0 & 1 \end{bmatrix} = \begin{bmatrix} 7.111 & 14.22 \\ 14.22 & 28.44 \end{bmatrix},$$

$$\mathbf{y}^{\mathrm{T}}\mathbf{B}^{(1)}\mathbf{y} = \begin{Bmatrix} 2.667 \\ 5.333 \end{Bmatrix}^{\mathrm{T}} \begin{bmatrix} 1 & 0 \\ 0 & 1 \end{bmatrix} \begin{Bmatrix} 2.667 \\ 5.333 \end{Bmatrix} = 35.56,$$

$$\mathbf{p}\,\mathbf{p}^{\mathrm{T}} = \begin{Bmatrix} 1.333 \\ 1.333 \end{Bmatrix} \begin{Bmatrix} 1.333 \\ 1.333 \end{Bmatrix}^{\mathrm{T}} = \begin{bmatrix} 1.778 & 1.778 \\ 1.778 & 1.778 \end{bmatrix},$$

$$\mathbf{p}^{\mathrm{T}}\mathbf{y} = \begin{Bmatrix} 1.333 \\ 1.333 \end{Bmatrix}^{\mathrm{T}} \begin{Bmatrix} 2.667 \\ 5.333 \end{Bmatrix} = 10.67.$$

After dividing each term in the matrices $\mathbf{B}^{(1)}\mathbf{y}\,\mathbf{y}^{\mathrm{T}}\mathbf{B}^{(1)}$ and $\mathbf{p}\,\mathbf{p}^{\mathrm{T}}$ above by 35.56 and 10.67, respectively, we have according to Eq. (4.4) for the second iteration:

$$\mathbf{B}^{(2)} = \begin{bmatrix} 1 & 0 \\ 0 & 1 \end{bmatrix} - \begin{bmatrix} 0.2000 & 0.4000 \\ 0.4000 & 0.8000 \end{bmatrix} + \begin{bmatrix} 0.1667 & 0.1667 \\ 0.1667 & 0.1667 \end{bmatrix} = \begin{bmatrix} 0.9667 & -0.2333 \\ -0.2333 & 0.3667 \end{bmatrix}.$$

The new search direction is therefore:

$$\mathbf{s} = -\mathbf{B}^{(2)}\nabla f(\mathbf{x})^{(2)} = \begin{bmatrix} 0.9667 & -0.2333 \\ -0.2333 & 0.3667 \end{bmatrix} \begin{Bmatrix} -1.333 \\ 1.333 \end{Bmatrix} = \begin{Bmatrix} 1.600 \\ -0.800 \end{Bmatrix},$$

or

$$s_1 = 1.600, \quad s_2 = -0.800.$$

It will be observed that this search direction is the same (aside from the actual numerical values of s_1 and s_2) as at the second iteration in the Fletcher–Reeves method in Example 4.2. This is because the function to be minimized was chosen to be a simple quadratic function, but in general, the DFP update will, of course, give a different result to the Fletcher–Reeves method. ∎

4.2 Line Search Methods

As will already have become clear, the line search is an essential component of the optimization methods described in Sect. 4.1 and, as we shall see, in methods for constrained optimization in Chap. 5. In a line search, we are looking for the minimum of the objective function along the line of the current search direction. Being repeated many times during the optimization, the line search is generally one of the most time-consuming parts of the whole procedure. For this reason, it is essential that it is done as efficiently as possible. In principle, we might proceed step-by-step along the search direction, but this could lead to an excessive number of function evaluations and would soon be inefficient with a step size small enough to locate the minimum accurately. The most common methods of performing the line search are by 'region elimination' or by 'polynomial interpolation'. Generally, region elimination followed by polynomial interpolation is found to be an effective approach.

4.2.1 Region Elimination and the Golden Section Method

We define a point on the line of the current search direction by coordinate x (note that x is conventionally used for this and does not refer to any of the design variables of the original optimization problem). By region elimination, we aim to narrow down the search for the minimum as quickly as possible. For this, we require upper and lower limits x_U and x_L on the interval in which the search is to be conducted. One of these will normally be the current point in the design space and the other a limit set to cover the range within which it is expected to find the minimum. We then require two intermediate points x_1 and x_2, as shown in the upper part of Fig. 4.2. Name the values of the objective function at these two points f_1 and f_2. It has to be assumed that the function has only a single minimum within the search interval. From the figure, we see immediately that if $f_1 < f_2$ the minimum lies between x_L and x_2. We can then eliminate the region between x_2 and x_U, by moving the upper limit x_U to x_2. A similar argument applies if $f_1 > f_2$, the region between x_L and x_1 then being eliminated. By selecting two new points between the current x_L and x_U (that is after moving x_U or x_L as appropriate), the process can be repeated and the search interval further reduced. This is continued until the search interval has been sufficiently reduced and either a satisfactory value of the minimum has already been obtained or the result is to be further refined by polynomial interpolation, as described later.

It is seen that each region elimination as described above requires the objective function to be evaluated at two new points x_1 and x_2 in the search interval. However, no mention has been made up to now of the preferred location of these two points. The need to evaluate the function at two new points, rather than just one, is ingeniously avoided in the 'golden section' method (see [11, 12, 13]). In this method, the aim is, if $f_1 < f_2$ as in Fig. 4.2, to reuse the previously calculated point x_1 in the next step (or reuse point x_2 if $f_1 > f_2$). We define x_1 and x_2 in terms of a constant τ by:

$$x_2 = \tau b,$$

and for a symmetrical arrangement of the two points:

$$x_1 = (1 - \tau)b,$$

Fig. 4.2 Golden section
method

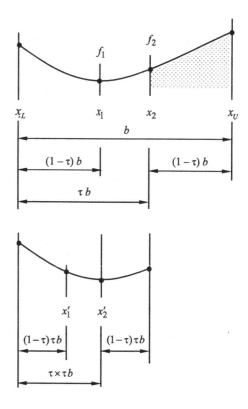

where b is the initial search interval, as shown in the upper part of Fig. 4.2. To reuse point x_1 in the next step (x_1 becomes the new x'_2 in the reduced search interval τb), we require:

$$x_1 = x'_2 = \tau \times \tau b,$$

as in the lower part of the figure. Equating the two forms of x_1:

$$\tau^2 b = (1 - \tau)b,$$

from which $\tau = 0.618034$. This number, or at least its reciprocal $1/\tau = 1.618034$, is a well-known number in history, with some special properties from which the name 'golden section' derives. This value of τ is used now for the location of the *one* new point at:

$$x'_1 = (1 - \tau)\tau b.$$

Otherwise, if $f_1 > f_2$, x_2 becomes the new x'_1 and we require a new point x'_2. However, we should note that the particular formulae for x_1, x_2, x'_1, x'_2 given above all refer to the *current* lower limit at each elimination. To refer all values to the same initial datum, general formulae are as follows:

$$
\begin{aligned}
\text{if } f_1 < f_2 : \quad & x_U' = x_2, \quad x_L' = x_L, \quad x_2' = x_1, \\
& x_1' = x_U' - \tau\,(x_U' - x_L'), \\
\text{if } f_1 > f_2 : \quad & x_U' = x_U, \quad x_L' = x_1, \quad x_1' = x_2, \\
& x_2' = x_L' + \tau(x_U' - x_L'),
\end{aligned}
\tag{4.6}
$$

where f_1 and f_2 refer to their values in the current elimination and x_1', x_2', x_U', x_L' are the new values for the next elimination. These formulae apply, of course, after an initial choice of x_U and x_L has been made, with corresponding x_1 and x_2. Using Eq. (4.6), the relative spacing of the two intermediate points is preserved, and the number of new function evaluations is halved. With a fraction $(1 - \tau)$ of the region eliminated each time, after five interval reductions, the search interval is reduced to about nine per cent of its original size and after ten interval reductions to less than one per cent.

4.2.2 Polynomial Interpolation

Alternatively, if a number of points along the search direction have already been evaluated, the minimum can be located by polynomial interpolation. This means simply fitting a polynomial function of appropriate degree to values of the original function at those points. Once the coefficients of the polynomial have been obtained, it is readily differentiated analytically to find the minimum along the curve. This does, of course, rely on the original function being smooth and continuous, at least in the vicinity of the minimum. Also, if it is highly nonlinear, the polynomial approximation may still be a poor match. For this reason, polynomial interpolation is commonly used to refine the result after a suitable number of steps of region elimination have been completed. Usually, either a second degree (quadratic) or third degree (cubic) approximation is satisfactory, at least if the available points are sensibly placed in the neighbourhood of the required minimum.

If we have three values of the objective function f_1, f_2, f_3 at points x_1, x_2, x_3 along the search direction, as shown in Fig. 4.3, the true curve between these points can be approximated by a parabola:

$$
f(x) = a_0 + a_1 x + a_2 x^2.
\tag{4.7}
$$

By substituting the three pairs of values $(f_1, x_1), (f_2, x_2), (f_3, x_3)$ in turn into Eq. (4.7), general formulae for the coefficients a_0, a_1, a_2 can be obtained as follows:

Fig. 4.3 Quadratic
interpolation

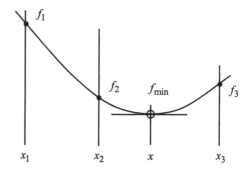

$$a_2 = \frac{(f_3 - f_1)/(x_3 - x_1) - (f_2 - f_1)/(x_2 - x_1)}{(x_3 - x_2)},$$

$$a_1 = \frac{(f_2 - f_1)}{(x_2 - x_1)} - a_2(x_1 + x_2),$$

$$a_0 = f_1 - a_1 x_1 - a_2 x_1^2.$$

Differentiating Eq. (4.7):

$$\frac{df}{dx} = a_1 + 2a_2 x$$

and putting this equal to zero for the minimum of $f(x)$ give:

$$x = -\frac{a_1}{2a_2}.$$

By substituting for a_1 and a_2 from above, we obtain an explicit formula for x:

$$x = x_1 + \frac{(f_3 - f_1)(x_2 - x_1)^2 - (f_2 - f_1)(x_3 - x_1)^2}{2[(f_3 - f_1)(x_2 - x_1) - (f_2 - f_1)(x_3 - x_1)]} \qquad (4.8)$$

at the minimum point of the parabolic approximation to the original function. The
minimum value of the function can then be calculated by this formula, or if preferred by means of the coefficients a_i above.

Extensive formulae for both quadratic interpolation and cubic interpolation (also
when a derivative df/dx of the function is available) are given by Vanderplaats
[13] and many other authors. By use of quadratic interpolation to complete the line
search after region elimination, it is of course possible to reduce the number of
region eliminations necessary and still obtain a satisfactory estimate of the minimum point. With cubic interpolation, requiring four points and not surprisingly
more complicated formulae, the number of region eliminations can be further
reduced or accuracy improved.

Example 4.4 Use the golden section method to find the minimum of the function:

$$f(x) = (x-1)^2 + 2x^4.$$

Refine the result by quadratic interpolation.

Again, we have a simple analytical function, the exact minimum of which is readily found by differentiation:

$$f_{\min} = 0.375 \quad \text{at} \quad x = 0.5.$$

Note that the presence of the second term in the function ensures that a parabolic curve cannot be a true fit to the function. Table 4.2 shows the sequence of steps in region elimination by the golden section method. By inspection of the function, it is found that the minimum must be between $x = 0$ and $x = 1$, so these values are used as initial lower and upper limits x_L and x_U. Intermediate values x_1 and x_2 are calculated as in Eq. (4.6). At each step, the larger of f_1 or f_2 is used to determine whether to replace x_L or x_U in the next row, as indicated in the last column of the table. After five region eliminations, the process is completed by quadratic interpolation. The three points selected, together with their corresponding function values extracted from the table, are:

$$x_1 = 0.4721, \quad f_1 = 0.3780,$$
$$x_2 = 0.4934, \quad f_2 = 0.3752,$$
$$x_3 = 0.5279, \quad f_3 = 0.3782.$$

These are shown in bold in Table 4.2. Substitution in Eq. (4.8) and then in the original function gives finally:

$$f_{\min} = 0.3750 \text{ at } x = 0.4995,$$

which is very close to the exact result. ∎

Table 4.2 Series of region eliminations by the golden section method in Example 4.4

x_L	x_U	x_1	x_2	f_1	f_2	$x_U - x_L$	Action
0.0000	1.0000	0.3820	0.6180	0.4245	0.4377	1.0000	Replace x_U by x_2
0.0000	0.6180	0.2361	0.3820	0.5898	0.4245	0.6180	Replace x_L by x_1
0.2361	0.6180	0.3820	0.4721	0.4245	0.3780	0.3820	Replace x_L by x_1
0.3820	0.6180	0.4721	**0.5279**	0.3780	**0.3782**	0.2361	Replace x_U by x_2
0.3820	0.5279	0.4377	0.4721	0.3896	0.3780	0.1459	Replace x_L by x_1
0.4377	0.5279	**0.4721**	**0.4934**	**0.3780**	**0.3752**	0.0902	

4.3   Spreadsheet Program

While Solver is used for the most part throughout this book, a program for the Hooke and Jeeves method is included here to demonstrate that numerical optimization does not necessarily require very complex programming; indeed, it will be seen that it can be programmed in rather few lines of code. The method is used again in the next chapter to illustrate the use of the penalty function method, since this is not included in the methods available in Solver. The Hooke and Jeeves method, referred to earlier in Sect. 4.1, was widely used as a general-purpose optimization method for many years after its first introduction, mostly in combination with a penalty function to accommodate constraints. Like most zero-order methods, the Hooke and Jeeves method has the advantage of being simple and robust, this of course at the cost of increased computational time. With modern computers, for a problem with no excessive number of variables and if the function to be minimized is itself not too time-consuming, the Hooke and Jeeves method can still prove effective.

4.3.1   'Hooke and Jeeves Method'

The spreadsheet uses the Hooke and Jeeves method to find the minimum of an unconstrained function $f(\mathbf{x})$ to be defined by the user. The spreadsheet is shown in Fig. 4.4.[1]

The Hooke and Jeeves procedure is as follows:

1. Choose suitable step sizes δ_i for each variable x_i.
2. *Exploratory move*: from an initial base point, increase x_1 by δ_1. If $f(\mathbf{x})$ is reduced, the move is a success, and the move is retained. Otherwise, retract the move and decrease x_1 by δ_1. If this move is a success, it is retained; otherwise, it is retracted. Repeat this procedure for each variable x_i in turn from the point reached in the last move. If any of these moves has been a success, a new base point has been found. Otherwise, the sequence of exploratory moves has been a failure.
3. *Reduction in step size*: if the sequence of exploratory moves has been a failure, halve the step size for all variables (or reduce it by some other factor) and perform a new sequence of exploratory moves.
4. *Pattern move*: if a new base point has been found after any sequence of exploratory moves, move along the line from the previous to the new base point by an amount equal to the difference between the new and previous base points.

---

[1]This spreadsheet, and those in the following chapters, contains macros. Depending on the chosen security settings, a Security Warning: 'Macros have been disabled' may appear on the Message Bar. Click Enable Content to continue with the spreadsheet.

Hooke and Jeeves Method

- *Insert data below, and initial step sizes and initial values of the variables in the columns to the right.*
- *Press function key f 9 to perform the optimization.*

Number of variables = 2

Required number of reductions in step size = 10

Maximum number of exploratory moves at each step size = 100

variables	initial step	initial values	optimum values
x_1	1.00	2.0000	0.4287
x_2	1.00	1.0000	0.1426
x_3			
x_4			
x_5			
x_6			
x_7			
x_8			
x_9			
x_{10}			
x_{11}			
x_{12}			
x_{13}			
x_{14}			
x_{15}			
x_{16}			
x_{17}			

Initial value of the function = 6

Make no changes to cells within this border!

Fig. 4.4 Spreadsheet 'Hooke and Jeeves Method'

Perform a new sequence of exploratory moves from this point. If any of these moves is a success, a new base point has again been found. Perform a new pattern move from this point. Otherwise, reject the pattern move, return to the last base point and perform a new sequence of exploratory moves from that point.
5. Continue this procedure until the solution is sufficiently converged.

The Hooke and Jeeves procedure is programmed in Visual Basic in the function HJ in a module in the same workbook as the spreadsheet. The function to be minimized has to be created by the user in Visual Basic in the function FN, in the same module, in terms of the variables in the spreadsheet. Access to both functions is by clicking Visual Basic in the Code group on the Developer tab. Function HJ is referred to directly in the spreadsheet, while FN is called from HJ. The spreadsheet itself is used only for input of data and display of results, and for certain control parameters. Any specific data required for function FN can most easily be included in that function or otherwise accessed from the spreadsheet by including the extra data in the arguments of HJ and FN. If the maximum of a function is required, rather than the minimum, this is of course just a matter of reversing its sign. The simple quadratic function in Table 4.3 is currently entered in function FN to demonstrate use of the method.

The spreadsheet is currently set up for up to 20 variables. Chosen initial values of the variables should be entered, with their corresponding initial step sizes, in order in the appropriate columns. For less than 20 variables, the remaining entries in the columns can be left blank. The actual number of variables has to be entered in the spreadsheet. For best performance, it is recommended that the initial step sizes should be related to the magnitude of the variables. A sequence of exploratory and pattern moves is repeated until no further reduction in the function value is found. At that stage, the step size is halved, and the sequence of exploratory and pattern moves repeated. The number of step size reductions to be made has to be entered in the spreadsheet. This depends, of course, on the accuracy required. The maximum

Table 4.3 Minimum of $f(\mathbf{x})$ by the Hooke and Jeeves method

$f(\mathbf{x}) = x_1^2 + x_1 x_2 + 2x_2^2 - x_1 - x_2 + 1$

Step size (both variables)	x_1	x_2	$f(\mathbf{x})$
	2	1	6
1	1	0	1
0.5	0.5000	0	0.7500
0.25	0.5000	0	0.7500
0.125	0.3750	0.1250	0.7188
0.0625	0.4375	0.1250	0.7148
0.03125	0.4375	0.1250	0.7148
0.01563	0.4219	0.1406	0.7144
0.00781	0.4297	0.1406	0.7143
0.00391	0.4297	0.1406	0.7143
0.00195	0.4277	0.1426	0.7143
0.00098	0.4287	0.1426	0.7143

Table 4.4 Data entry for spreadsheet program 'Hooke and Jeeves Method'

Parameters	
Number of variables	Enter the value in cell D11 (maximum 20)
Required number of reductions in step size	Enter the value in cell D14 (determines accuracy of the result)
Maximum number of sequences of exploratory moves at any step size	Enter the value in cell D18 (see Sect. 4.3.1)
Initial step size for variables x_i	Enter values in order in column G, starting at cell G5 (remaining cells may be left blank)
Variables	
Initial values of variables x_i	Enter values in order in column H, starting at cell H5 (remaining cells may be left blank)
Create the function to be minimized in function FN (to access the function click visual basic on the Developer tab to open the visual basic editor)	

number of times to perform the sequence of exploratory moves (whether or not followed by a pattern move) at any step size also has to be entered (this is currently set at 100). This is to avoid the possibility of an endless loop while running the program if no minimum of the function actually exists, or to avoid an excessive number of exploratory and pattern moves if the starting point is too far from the optimum. In this case, the spreadsheet shows the values of the variables and the function when the procedure was terminated. Parameters and variables to be entered are listed in Table 4.4. After entering the required data, optimization is performed by pressing function key f 9 on the keyboard (note that 'Manual Workbook Calculation' has been selected in Options on the spreadsheet).

After optimization, the optimum values of the variables are returned to the spreadsheet, together with the minimum value of the function and the number of function evaluations necessary to reach the optimum. For the quadratic function already entered to demonstrate the program, with initial values $x_1 = 2$, $x_2 = 1$ and initial step sizes $s_1 = s_2 = 1$, it requires 101 function evaluations to reach the optimum to within an accuracy of ± 0.001. Values of x_1, x_2 and corresponding values of the function on completion of each step size are given in Table 4.3 to illustrate convergence.

Note that cells I5:I26 in the column headed 'optimum values' contain an array formula (enclosed in braces) for return of the results of the optimization to the spreadsheet. *No changes may be made to cells in this array unless the array formula is first deleted.*

4.4 Summary

Problems of unconstrained optimization—to find the minimum (or maximum) of a function in the absence of constraints—can in principle be solved by a direct search method, in which a sequence of values of the objective function is calculated until a solution sufficiently close to the minimum is found. Such methods are termed zero-order methods. The Hooke and Jeeves method is an example of this. However, zero-order methods, while simple and robust, become impractical with a large number of variables, or when the objective function itself is expensive to calculate. More powerful methods first find a best search direction in the design space and then perform a line search for the minimum along this line. The simplest method for finding the search direction is by steepest descent—the fastest 'downhill' direction —requiring calculation of first derivatives, or the gradient, of the objective function. Such methods are termed therefore first-order methods. In numerical optimization, first derivatives are normally calculated by finite difference. However, this inevitably leads to an iterative process because the 'best' search direction can only be based on derivatives at the current point in the design space and cannot 'see' the true minimum of the objective function. When the minimum along the current search direction has been found, the gradient has to be recalculated and the line search repeated. This is continued until a sufficiently accurate optimum has been found. More sophisticated first-order methods, such as the Fletcher–Reeves method, make use of both the current and previous iterations to deduce a search direction which is more nearly directed towards the optimum. Other methods, referred to as quasi-Newton or second-order methods, aim to progressively build up an approximation to the inverse of the Hessian matrix of second derivatives of the objective function, improving the search direction at each iteration, with faster convergence at the cost of extra complexity.

The line search is an essential part of all methods based on first finding a search direction. While the line search could be performed simply by proceeding step-by-step along the search direction, this is unlikely to be efficient. Usual methods for the line search are by region elimination or by polynomial interpolation. In region elimination, such as by the golden section method, the range within which the minimum is to be found is progressively reduced until it has been located to within specified limits. By polynomial interpolation, a second- or higher-order function is fitted to points along the search direction in the vicinity of the optimum, and this function is differentiated for the minimum. Standard formulae are available for this. For example, in parabolic interpolation, three points along the search direction are required to estimate the minimum. If necessary, the process can then be repeated with new points closer to the previously found minimum. Golden section elimination followed by parabolic interpolation is generally found to be a satisfactory method.

Finally, it is important to point out that the methods described here can only find a local minimum, generally the one closest to the starting point, should more local minima exist. The way around this would be to repeat the optimization from more starting points until satisfied that the true optimum had been found, or otherwise to resort to a method such as the genetic algorithm in Chap. 8, which retains more possible solutions during its search for the optimum. Again, the discussion throughout this chapter refers only to finding the minimum of a function, whether constrained or unconstrained, but with little change, the methods are equally applicable to finding a maximum.

Exercises

4.1 Repeat the first few steps of the problem in Example 4.1 for the same function $f(\mathbf{x})$, from a different starting point.
Follow the same analytical procedure as in the example. Plot values of the variables at each step in a figure similar to Fig. 4.1.

4.2 Perform the first few steps of the Fletcher–Reeves method to minimize the function:

$$f(\mathbf{x}) = x_1^2 + x_2^4 + 1.$$

Follow the same analytical procedure as in Example 4.2. Choose a suitable initial point to start the procedure.

4.3 Repeat Exercise 4.2 using the DFP update formula in Eq. (4.4).
Make an Excel spreadsheet to calculate the update formula. Use the MMULT function in Excel to perform the matrix multiplication.

4.4 Repeat Exercise 4.3 using the BFGS update formula in Eq. (4.5).
Extend the spreadsheet in Exercise 4.3 to calculate the BFGS update formula.

4.5 Use the golden section method to find the minimum of the function:

$$f(x) = 2x^3 - 3x + 2.$$

Use initial lower and upper bounds $x_L = 0$, $x_U = 1$ to start region elimination. Continue until the search interval has been reduced to about 10 per cent of its initial value.

4.6 Complete the minimization of the function in Exercise 4.5 by parabolic interpolation.
Use the standard formulae for parabolic interpolation in Sect. 4.2.2. Compare the result with the exact minimum found by differentiation of the original function.

4.7 Perform the first few steps of the procedure in the spreadsheet 'Hooke and Jeeves Method' by hand and compare the results with those in Table 4.3.
Take the same quadratic function with the same initial values and step size. Observe the working of the series of 'If' and 'ElseIf' statements enabling a pattern move towards the end of function HJ.

4.8 Use the spreadsheet 'Hooke and Jeeves Method' to repeat the optimization of the quadratic function in Table 4.3 with different initial values and step sizes.
Values of the variables and minimum of the function on completion of each step size can be seen by progressively increasing the number of reductions in step size entered on the spreadsheet. Observe the number of function evaluations necessary for the same accuracy with different initial values and step size.

4.9   Use the spreadsheet 'Hooke and Jeeves Method' to optimize the angles α
and β of the seven-bar truss in Fig. 1.15 of Chap. 1.
Formulae for analysis of the truss are given in Table 1.5 of Chap. 1.
Replace the Visual Basic code in function FN with some lines of code to
calculate the volume of the truss in terms of design variables $x_1 = D$ and
$x_2 = H$. Variable FN in the function FN must be the volume of the truss
for any x_1 and x_2. Choose convenient values for the load on the truss, its
span and the allowable stress. Express the result in terms of angles α
and β.

References

1. Hooke R, Jeeves TA (1961) 'Direct search' solution of numerical and statistical problems. J Assn Comput Mach 8:212–229
2. Walsh GR (1975) Methods of optimization. John Wiley & Sons, London
3. Powell MJD (1964) An efficient method for finding the minimum of a function of several variables without calculating derivatives. Comput J 7:303–307
4. Fletcher R, Reeves CM (1964) Function minimisation by conjugate gradients. Comput J 7(2):149–154
5. Davidon WC (1959) Variable metric method for minimization. Argone National Laboratory, ANL-5990 Rev., University of Chicago
6. Fletcher R, Powell MJD (1963) A rapidly convergent method for minimization. Comput J. 6(2):163–168
7. Broydon CG (1970) The convergence of a class of double rank minimization algorithms, parts I and II. J Inst Math Appl 6:76–90, 222–231
8. Fletcher R (1970) A new approach to variable metric algorithms. Comput J 13:317–322
9. Goldfarb D (1970) A family of variable metric methods derived by variational means. Math Comput 24:23–36
10. Shanno DF (1970) Conditioning of quasi-newton methods for function minimization. Math Comput 24:647–656
11. Kiefer J (1953) Sequential minimax search for a maximum. Proc Am Math Soc 4:502–506
12. Reklaitis GV, Ravindran A, Ragsdell KM (1983) Engineering optimization. John Wiley and Sons, New York
13. Vanderplaats GN (1984) Numerical optimization techniques for engineering design. McGraw-Hill, New York

Chapter 5
Numerical Methods for Constrained Optimization

Abstract Methods for constrained optimization described in this chapter can be broadly classified as constraint-following methods or penalty function methods. The gradient projection method and the generalized reduced gradient method are both constraint-following methods, on the basis that the optimum will lie on some or many constraints, and the aim is therefore to follow the constraints as closely as possible around the design space. In the gradient projection method, only those constraints currently active are included at any stage. The best search direction is found on the intersection of those constraints. Due to constraint nonlinearity, constraint gradients have to be re-evaluated at each step, and the process continued. In the generalized reduced gradient method, one of the methods in Solver, instead of an active constraint strategy surplus variables are added to convert inequality constraints into equalities. A search direction is then obtained from the reduced gradient in a set of independent variables. Again, constraint gradients have to be re-evaluated at each step. In a penalty function method, terms containing the constraint functions are added to the objective function to convert it in effect into an unconstrained problem, the aim being to avoid constraints or to penalize constraint violation. By the increase or decrease of a penalty parameter, the solution converges to the optimum of the constrained problem. A spreadsheet program for the penalty function method is based on the Hooke and Jeeves method in the previous chapter.

The numerical methods we studied in the previous chapter referred only to unconstrained optimization, in other words the search for the minimum of some function in the absence of constraints and with no other limits on the values of the variables. Unconstrained optimization forms the basis of the methods for constrained optimization in the present chapter, even though some substantial further development is required. The objective function and the constraints might be analytical functions but in practice are more likely to be the result of a more complex numerical analysis, while in structural design we are largely concerned with inequality rather than equality constraints and constraints that are in most cases nonlinear. The methods described in this chapter are aimed particularly at this class of problem, for which a numerical optimization procedure becomes almost

© Springer International Publishing AG 2017
A. Rothwell, *Optimization Methods in Structural Design*, Solid Mechanics and Its Applications 242, DOI 10.1007/978-3-319-55197-5_5

inevitable. The concept of a search direction and the subsequent line search were introduced in the previous chapter. The principal difference in constrained optimization is that the search direction now has to conform to the presence of constraints, that is, it depends not only on the objective function but also on the different constraint functions if violation of constraints is to be avoided. This is with the considerable added difficulty that it mostly is not known in advance which of the often many inequality constraints will prove to be active at the optimum.

Methods for constrained optimization currently available in Solver are as follows:

– the generalized reduced gradient method (GRG nonlinear),
– linear programming (simplex LP),
– the genetic algorithm (evolutionary).

The first is the one most generally applicable to structural design problems and the one we have been using up to now. It is one of a class of so-called constraint-following methods described in Sect. 5.1. Linear programming applies, as its name suggests, to problems in which both the objective function and all constraints are linear functions of the design variables. While linear programming is a very efficient method, such problems are less common in structural design, and we leave it to the many texts in which linear programming is explained in detail (see [3, 4, 7, 8] and many general texts on optimization methods). The genetic algorithm has been developed in particular for optimization problems with discontinuities or other irregularities, problems with discrete variables and those where local minima exist in a search for the true optimum. This is the case in the design of composite laminates, and further discussion of the genetic algorithm is deferred therefore until Chap. 8.

The penalty function method is an alternative approach to constrained optimization. By the addition of a penalty term to the objective function to account for constraints, the original constrained problem is converted into an unconstrained one. Although not available in Solver (and in view of the other methods provided might well be considered unnecessary), it is an intuitive and widely used method with the advantage that it can readily be implemented in an existing unconstrained optimization routine. The penalty function method is discussed in Sect. 5.2.

Finally, it might again be pointed out that while we refer in this chapter only to the minimum of a function, not the maximum, all optimization methods can, of course, be applied equally to either problem with no significant change.

5.1   Constraint-Following Methods

Suppose that with a conventional method for the minimization of an unconstrained function, we were to proceed from a feasible point in the design space until a constraint is first encountered. The process would simply stop at that point, because no mechanism has been provided by which the search direction can be modified to take account of the constraint. We shall consider first methods which 'follow the constraints' around the design space. On the basis that in a constrained optimization, the optimum will lie on at least one constraint, the search can then be confined to remain on that constraint while exploring further the design space. Additional constraints may be encountered as the process continues, while earlier constraints may be rejected as no longer active. This is the basis of the gradient projection method in Sect. 5.1.1. While equality constraints can be accommodated in this method, it is generally more suited to inequality constraints. The generalized reduced gradient method in Sect. 5.1.2 adopts a different strategy. Instead of identifying the currently active constraints, additional variables are introduced so that all constraints remain satisfied during the line search. In this way, the original problem is reduced to an unconstrained one, subject only to side constraints on the additional variables. This method is equally suited to both equality and inequality constraints. Other methods are referred to more briefly in Sect. 5.1.3.

Both methods above employ linearization of constraints, that is, they have in common that constraints are represented by a linear approximation at the current point in the optimization. In fact, the theoretical derivation in the following two sections is essentially for linearly constrained problems. A correction procedure becomes necessary to return to the true constraint boundary when the constraints are nonlinear. Although it is the generalized reduced gradient method that is used in Solver, we discuss first the gradient projection method in the next section because this is a more intuitive and visually appealing method and shares much in common with the generalized reduced gradient method.

5.1.1 Gradient Projection Method

The gradient projection method [10, 11] is a direct constraint-following method, based on an 'active constraint strategy' for an inequality constrained problem in which only those constraints currently active are included at any stage. It might be seen as the forerunner of the generalized reduced gradient method to be discussed in the following section. We consider first a problem with only linear constraints. Figure 5.1 shows a three-dimensional design space with a single constraint represented by the larger triangle in the figure. Starting from a feasible point in the design space, once this constraint has been encountered, the further search is in the steepest descent direction on this plane. In a multidimensional design space, the search direction is confined to the intersection of the currently active constraints (a hyperplane impossible to visualize!).

To determine the steepest descent direction on the condition that we remain on the active constraints, we consider first a small increment dg_j in a constraint $g_j(\mathbf{x})$:

$$dg_j = \frac{\partial g_j}{\partial x_1} dx_1 + \frac{\partial g_j}{\partial x_2} dx_2 + \cdots + \frac{\partial g_j}{\partial x_n} dx_n,$$

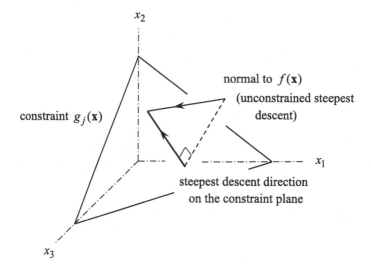

Fig. 5.1 Projection of the gradient of the objective function on the plane of a constraint

where $dx_1 = s_1, dx_2 = s_2, \ldots, dx_n = s_n$ are the increments in variables **x** corresponding to some movement in a search direction **s**. In terms of the gradient ∇g_j of the constraint, we can express the above formula in the usual way as:

$$dg_j = \mathbf{s}^T \nabla g_j.$$

We require $dg_j = 0$ to remain on the constraint $g_j(\mathbf{x})$. Therefore, for a *feasible* search direction **s** remaining on the intersection of all currently active constraints, we require

$$\mathbf{s}^T \nabla g_j = 0$$

for each active constraint (a set of linear equations equal to the number of active constraints). Any equality constraints must, of course, be included in the active constraints, but then, our initial feasible point should be one that satisfies all such equality constraints. If we define a matrix **N** as one whose columns are the gradients of the r active constraints, numbered for convenience $g_1, \ldots, g_r$:

$$\mathbf{N} = \begin{bmatrix} \frac{\partial g_1}{\partial x_1} & \cdots & \frac{\partial g_r}{\partial x_1} \\ \vdots & \ddots & \vdots \\ \frac{\partial g_1}{\partial x_n} & \cdots & \frac{\partial g_r}{\partial x_n} \end{bmatrix}, \tag{5.1}$$

we can write the above conditions for a feasible search direction more compactly as:

$$\mathbf{s}^{\mathrm{T}}\mathbf{N} = 0. \tag{5.2}$$

Similarly, a small increment in the objective function $f(\mathbf{x})$ can be written as:

$$df = \frac{\partial f}{\partial x_1} dx_1 + \frac{\partial f}{\partial x_2} dx_2 + \cdots + \frac{\partial f}{\partial x_n} dx_n,$$
$$\text{or}: \ df = \mathbf{s}^{\mathrm{T}}\nabla f.$$

For a *usable* search direction, we require simply that $df < 0$, but for the steepest descent direction, we have to minimize df (i.e. for the largest decrease in $f(\mathbf{x})$), subject to the constraint conditions in Eq. (5.2), at a nominal unit distance $\mathbf{s}^{\mathrm{T}}\mathbf{s} = 1$ from the current point in the design space. The optimization problem can now be expressed as:

$$\begin{aligned} \text{minimize:} \quad & df = \mathbf{s}^{\mathrm{T}}\nabla f \\ \text{subject to:} \quad & \mathbf{s}^{\mathrm{T}}\mathbf{N} = 0 \\ \text{and:} \quad & \mathbf{s}^{\mathrm{T}}\mathbf{s} = 1. \end{aligned}$$

This is a classical, equality constrained optimization problem. We write the Lagrangian function:

$$F = \mathbf{s}^{\mathrm{T}}\nabla f - \mathbf{s}^{\mathrm{T}}\mathbf{N}\lambda - \mu(\mathbf{s}^{\mathrm{T}}\mathbf{s} - 1),$$

where λ is the set of Lagrange multipliers corresponding to each active constraint in matrix $\mathbf{N}$ and μ is the single Lagrange multiplier referring to the condition $\mathbf{s}^{\mathrm{T}}\mathbf{s} = 1$. The condition for a minimum of $f(\mathbf{x})$ is:

$$\frac{\partial F}{\partial \mathbf{s}} = \nabla f - \mathbf{N}\lambda - 2\mu\mathbf{s} = 0. \tag{5.3}$$

Premultiplying by $\mathbf{N}^{\mathrm{T}}$ and observing that Eq. (5.2) can be rewritten $\mathbf{N}^{\mathrm{T}}\mathbf{s} = 0$, we obtain

$$\mathbf{N}^{\mathrm{T}}\nabla f - \mathbf{N}^{\mathrm{T}}\mathbf{N}\lambda = 0$$
$$\text{or}: \ \lambda = (\mathbf{N}^{\mathrm{T}}\mathbf{N})^{-1}\mathbf{N}^{\mathrm{T}}\nabla f. \tag{5.4}$$

Substituting in Eq. (5.3) and solving for $\mathbf{s}$, we obtain finally

$$\mathbf{s} = \frac{1}{2\mu}\left[\mathbf{I} - \mathbf{N}(\mathbf{N}^{\mathrm{T}}\mathbf{N})^{-1}\mathbf{N}^{\mathrm{T}}\right]\nabla f,$$

where $\mathbf{I}$ is the identity matrix. Since $\mathbf{s}$ is simply the set of directions along which to perform a line search, the factor $\frac{1}{2\mu}$ can be ignored. The search direction $\mathbf{s}$ above is commonly written as:

$$\mathbf{s} = -\mathbf{P}\,\nabla f,$$

where

$$\mathbf{P} = \left[\mathbf{I} - \mathbf{N}(\mathbf{N}^{\mathrm{T}}\mathbf{N})^{-1}\mathbf{N}^{\mathrm{T}}\right] \tag{5.5}$$

is the so-called projection matrix. This recognizes that $\mathbf{s}$ is actually the projection of the normal to the objective function on the intersection of the currently active constraints, as already illustrated in three variables in Fig. 5.1.

We now have a means of calculating a search direction which is both usable and feasible, in the direction of steepest descent on the currently active constraints. We perform a line search on this search direction until either a minimum is found or a constraint previously inactive is encountered, as follows:

1. If a minimum is found, the Lagrange multipliers are evaluated from Eq. (5.4) to validate the set of active constraints. If all the Lagrange multipliers are positive, the correct set of active constraints has been identified, and we have found the optimum. (The role of Lagrange multipliers in identifying the currently active constraints was discussed earlier in Sect. 3.3.)
2. If a previously inactive constraint is encountered, this is added to the current set of active constraints, and a new search direction $\mathbf{s}$ is calculated. Provided $\|\mathbf{s}\| \neq 0$, we proceed with the line search. However, if $\|\mathbf{s}\| = 0$ (to within a small margin to allow for numerical error), this means that there is no search direction in which further improvement can be made. We then have a potential solution to the problem, and as above, the Lagrange multipliers are evaluated. Again, if all the Lagrange multipliers are positive, then the correct set of active constraints has been identified, and we have found the optimum.
3. Otherwise, any constraints with negative Lagrange multipliers in (1) or (2) above are removed one by one (starting with the one with the most negative Lagrange multiplier), a new search direction is calculated, and the procedure is repeated.

The method is illustrated in the simple, two-dimensional design space in Fig. 5.2. Starting from a feasible point A, the first step is an unconstrained, steepest descent until constraint $g_1 = 0$ is encountered at point B. Here, we calculate a new search direction and proceed in the downhill direction along constraint $g_1 = 0$ until a second constraint $g_2 = 0$ is met at point C. Recalculating the search direction at this point, we now find $\|\mathbf{s}\| = 0$ (This is, of course, inevitable since in two dimensions the problem is now fully constrained.). The Lagrange multipliers are evaluated at point C, where we find λ_1 negative and λ_2 positive. Constraint g_1 is therefore removed from the active set, a new search direction is calculated, and we continue down constraint g_2 to point D, where again we find $\|\mathbf{s}\| = 0$ and the

Fig. 5.2 Constraint selection in the gradient projection method

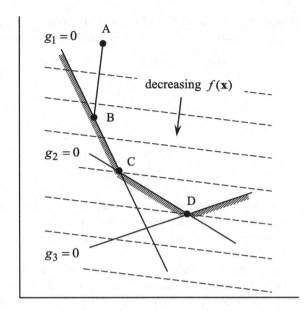

Lagrange multipliers are evaluated. Depending on the signs of these, point D may or may not be the optimum. It can already be seen in the figure that point D is indeed the optimum, but in terms of the Lagrange multipliers, this is detected by both λ_2 and λ_3 being positive. In the figure, the objective function $f(\mathbf{x})$ is shown as linear, but had it been some other nonlinear function, a minimum might alternatively have been found along either constraint g_1 or g_2 during the line search.

The basic assumption of the gradient projection method is that the constraints are linear. As already stated, in structural design we are mostly dealing with nonlinear constraints. This can be handled by the so-called linearization of constraints, that is, by treating them as linear constraints at the current point, calculating the constraint gradients and proceeding as before. Of course, as we proceed along the search direction, we move away from nonlinear constraints. A 'restoration move' is then necessary to return to the constraint boundary. For this, we move in a direction normal to the constraint surface, according to the formula:

$$\mathbf{x}' - \mathbf{x} = -\mathbf{N}(\mathbf{N}^{\mathrm{T}}\mathbf{N})^{-1}\mathbf{g}(\mathbf{x}),$$

where $\mathbf{x}' - \mathbf{x}$ is an estimate of the correction necessary to restore the solution to the constraint boundary and vector $\mathbf{g}(\mathbf{x})$ contains the values of the currently active constraints. Because of the nonlinearity, the above formula may have to be used iteratively until sufficiently small constraint values are obtained. At this point, the gradients have to be recalculated, and the procedure is continued. Usually, 'move limits' are imposed to limit the extent of movement in the search direction, avoiding excessive departures from the constraints so that the restoration process converges satisfactorily. The need for relatively tight move limits with highly nonlinear constraints, and the increased number of iterations necessary to return to the constraint boundary can contribute substantially to the time taken to reach an optimum.

Example 5.1 Use the gradient projection method to minimize the function:

$$f(\mathbf{x}) = x_1^2 + x_2^2 + x_3^2,$$

subject to constraints:

$$g_1(\mathbf{x}) = 3x_1 + 2x_2 + x_3 - 6 \geq 0,$$
$$g_2(\mathbf{x}) = x_3 - x_2 \geq 0.$$

The matrix operations in the gradient projection method rule out any but a simple example if extensive arithmetic is to be avoided. The present example serves to illustrate the principal steps in the procedure, in a problem in which the two constraints are both linear.

If we could be sure that constraint g_2 will be active at the optimum, it would be sensible to use this to eliminate variable x_3 from the problem and to continue with only two variables. For the purpose of this example, we do not do this. However, to simplify the example, we choose an initial feasible point:

$$x_1 = x_2 = x_3 = 2$$

at which constraint g_2 is already active. In this way, we avoid the first unconstrained steepest descent otherwise necessary to locate one or other constraint. At this point,

$$f(\mathbf{x}) = 12, \quad g_1 = 6, \quad g_2 = 0.$$

To find the search direction on this single active constraint, we proceed as follows. Inserting derivatives of constraint g_2 into matrix $\mathbf{N}$ in Eq. (5.1), we have

$$\mathbf{N} = \begin{bmatrix} 0 \\ -1 \\ 1 \end{bmatrix},$$

with transpose:

$$\mathbf{N}^T = \begin{bmatrix} 0 & -1 & 1 \end{bmatrix}.$$

Multiplying the two matrices:

$$\mathbf{N}^T\mathbf{N} = 2,$$

(a single numerical value) with inverse:

$$(\mathbf{N}^T\mathbf{N})^{-1} = 0.5.$$

Post-multiplying by $\mathbf{N}^T$:

$$(\mathbf{N}^T\mathbf{N})^{-1}\mathbf{N}^T = \begin{bmatrix} 0 & -0.5 & 0.5 \end{bmatrix}$$

and premultiplying by $\mathbf{N}$:

$$\mathbf{N}(\mathbf{N}^T\mathbf{N})^{-1}\mathbf{N}^T = \begin{bmatrix} 0 & 0 & 0 \\ 0 & 0.5 & -0.5 \\ 0 & -0.5 & 0.5 \end{bmatrix}.$$

The projection matrix in Eq. (5.5) is then:

$$\mathbf{P} = \mathbf{I} - \mathbf{N}(\mathbf{N}^T\mathbf{N})^{-1}\mathbf{N}^T = \begin{bmatrix} 1 & 0 & 0 \\ 0 & 0.5 & 0.5 \\ 0 & 0.5 & 0.5 \end{bmatrix}.$$

With the gradient of the objective function at the initial point:

$$\nabla f = \begin{bmatrix} 4 \\ 4 \\ 4 \end{bmatrix}$$

we have a search direction on the constraint g_2:

$$\mathbf{s} = -\mathbf{P}\nabla f = \begin{bmatrix} -4 \\ -4 \\ -4 \end{bmatrix}.$$

By a line search along the line:

$$x_1 = 2 - 4\alpha,$$
$$x_2 = 2 - 4\alpha,$$
$$x_3 = 2 - 4\alpha$$

(i.e. from the point $x_1 = x_2 = x_3 = 2$, with the above components of the search direction and line search parameter α), we meet constraint g_1 at the point:

$$\alpha = \frac{1}{4},$$

$$x_1 = x_2 = x_3 = 1,$$

$$f(\mathbf{x}) = 3.$$

It is easily verified that a minimum of the objective function does not exist between the initial point and this new point.

We continue by determining a search direction from the new point. Inserting now derivatives of the two constraints into the **N** matrix, we have

$$\mathbf{N} = \begin{bmatrix} 3 & 0 \\ 2 & -1 \\ 1 & 1 \end{bmatrix},$$

with transpose:

$$\mathbf{N}^{\mathrm{T}} = \begin{bmatrix} 3 & 2 & 1 \\ 0 & -1 & 1 \end{bmatrix}.$$

Multiplying the two matrices:

$$\mathbf{N}^{\mathrm{T}}\mathbf{N} = \begin{bmatrix} 14 & -1 \\ -1 & 2 \end{bmatrix}$$

and taking the inverse:

$$(\mathbf{N}^{\mathrm{T}}\mathbf{N})^{-1} = \begin{bmatrix} 0.07407 & 0.03704 \\ 0.03704 & 0.51852 \end{bmatrix}.$$

(Matrix inversion and other matrix operations can all be done by hand in this example, but are no doubt more easily performed by means of the matrix functions in Excel.)

Post-multiplying now by $\mathbf{N}^{\mathrm{T}}$:

$$(\mathbf{N}^{\mathrm{T}}\mathbf{N})^{-1}\mathbf{N}^{\mathrm{T}} = \begin{bmatrix} 0.2222 & 0.1111 & 0.1111 \\ 0.1111 & -0.4444 & 0.5556 \end{bmatrix}$$

and premultiplying by **N**:

$$\mathbf{N}(\mathbf{N}^{\mathrm{T}}\mathbf{N})^{-1}\mathbf{N}^{\mathrm{T}} = \begin{bmatrix} 0.6667 & 0.3333 & 0.3333 \\ 0.3333 & 0.6667 & -0.3333 \\ 0.3333 & -0.3333 & 0.6667 \end{bmatrix}.$$

The projection matrix in Eq. (5.5) is then:

$$\mathbf{P} = \mathbf{I} - \mathbf{N}(\mathbf{N}^\mathsf{T}\mathbf{N})^{-1}\mathbf{N}^\mathsf{T} = \begin{bmatrix} 0.3333 & -0.3333 & -0.3333 \\ -0.3333 & 0.3333 & 0.3333 \\ -0.3333 & 0.3333 & 0.3333 \end{bmatrix}.$$

With the gradient of the objective function at the new point:

$$\nabla f = \begin{bmatrix} 2 \\ 2 \\ 2 \end{bmatrix}$$

we have a new search direction:

$$\mathbf{s} = -\mathbf{P}\,\nabla f = \begin{bmatrix} 0.6667 \\ -0.6667 \\ -0.6667 \end{bmatrix}.$$

For convenience, we scale the above components of the search direction to give

$$s_1 = 1, \quad s_2 = -1, \quad s_3 = -1.$$

With both constraints active, there can be no more constraints to be encountered along the new search direction.

We search now for a minimum along the line:

$$x_1 = 1 + \alpha,$$
$$x_2 = 1 - \alpha,$$
$$x_3 = 1 - \alpha,$$

(from the point $x_1 = x_2 = x_3 = 1$). In general, this would have to be done numerically, but in the present example, we can do this by simple differentiation. Substituting the above expressions for x_1, x_2 and x_3 into the formula for the objective function, we have

$$f(\mathbf{x}) = (1 + \alpha)^2 + 2(1 - \alpha)^2,$$

with a minimum:

$$\alpha = \frac{1}{3},$$

$$x_1 = \frac{4}{3}, \quad x_2 = x_3 = \frac{2}{3},$$

$$f_{\min}(\mathbf{x}) = \frac{8}{3},$$

which is the required solution. By substituting the above values of x_1, x_2 and x_3 in the constraint formulae, it is seen that the above point does indeed lie on both constraints and that the last move has, as expected, followed the intersection of the two constraints to the minimum. The Lagrange multipliers can be evaluated by Eq. (5.4) to verify that both constraints are active at this minimum point.

The constraints in this example were expressly chosen to be linear to avoid the additional arithmetic of the repeated restoration moves that would be necessary to return to the constraint boundary with nonlinear constraints. If the constraints are significantly nonlinear, this can greatly increase the burden of numerical work.

At the start of this example, it was pointed out that the second constraint might have been used to eliminate one variable and simplify the problem. This is in fact the principle behind the reduced gradient method in the following section, although this is then done implicitly rather than explicitly, since elimination of variables is generally not possible by simple algebra.   ■

Example 5.2 Use the gradient projection method to minimize the function:

$$f(\mathbf{x}) = x_1^2 + x_2^2 + x_3^2,$$

subject to constraints:

$$g_1(\mathbf{x}) = 3x_1 + 2x_2 + x_3 - 6 \geq 0,$$
$$g_2(\mathbf{x}) = x_3 - x_2 \geq 0,$$
$$g_3(\mathbf{x}) = 5 - 4x_1 \geq 0.$$

A third constraint has been added to the problem in Example 5.1, to change the solution from one in which a minimum is found along the final search direction to a fully constrained minimum at which all three constraints become active. We follow Example 5.1 up to the point where constraints g_1 and g_2 are both active. Then, proceeding along the same search direction from that point, we encounter the third constraint g_3 *before* finding a minimum of the function. This is the point:

$$x_1 = 1.25, \quad x_2 = x_3 = 0.75,$$
$$f(\mathbf{x}) = 2.688, \quad g_1 = g_2 = g_3 = 0.$$

This is where we take up the example.

Inserting derivatives of the three constraints into matrix $\mathbf{N}$ in Eq. (5.1), we have

$$\mathbf{N} = \begin{bmatrix} 3 & 0 & -4 \\ 2 & -1 & 0 \\ 1 & 1 & 0 \end{bmatrix},$$

with transpose:

$$\mathbf{N}^\mathrm{T} = \begin{bmatrix} 3 & 2 & 1 \\ 0 & -1 & 0 \\ -4 & 0 & 0 \end{bmatrix}.$$

Multiplying the two matrices:

$$\mathbf{N}^\mathrm{T}\mathbf{N} = \begin{bmatrix} 14 & -1 & -12 \\ -1 & 2 & 0 \\ -12 & 0 & 16 \end{bmatrix}$$

and taking the inverse:

$$(\mathbf{N}^\mathrm{T}\mathbf{N})^{-1} = \begin{bmatrix} 0.2222 & 0.1111 & 0.1667 \\ 0.1111 & 0.5556 & 0.0833 \\ 0.1667 & 0.0833 & 0.1875 \end{bmatrix}.$$

Post-multiplying now by $\mathbf{N}^\mathrm{T}$:

$$(\mathbf{N}^\mathrm{T}\mathbf{N})^{-1}\mathbf{N}^\mathrm{T} = \begin{bmatrix} 0 & 0.3333 & 0.3333 \\ 0 & -0.3333 & 0.6667 \\ -0.2500 & 0.2500 & 0.2500 \end{bmatrix}$$

and premultiplying by $\mathbf{N}$:

$$\mathbf{N}(\mathbf{N}^\mathrm{T}\mathbf{N})^{-1}\mathbf{N}^\mathrm{T} = \begin{bmatrix} 1 & 0 & 0 \\ 0 & 1 & 0 \\ 0 & 0 & 1 \end{bmatrix}.$$

The projection matrix in Eq. (5.5) is therefore:

$$\mathbf{P} = \mathbf{I} - \mathbf{N}(\mathbf{N}^\mathrm{T}\mathbf{N})^{-1}\mathbf{N}^\mathrm{T} = \begin{bmatrix} 0 & 0 & 0 \\ 0 & 0 & 0 \\ 0 & 0 & 0 \end{bmatrix},$$

or $\|\mathbf{s}\| = 0$ at the given point. We now have to evaluate the Lagrange multipliers to verify that this point (fully constrained with three active constraints and three variables) is a valid optimum. The gradient of the objective function at the current point ($x_1 = 1.25$, $x_2 = x_3 = 0.75$) is

$$\nabla f = \begin{bmatrix} 2.5 \\ 1.5 \\ 1.5 \end{bmatrix}.$$

By Eq. (5.4), the Lagrange multipliers are

$$\boldsymbol{\lambda} = (\mathbf{N}^{\mathrm{T}}\mathbf{N})^{-1}\mathbf{N}^{\mathrm{T}}\nabla f = \begin{bmatrix} 1 \\ 0.5 \\ 0.125 \end{bmatrix}.$$

Since all three Lagrange multipliers are positive, we confirm that the point at which all three constraints are active is indeed the optimum. Had any of the Lagrange multipliers been negative, we would have to remove a constraint (from Example 5.1, this would obviously be the constraint that we added) and continue with the line search for the optimum. As a result of the added constraint, the minimum function value has increased from $f_{\min} = 2.667$ (in Example 5.1) to $f_{\min} = 2.688$. ∎

5.1.2 Generalized Reduced Gradient Method

The essential difference between the generalized reduced gradient method (see [1, 6]) and the gradient projection method in the previous section is that a number of variables equal to the number of constraints are selected to satisfy the constraints before calculating a 'reduced gradient' in the remaining variables. In this way, the problem is reduced to an unconstrained one. The method is in principle for linear constraints. A correction process is introduced later to accommodate nonlinear constraints. The word 'generalized' in the name of the method refers to the development of an earlier reduced gradient method to include now inequality and equality constraints. In the sense that constraints remain satisfied during the line search, this can again be seen as a constraint-following method.

To begin, let us consider a problem with only equality constraints and assume further that these are linear. If there are p such constraints:

$$h_k(\mathbf{x}) = 0, \quad k = 1, \ldots, p,$$

we select any p variables, for convenience numbering these $x_1, \ldots, x_p$. These become dependent variables to satisfy the constraints, the remaining variables $x_{p+1}, \ldots, x_n$ being independent variables for optimization. We may recall that in a similar way, design variables were identified as dependent and independent in the Lagrange multiplier method in Sect. 3.2, while elimination of variables to reduce the size of the problem was already referred to in Sect. 3.1.1, in the context of optimality criteria. Taking the transpose of matrix $\mathbf{N}$ defined in Eq. (5.1), now with inequality constraints $g_j(\mathbf{x})$ replaced by equality constraints $h_k(\mathbf{x})$, the condition that the solution remains feasible in Eq. (5.2), that is, remains on the constraints, can be rewritten as:

$$\mathbf{dh(x)} = \mathbf{N}^T\mathbf{dx} = \mathbf{0}. \tag{5.6}$$

Expanding the matrix in the above equation:

$$\mathbf{N}^T dx = \begin{bmatrix} \dfrac{\partial h_1}{\partial x_1} & \cdots & \dfrac{\partial h_1}{\partial x_p} & \vdots & \dfrac{\partial h_1}{\partial x_{p+1}} & \cdots & \dfrac{\partial h_1}{\partial x_n} \\ \vdots & \ddots & \vdots & \vdots & \vdots & \ddots & \vdots \\ \dfrac{\partial h_p}{\partial x_1} & \cdots & \dfrac{\partial h_p}{\partial x_p} & \vdots & \dfrac{\partial h_p}{\partial x_{p+1}} & \cdots & \dfrac{\partial h_p}{\partial x_n} \end{bmatrix} \begin{bmatrix} dx_1 \\ \vdots \\ dx_p \\ dx_{p+1} \\ \vdots \\ dx_n \end{bmatrix}. \tag{5.7}$$

Note how the matrix has been partitioned into a sub-matrix with derivatives with respect to the p dependent variables and a second sub-matrix in the remaining independent variables. If we name the two sets of variables:

$$\mathbf{u} = x_1, \ldots, x_p,$$
$$\mathbf{v} = x_{p+1}, \ldots, x_n,$$

we can express Eq. (5.6) as:

$$\mathbf{N}_1 d\mathbf{u} + \mathbf{N}_2 d\mathbf{v} = \mathbf{0},$$

where $\mathbf{N}_1$ refers to the first partition of $\mathbf{N}^T$ and $\mathbf{N}_2$ to the second. Premultiplying by $\mathbf{N}_1^{-1}$, we obtain

$$d\mathbf{u} = -\mathbf{N}_1^{-1}\mathbf{N}_2 d\mathbf{v}. \tag{5.8}$$

This formula enables us to update the dependent variables $\mathbf{u} = x_1, \ldots, x_p$ with changes in the independent variables, to ensure that constraints remain satisfied. In terms of both the dependent and independent variables, an increment df in the objective function $f(\mathbf{x})$ can be written as:

$$df = \nabla_v^T f \, d\mathbf{v} + \nabla_u^T f \, d\mathbf{u},$$

where subscript v refers to the gradient of $f(\mathbf{x})$ with respect to variables $\mathbf{v}$ and subscript u with respect to variables $\mathbf{u}$. Substituting now for $d\mathbf{u}$ from above, we obtain

$$df = \nabla_v^T f \, d\mathbf{v} - \nabla_u^T f \left[\mathbf{N}_1^{-1}\mathbf{N}_2 \right] d\mathbf{v},$$

or finally:

$$\mathbf{G}_r = \frac{\partial f}{\partial \mathbf{v}} = \nabla_v f - \left[\mathbf{N}_1^{-1}\mathbf{N}_2\right]^T \nabla_u f. \tag{5.9}$$

$\mathbf{G}_r$ is termed the 'reduced gradient' of $f(\mathbf{x})$ and is the gradient in a reduced set of independent variables. This is used now for the search direction:

$$\mathbf{s} = -\mathbf{G}_r.$$

Note that the line search for a minimum along this search direction requires recalculation of the dependent variables from Eq. (5.8) at each step to evaluate the objective function, which remains, of course, a function of both the dependent and independent variables. When a minimum has been found in the line search, the reduced gradient is re-evaluated at the new current point and the line search repeated.

If all constraints are in fact equality constraints, then with elimination of constraints and use of the reduced gradient, the optimization problem has been reduced to an unconstrained one which can be solved by the usual methods. For inequality constraints, an active constraint strategy could still be used, as in the gradient projection method previously. However, in the generalized reduced gradient method, this is usually not done, principally because as the set of active constraints changes, so does the partitioning of the matrix $\mathbf{N}^T$ into sub-matrices $\mathbf{N}_1$ and $\mathbf{N}_2$, for which we again have to select a suitable set of dependent variables. Instead of an active constraint strategy, inequality constraints are converted into equality constraints by the use of additional, so-called surplus variables $x_{n+1}, \ldots, x_{n+m}$, where m is the number of inequality constraints (although not discussed in this book, use of slack and surplus variables is a feature of linear programming methods). Inequality constraints $g_j(\mathbf{x}) \geq 0$ become

$$g_j(\mathbf{x}) - x_{j+n} = 0, \quad j = 1, \ldots, m$$

subject to:

$$x_{j+n} \geq 0.$$

This last condition is in effect the selection of active constraints, since for any $x_{j+n} = 0$ the corresponding constraint is active. This procedure has the computational advantage that all constraints are treated consistently—no explicit constraint selection has to be made and no change in partitioning of sub-matrices $\mathbf{N}_1$ and $\mathbf{N}_2$. Any actual equality constraints are simply included in the list of constraints but do not, of course, require more surplus variables. The reduced gradient $\mathbf{G}_r$ in Eq. (5.9) is now calculated for the complete set of independent variables, that is, including surplus variables, and used for the search direction in these variables. Lower bounds on surplus variables are taken care of in the following way. When, during the line search, a surplus variable is reduced to zero, the line search is stopped at that point and the reduced gradient recalculated. Any components of the new search direction that would cause a surplus variable to become negative are omitted. Changes in dependent variables, corresponding to changes in the independent variables during the line search, are obtained by Eq. (5.8). A zero entry is made in $\mathbf{dv}$ in Eq. (5.8) for those independent variables for which the corresponding components of the search direction have been omitted.

As already said, the basis of the generalized reduced gradient method is that all constraints are linear. In general, this is unlikely to be so, and an approach similar to that in the gradient projection method can be employed, in which constraints are represented by equivalent linear constraints at the current point. The inclusion of nonlinear constraints means that during the subsequent line search, we shall gradually move away from the true constraint surface. As this occurs, a correction is necessary to restore constraints to their proper zero values (note that by inclusion of surplus variables, all constraints are now equalities). In the generalized reduced gradient method, this is done by correcting the dependent variables, keeping the independent variables unchanged. Making use of Eq. (5.8), new values $\mathbf{du}$ are

$$\mathbf{du} = \mathbf{N}_1^{-1}\{-\mathbf{h}(\mathbf{x}) - \mathbf{N}_2\mathbf{dv}\},$$

where $\mathbf{h}(\mathbf{x})$ are the values of the constraints to be corrected. These are added to the current values of the dependent variables. This is repeated until values of $\mathbf{h}(\mathbf{x})$ sufficiently close to zero are obtained, after which the next line search is performed.

While we have discussed the principle of the generalized reduced gradient method in sufficient detail, it is hoped, to follow its working, some important technical details remain for an efficient computer implementation. These include the choice of dependent variables to avoid a singular matrix, means of accelerating convergence both in the line search and in a restoration move, and how to proceed if the initial point is infeasible. However, it goes beyond the scope of this book to discuss these here ([12] gives much useful information on these aspects.). It will be clear that the methods in this and in the previous section, while generally very efficient, require a substantial programming effort and can scarcely be illustrated adequately in simple examples. Further, it should be realized that the generalized reduced gradient method, as described here, has been the result of some continuous development, different flavours exist, and extra features are commonly added with each new implementation. This no doubt also applies to the current version of Solver in Excel. As implemented, Solver also includes a useful option to restrict variables to integer values.

Example 5.3 Calculate the reduced gradient at an initial point:

$$x_1 = 1.5, \quad x_2 = 0.75, \quad x_3 = 1$$

in the generalized reduced gradient method for the problem in Example 5.1.

We first subtract surplus variables x_4 and x_5 from the two inequality constraints in Example 5.1 to convert these to equalities:

$$h_1(\mathbf{x}) = 3x_1 + 2x_2 + x_3 - x_4 - 6 = 0,$$
$$h_2(\mathbf{x}) = x_3 - x_2 - x_5 = 0,$$

with lower limits:

$$x_4 \geq 0, \quad x_5 \geq 0.$$

Values of x_4 and x_5 to satisfy the equality constraints at the initial point are

$$x_4 = 1, \quad x_5 = 0.25.$$

Next, we select dependent and independent variables. Since there are no specified limits on the original variables x_1, x_2, x_3, we choose $\mathbf{u} = x_1, x_2$ as dependent variables. We may anticipate that both x_4 and x_5 will reach their lower limits, that is, if both of the original inequality constraints are active at the optimum, and limits on these variables are more easily dealt with during optimization of the objective function in the independent variables $\mathbf{v} = x_3, x_4, x_5$.

Differentiating constraints $h_1(\mathbf{x})$ and $h_2(\mathbf{x})$, appropriate derivatives are inserted in the matrices $\mathbf{N}_1$ and $\mathbf{N}_2$ in Eq. (5.7):

$$\mathbf{N}_1 = \begin{bmatrix} 3 & 2 \\ 0 & -1 \end{bmatrix},$$

$$\mathbf{N}_2 = \begin{bmatrix} 1 & -1 & 0 \\ 1 & 0 & -1 \end{bmatrix}.$$

The inverse of $\mathbf{N}_1$ is

$$\mathbf{N}_1^{-1} = \begin{bmatrix} 0.3333 & 0.6667 \\ 0 & -1 \end{bmatrix}$$

and post-multiplying by $\mathbf{N}_2$:

$$\mathbf{N}_1^{-1}\mathbf{N}_2 = \begin{bmatrix} 1 & -0.3333 & -0.6667 \\ -1 & 0 & 1 \end{bmatrix}.$$

The objective function in Example 5.1 is

$$f(\mathbf{x}) = x_1^2 + x_2^2 + x_3^2,$$

which has a gradient with respect to the dependent variables:

$$\nabla_u f = \begin{bmatrix} 3 \\ 1.5 \end{bmatrix},$$

and with respect to the independent variables:

$$\nabla_v f = \begin{bmatrix} 2 \\ 0 \\ 0 \end{bmatrix},$$

at the initial point. Therefore,

$$\left[\mathbf{N}_1^{-1}\mathbf{N}_2\right]^T \nabla_u f = \begin{bmatrix} 1.5 \\ -1 \\ -0.5 \end{bmatrix},$$

and from Eq. (5.9), the reduced gradient is

$$\mathbf{G_r} = \begin{bmatrix} 0.5 \\ 1.0 \\ 0.5 \end{bmatrix}.$$

Having calculated the reduced gradient, we can continue now with the first-line search. The negative of the reduced gradient is the initial search direction in the independent variables:

$$\mathbf{s} = \begin{bmatrix} dx_3 \\ dx_4 \\ dx_5 \end{bmatrix} = \begin{bmatrix} -0.5 \\ -1 \\ -0.5 \end{bmatrix}.$$

The line search is therefore along the line:

$$x_3 = 1 - 0.5\alpha,$$
$$x_4 = 1 - \alpha,$$
$$x_5 = 0.25 - 0.5\alpha,$$

from the chosen initial point, with a line search parameter α. For the dependent variables, we have from Eq. (5.8):

$$\mathbf{du} = \begin{bmatrix} dx_1 \\ dx_2 \end{bmatrix} = -\begin{bmatrix} 1 & -0.3333 & -0.6667 \\ -1 & 0 & 1 \end{bmatrix} \begin{bmatrix} -0.5 \\ -1 \\ -0.5 \end{bmatrix} = \begin{bmatrix} -0.1667 \\ 0 \end{bmatrix}$$

(where the column matrix contains the components of the search direction in the independent variables). We can also express x_1 in terms of α (x_2 does not change because of the zero above):

$$x_1 = 1.5 - 0.1667\,\alpha,$$
$$x_2 = 0.75.$$

By setting surplus variables x_4 and x_5 to zero in the line search formulae, we can calculate directly values of α at which these reduce to their lower limit. With both surplus variables reducing, variable x_5 is the first to reach its lower limit, at $\alpha = 0.5$, where

$$x_1 = 1.417, \quad x_2 = 0.75, \quad x_3 = 0.75,$$
$$x_4 = 05, \quad x_5 = 0.$$

No minimum of the objective function along the search direction is found up to this point. By substituting the above values of the variables in the two equality constraints, we can confirm that they do indeed remain satisfied at the new point.

The objective function has been reduced from 3.8125 at the initial point to 3.132 at the current point, while we know from Example 5.1 that it is 2.667 at the optimum. For the following step, we have to recalculate the reduced gradient at the new current point and proceed with a second-line search towards the optimum, in a similar manner to above. Note that with the linear constraints in the present example, the matrices N_1 and N_2 are unchanged, reducing the subsequent calculation. ∎

5.1.3 Other Methods for Constrained Optimization

While the methods in the previous two sections have been and remain widely used, as well as the penalty function methods in Sect. 5.2, not surprisingly a variety of other methods that might broadly be termed constraint-following methods have been developed. In particular, we should refer to the method of feasible directions and to sequential quadratic programming. Also, with the increasing power of computers, some zero-order methods, including the genetic algorithm described later in Chap. 8, have become increasingly useful in certain types of problem.

The method of feasible directions [13] is specifically aimed at optimization problems with nonlinear, inequality constraints. If a search direction is chosen tangent to the active constraints (as in the gradient projection method) and the constraints are nonlinear, this immediately leads to a departure from the true constraints, for which move limits and a correction procedure are necessary. In the method of feasible directions, the search direction is allowed to point into the feasible domain at some angle to the constraint boundary. This is illustrated in Fig. 5.3, showing the current point located on a nonlinear constraint $g(\mathbf{x}) = 0$ and a weight line, or line of constant value of the objective function $f(\mathbf{x})$, through that point (the weight line is shown linear in the figure, but this need not be so). A feasible search direction $\mathbf{s}$ must point to the space *above* the constraint line

Fig. 5.3 Feasible directions method

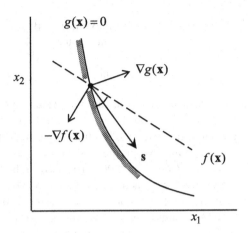

$(g(\mathbf{x}) \geq 0)$, while for a usable search direction it must point to the space *below* the weight line ($f(\mathbf{x})$ reducing). This leaves the sector shown in the figure available for a feasible–usable search direction. A search direction chosen within this sector enables more progress to be made before again encountering a constraint.

For a suitable search direction $\mathbf{s}$, we set up the following optimization problem:

$$\begin{aligned}
\text{maximize} \quad & \beta \quad (\beta \geq 0) \\
\text{subject to}: \quad & \mathbf{s}^\mathrm{T} \nabla g(\mathbf{x}) \geq \beta, \\
& \mathbf{s}^\mathrm{T} \nabla f(\mathbf{x}) \leq -\beta, \\
\text{and} \quad & |s_i| \leq 1.
\end{aligned}$$

This makes a compromise between the extent to which the search direction is allowed to depart from the constraint boundary and the reduction in function $f(\mathbf{x})$. The result is a search direction which roughly bisects the feasible–usable sector in Fig. 5.3 (provided of course that both the constraint and the objective function have been normalized).

A so-called push-off factor $\theta (\theta \geq 0)$ is introduced into the first constraint to control the extent to which the search direction is angled away from the constraint boundary:

$$\mathbf{s}^\mathrm{T} \nabla g(\mathbf{x}) \geq \theta \beta.$$

A value $\theta = 0$ allows a search direction along the tangent to the constraint, while larger values of θ apply to increasingly nonlinear constraints. Furthermore, we have up to now considered only a single constraint. We should include both active and near-active constraints g_j for better convergence. The optimization problem, written in its usual form, then becomes

$$\begin{aligned}
\text{maximize} \quad & \beta \, (\beta \geq 0) \\
\text{subject to :} \quad & -\mathbf{s}^T \nabla g_j(\mathbf{x}) + \theta_j \beta \leq 0, \\
& \mathbf{s}^T \nabla f(\mathbf{x}) + \beta \leq 0, \\
\text{and} \quad & |s_i| \leq 1.
\end{aligned}$$

This is in fact a linear optimization problem, usually solved by linear programming. Note the form of the third constraint, to impose bounds on the search direction. It is expressed in this way (rather than the usual $\mathbf{s}^T\mathbf{s} = 1$) to preserve the linear optimization problem. Values of the push-off factor θ_j can be defined for each constraint. Commonly, $\theta_j = 1$ is used in the first place, but it can also be related to the values of near-active constraints to avoid these becoming repeatedly active and inactive (see [12]). For a linear constraint, $\theta_j = 0$.

If we assume that we have already located a point on a constraint boundary, we perform the optimization problem above for the components s_i of the search direction. We proceed along this direction until either a minimum is found or the same or a new constraint is encountered. In the first case, we perform an unconstrained optimization to return to the constraint boundary, and in the second case, a new search direction is computed directly. This is continued until we have approached closely enough to the required optimum.

In a development of the method above, known as a 'robust feasible direction method' [12], the same partitioning in dependent and independent variables is used as in the generalized reduced gradient method, but without the addition of surplus variables. Instead of surplus variables, the active constraint strategy of the feasible directions method is retained. The search direction is based on formulae for the feasible directions method, but with push-off factors set to zero so that the search direction is now tangent to the active constraints. The line search is in the independent variables, with updating of the dependent variables as in the generalized reduced gradient method. Other than in that method, only the gradients of the currently active constraints have to be computed, and with no surplus variables, the number of variables is not increased.

Sequential quadratic programming [9] employs a more complex procedure for determining a search direction than in the methods described up to now. It is based on a quadratic approximation to the Lagrangian function and linearized constraints. Special methods exist for solving the resulting quadratic programming problem from which the search direction is deduced. Lagrange multipliers are updated in successive iterations. Different formulations of the method exist, and the theoretical background is quite complex; therefore, we shall not go into this further here. Arora [2] provides an extensive further discussion of sequential quadratic programming methods. The improved efficiency of sequential quadratic programming can more than compensate for the extra effort required in programming the method for the computer.

5.1.4 Substitution of Variables

The methods for constrained optimization discussed in this chapter up to now all depend on linearization of nonlinear constraints, as approximation to the true constraints at the current point. Move limits are usually necessary to limit how far the solution is allowed to depart from the true constraint boundary. Clearly, the chosen extent of such move limits depends on the degree of nonlinearity of the constraints. Linearity can often be improved by appropriate substitution of variables. Commonly, this is done by use of so-called inverse variables, that is, by replacing simple variables such as cross-sectional area A or thickness t by $\frac{1}{A}$ or $\frac{1}{t}$. The reasoning behind this is as follows. Stress is load over area, and if we examine any formula for the stress in a structure, or for its deflection, we see that quantities such as A and t appear in the denominator of the formula or in any case to a higher power in the denominator than in the numerator. Therefore, use of inverse variables leaves stress and deflection constraints more directly related to these variables. However, the same is not necessarily true for other constraints, such as buckling constraints. An effect of using inverse variables is that the feasible region is moved below to the constraints in the design space, rather than above as in all the figures up to now.

Throughout this chapter, it is assumed that we do not know in advance which of the inequality constraints will be active at the optimum, these being identified during the optimization process. However, if we can be sure that certain constraints will be active at the optimum, these can be treated as equalities and we may be able to solve for the same number of variables or, better said, express these in terms of the remaining variables. Provided that this is algebraically possible, or not otherwise impractical, this reduces the number of variables participating as such in the optimization. This is also, of course, the basis of a fully stressed design. For the statically determinate truss structures in Chap. 1, stress constraints in individual members are used to solve for the required areas of those members, eliminating those variables from the optimization and leaving, for example, only those variables relating to the shape of the truss, or other variables referring to the shape of cross section of the members. For a statically indeterminate truss, under certain conditions, this results in the familiar iterative process leading to an optimum layout of truss.

Reduction in the number of variables for large problems, by selection of 'master' and 'slave' variables, with only the master variables entering into the actual optimization, is discussed in Chap. 9.

5.2 Penalty Function Methods

The constraint-following methods in Sects. 5.1.1 and 5.1.2 reduce the problem, by selection of active constraints or by use of surplus variables, to what is in effect an equality constrained problem. Penalty function methods, on the other hand, convert the constrained problem into a fully unconstrained one. This is achieved by adding a suitable penalty term to the objective function, either to avoid or to penalize constraint violations. The unconstrained problem is then solved by any of the previous methods. As the magnitude of the penalty term is decreased (or increased, depending on the type of penalty function) and the unconstrained optimization repeated, the solution approaches the optimum of the original constrained problem. For this reason, these methods are also referred to as 'sequential unconstrained optimization techniques' [5]. An extensive review of these methods is provided by Vanderplaats [12]. They provide a simple, intuitive way of including constraints in an optimization problem and are easy to implement. Penalty function methods can be applied to both equality and inequality constrained problems and do not require linearization of nonlinear constraints. However, it remains an iterative process requiring repeated unconstrained optimization, and penalty function methods are generally less efficient than the more sophisticated constraint-following methods. Many different forms of penalty function have been proposed over the years. Three of the most commonly used penalty functions are discussed in the following sections.

5.2.1 Interior Penalty Function

The simplest form of penalty function, first for an inequality constrained problem, is the so-called interior penalty function. This takes the form:

$$F(\mathbf{x}) = f(\mathbf{x}) + r\varphi(\mathbf{x}), \tag{5.10}$$

where $f(\mathbf{x})$ is the function for which we seek a constrained minimum, r is a penalty parameter and

$$\varphi(\mathbf{x}) = \sum_{j=1}^{m} \frac{1}{g_j(\mathbf{x})}$$

is the penalty term containing the m inequality constraints $g_j(\mathbf{x})\varphi \to \infty$. Since if any of the constraints are precisely satisfied ($g_j = 0$), this function in effect puts up a barrier along the constraint boundary, with $F(\mathbf{x})$ increasing rapidly as the boundary is approached. None of the constraints may be negative, as this would lead to negative values of φ and cause the solution not to converge, so this form of penalty function is necessarily restricted to the feasible region. Away from the constraint boundary, the penalty term gradually decays, and function $F(\mathbf{x})$ approaches the original function $f(\mathbf{x})$.

Provided that parameter $r > 0$, we have applied a penalty to the original function $f(\mathbf{x})$ for approaching the constraint boundary. The minimum of $F(\mathbf{x})$ is not, of course, the constrained minimum of $f(\mathbf{x})$, but approaches the required optimum as

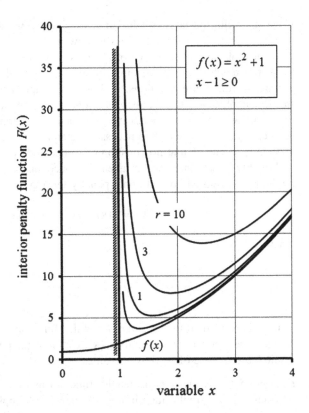

Fig. 5.4 Effect of reducing values of penalty parameter r on the form of the interior penalty function

$$f(x) = x^2 + 1$$
$$x - 1 \geq 0$$

r is reduced. We take therefore a sequence of reducing values of r and repeat the unconstrained optimization for each r, continuing each time from the previous solution. The effect of reducing r on the interior penalty function is illustrated in Fig. 5.4 for an elementary, one-dimensional problem. By reducing r, the minimum of $F(x)$ is gradually drawn into the constrained minimum of function $f(x)$ at $x = 1$, with smaller changes in variable x at each step. We also see in the figure how the penalty function $F(x)$ approaches the original function $f(x)$ with increasing x away from the constraint and that $F(x)$ increases rapidly close to the constraint.

While Fig. 5.4 illustrates a one-dimensional problem, a similar plot would be obtained if plotted on the line along which the solution approaches the optimum in a multidimensional problem. Since for small values of r the solution is close to the optimum, the need for a *sequence* of decreasing r values may not be obvious at this point. In simple terms, the effect of this is to enable the process to steer clear of all constraints in the beginning and in the end to approach the optimum from a more favourable direction in which it is then possible to locate the optimum with the required accuracy. If we start with too small a value of r, we may approach too closely to one or more constraints before we are in the neighbourhood of the optimum, and the rapid rise in $F(\mathbf{x})$ can lead to a poorly conditioned numerical minimization of the penalty function or even failure to converge. On the other hand, if we start with too large a value of r, we find an initial point far distant from the required optimum, and the process is slowed down.

Choice of the best sequence of r values is problem dependent and has been the subject of much research, the aim of course being to find the optimum in the least number of function evaluations. It should be emphasized that the interior penalty function applies *only* to the feasible region. If any constraint becomes negative, then the 'penalty' turns into an advantage, that is, it becomes negative, at least with respect to that constraint. It is necessary, therefore, to apply some safeguard in the procedure to ensure that no constraint can become negative at any stage.

Equality constraints can be taken into account in the interior penalty function method by means of an additional penalty term $\varphi'(\mathbf{x})$ in the penalty function:

$$F(\mathbf{x}) = f(\mathbf{x}) + r\varphi(\mathbf{x}) + r'\varphi'(\mathbf{x}),$$

where

$$\varphi'(\mathbf{x}) = \sum_{k=1}^{p} [h_k(\mathbf{x})]^2$$

(a term borrowed from the exterior penalty function method in the next section) contains the p equality constraints $h_k(\mathbf{x})$. Unlike r, the penalty parameter r' requires a suitable sequence of *increasing* values.

Example 5.4 Use the interior penalty function method to optimize the radius and thickness of the circular tube in compression in Example 2.1.

To illustrate the interior penalty function method and to compare the result with the theoretical optimum in Example 2.1, we take only Euler and local buckling constraints into account. The minimum thickness limitation in that example is therefore ignored. Table 5.1 shows the sequence of values of radius R, thickness

Table 5.1 Sequence of iterations by the interior penalty function method for the circular tube in Example 2.1, with initial penalty parameter $r = 250$ reduced by a factor of 10 at each iteration ($P = 10,000$ N, $L = 1000$ mm, $E = 72,000$ N/mm^2)

Iteration	Radius R (mm)	Thickness t (mm)	Area A (mm^2)
0	20	1	125.66
1	40.07	0.5249	132.17
2	33.87	0.3169	67.43
3	30.62	0.2341	45.03
4	29.30	0.2051	37.75
5	28.84	0.1956	35.44
6	28.69	0.1926	34.71
7	28.64	0.1916	34.48
8	28.62	0.1913	34.40
9	28.62	0.1912	34.38
10	28.62	0.1912	34.38
20	28.62	0.1911	34.37

t and cross-sectional area *A* generated in each iteration. These were calculated by the spreadsheet 'penalty function method' in Sect. 5.3.1. Initial values of *R* and *t* were chosen to be close to the constraint boundary. An initial value of the penalty parameter $r = 250$ was chosen, reducing by a factor of 10 at each iteration. Within five iterations, the optimum is approached quite closely, while a sufficiently accurate result for most purposes is obtained after only ten iterations. The result agrees with the theoretical optimum in Example 2.1. It will be observed in Table 5.1 that the procedure first approaches the optimum quite quickly, but is much slower in later iterations in reaching an accurate result. This is an inevitable characteristic of an interior penalty function and of the exterior penalty function in the next section. ■

5.2.2 Exterior Penalty Function

The need to ensure that constraints do not become negative in the interior penalty function method can be avoided by use of an exterior penalty function. This again takes the form:

$$F(\mathbf{x}) = f(\mathbf{x}) + r\varphi(\mathbf{x}),$$

where the penalty term $\varphi(\mathbf{x})$ is now:

$$\varphi(\mathbf{x}) = \sum_{j=1}^{m} \left\{ \max\left[0, -g_j(\mathbf{x})\right] \right\}^2.$$

The effect of the above formula is that only *violated* constraints ($g_j < 0$) are included in the penalty term. As a consequence, this new function penalizes constraint violations, leaving it unaffected by constraints that are satisfied. The square in the penalty term is to avoid a discontinuity at the constraint boundary and to make constraint violation progressively more severe. The exterior penalty function applies to both the feasible and infeasible regions, but now the solution approaches the optimum from the infeasible region. By progressively *increasing* the value of *r* in the penalty function and repeating the unconstrained optimization, the solution is pushed towards the constrained optimum until a sufficiently accurate solution can be obtained. This is illustrated in Fig. 5.5 for the same one-dimensional problem as previously for the interior penalty function in Fig. 5.4.

Clearly, a disadvantage of the exterior penalty function method is that unlike the interior penalty function, if the optimization is stopped before it is finally converged, the solution will be to some degree on the infeasible side of the constraints. However, because the method is valid in both the feasible and infeasible regions, no special measures are necessary to prevent the solution from crossing the constraint boundary (as in the interior penalty function method). Similar remarks as before apply with regard to the need for a suitable sequence of *r* values.

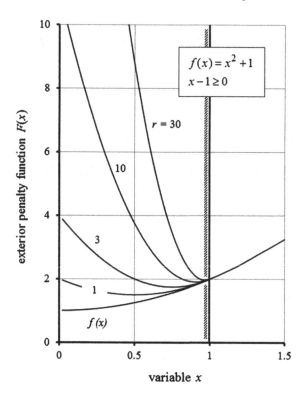

Fig. 5.5 Effect of increasing values of penalty parameter r on the form of the exterior penalty function

The exterior penalty function method is readily modified to accommodate equality constraints by means of an additional penalty term $\varphi'(\mathbf{x})$:

$$F(\mathbf{x}) = f(\mathbf{x}) + r\varphi(\mathbf{x}) + r'\varphi'(\mathbf{x}),$$

where

$$\varphi'(\mathbf{x}) = \sum_{k=1}^{p} [h_k(\mathbf{x})]^2$$

contains the p equality constraints $h_k(\mathbf{x}) = 0$. For these constraints, this term has the effect of pushing the solution towards the constraints from either side. A suitable increasing sequence of both r and r' is now necessary. As we have seen, the same additional term $\varphi'(\mathbf{x})$ is used in the interior penalty function method to include equality constraints.

For the proper working of both types of penalty function, it is advisable first to normalize the constraints. By ensuring that all constraints have an initial value somewhere in the region of unity, all constraints can contribute to the penalty term on the same basis. For example, a stress constraint:

Table 5.2 Sequence of iterations by the exterior penalty function method for the circular tube in Example 2.1, with initial penalty parameter $r = 25$ increased by a factor of 10 at each iteration ($P = 10,000$ N, $L = 1000$ mm, $E = 72,000$ N/mm^2)

Iteration	Radius R (mm)	Thickness t (mm)	Area A (mm^2)
0	15	0.1	9.42
1	27.26	0.1653	28.31
2	28.51	0.1889	33.84
3	28.61	0.1909	34.32
4	28.62	0.1911	34.36
5	28.62	0.1911	34.37
10	28.62	0.1911	34.37

$$g_j = \sigma_0 - \sigma \geq 0$$

can better be written as:

$$1 - \frac{\sigma}{\sigma_0} \geq 0$$

by dividing by the allowable stress σ_0 (or some other appropriate constant). The same applies, of course, to equality constraints.

Example 5.5 Repeat Example 5.4 using the exterior penalty function.

The spreadsheet 'penalty function method' was modified for an exterior penalty function, as described later in Sect. 5.3.1. Table 5.2 shows the sequence of values of radius R, thickness t and cross-sectional area A generated in each iteration (again no minimum thickness requirement is applied). An initial value of the penalty parameter $r = 25$ was chosen, increasing by a factor of 10 at each iteration. A sufficiently accurate solution, again agreeing with the theoretical optimum, is obtained after only three iterations.

The factor of 10 for the penalty parameter in both this and the previous example is quite large and may well have to be reduced in other problems when a conventional gradient-based method is used for the unconstrained optimization. The Hooke and Jeeves routine used for convenience in the spreadsheet, being a zero-order method not requiring gradient data, is less sensitive to the steepness of the contours of the penalty function, allowing therefore the larger reduction factor.   ∎

5.2.3   Augmented Lagrangian Penalty Function

As has been pointed out, a characteristic of both the interior and exterior penalty functions is the rapidly increasing steepness of the function as the optimum is approached. This inevitably makes the task of locating a minimum in the unconstrained optimization more difficult. The augmented Lagrangian penalty function avoids this problem by including Lagrange multipliers in the previous exterior penalty term. The aim is to define a function which moves progressively towards satisfaction of the Kuhn–Tucker conditions, rather than one which simply enables us to locate the minimum, as has been the objective up to now. As added advantage, values of the Lagrange multipliers are obtained at the same time as the required optimum values of the variables. For a problem with only inequality constraints, the augmented Lagrangian penalty function is

$$F(\mathbf{x}) = f(\mathbf{x}) + r \sum_{j=1}^{m} \left\{ \max \left[\frac{\lambda_j}{2r} - g_j(\mathbf{x}), 0 \right] \right\}^2, \tag{5.11}$$

where λ_j are the Lagrange multipliers. These are mostly taken to be zero in the first iteration. It is seen that the above function then reduces simply to the conventional exterior penalty function in this first step. However, we now minimize the function in Eq. (5.11) by updating the values of λ_j at each iteration, rather than by increasing the value of the penalty parameter r, as we did before. The updating formula for λ_j now has to be deduced.

If we differentiate function $F(\mathbf{x})$ analytically with respect to each x_i for a minimum, we obtain

$$\frac{\partial F}{\mathrm{d}x_i} = \frac{\partial f}{\partial x_i} - \sum_{j=1}^{m} \max \left[\lambda_j - 2rg_j, 0 \right] \frac{\partial g_j}{\partial x_i} = 0.$$

For an inequality constrained problem, the third of the Kuhn–Tucker conditions in Sect. 3.3.1 for a variable x_i reduces to:

$$\frac{\partial f}{\partial x_i} - \sum_{j=1}^{m} \lambda_j^* \frac{\partial g_j}{\partial x_i} = 0,$$

where the asterisk denotes the optimum value of the Lagrange multiplier. Compare now the two formulae above. We see that the two expressions agree if:

$$\max \left[\lambda_j - 2rg_j, 0 \right] = \lambda_j^*.$$

It is reasonable to assume then that:

$$\lambda_j' = \max \left[\lambda_j - 2rg_j, 0 \right], \tag{5.12}$$

where λ_j and g_j are their current values and will provide an improved estimate λ_j' of the Lagrange multipliers. In an iterative process, the numerical procedure is then to minimize the unconstrained function $F(\mathbf{x})$ in Eq. (5.11), substituting at each iteration updated values of λ_j' from Eq. (5.12). This will then converge to the required optimum of λ and $\mathbf{x}$. Note that as implied earlier, it may no longer be necessary to increase the value of r at each iteration, as in the conventional exterior penalty function method or in any case not to the same extent. This is why we do not get an increasingly steep increase in the penalty function as we approach the optimum. Convergence is generally significantly better than with either of the previous two penalty functions. Note also that Eq. (5.12) ensures that all the λ_j remain positive or zero, as required. To include equality constraints $h_k(\mathbf{x})$ in the augmented Lagrangian penalty function, we simply add a term:

$$\sum_{k=1}^{p} \left(-\lambda_k h_k + r h_k^2 \right)$$

to the penalty function in Eq. (5.11). The updating formula for λ_k becomes

$$\lambda_k' = \lambda_k - 2 r h_k.$$

Example 5.6 Repeat Example 5.5 using the augmented Lagrangian penalty function.

We choose the same initial values of radius $R = 15$ mm and thickness $t = 0.1$ mm as in the exterior penalty function method in the previous example, with a penalty parameter $r = 25$ equal to the initial value of r in that example. This is kept constant in each iteration. With the same data as in Example 2.1, and using the formulae for the Euler buckling load P_E and the local buckling load P_L from Sect. 2.1, the following initial constraint values can be calculated:

$$g_1 = (P_E/P) - 1 = -0.9247, \quad g_2 = (P_L/P) - 1 = -0.7263.$$

Both being well below zero, it is seen that the initial values of R and t are quite far into the infeasible region. For this example, Eq. (5.11) can be written as:

$$F = A + r\left\{ \max\left[\frac{\lambda_1}{2r} - g_1, 0 \right] \right\}^2 + r\left\{ \max\left[\frac{\lambda_2}{2r} - g_2, 0 \right] \right\}^2.$$

Starting with initial values of the Lagrange multipliers $\lambda_1 = \lambda_2 = 0$, we insert the appropriate formulae for the cross-sectional area A and for constraints g_1 and g_2 into the above equation. With variables R and t, function F is now minimized (using Solver or by the Hooke and Jeeves method in Sect. 4.3.1). This gives

$$R = 27.26 \text{ mm}, \quad t = 0.1653 \text{ mm},$$

$$g_1 = -0.2525, \quad g_2 = -0.2525.$$

It is seen that a substantial reduction in constraint violation has already taken place in this first iteration.

Table 5.3 Sequence of iterations by the augmented Lagrangian penalty function method for the circular tube in Example 2.1, with fixed penalty parameter $r = 25$ ($P = 10,000$ N, $L = 1000$ mm, $E = 72,000$ N/mm^2)

Iteration	Radius R (mm)	Thickness t (mm)	Area A (mm^2)	Lagrange multiplier λ_1	Lagrange multiplier λ_2
0	15	0.1	9.42	0	0
1	27.26	0.1653	28.31	12.62	12.62
2	28.74	0.1935	34.94	11.36	11.36
3	28.61	0.1910	34.32	11.46	11.46
4	28.62	0.1912	34.37	11.46	11.46
5	28.62	0.1911	34.37	11.46	11.46

The update formula for the Lagrange multipliers in Eq. (5.12) is

$$\lambda_j' = \max\left[\lambda_j - 2rg_j, 0\right],$$

where λ_j and g_j refer to their current values for each constraint. In both cases, the first term in the brackets above becomes

$$0 - 2 \times 25 \times (-0.2525) = 12.62.$$

This is greater than zero, so λ_1' and λ_2' for the next iteration are both 12.62 (it turns out in this example that λ_1 and λ_2 remain equal throughout). The updated Lagrange multipliers are now used for the next iteration. The complete sequence of iterations is shown in Table 5.3. This was made with the spreadsheet 'penalty function method' suitably modified for the augmented Lagrangian penalty function method. We see that a solution accurate to within 0.1% is reached in three iterations. However, we should understand that this simple example does not show the full advantage of the Lagrange multiplier method over other penalty function methods in larger problems.   ■

5.3 Spreadsheet Program

The spreadsheet illustrates use of a penalty function for constrained optimization. Since Solver does not itself include such an option, it is convenient to make further use of the program for the Hooke and Jeeves method in Sect. 4.3.1, with the aim of making the minimum of changes to that program. As initially set up, the spreadsheet uses an interior penalty function, but it is readily modified for another form of penalty function. A single new function is introduced into the program, responsible for progressive reduction of the penalty parameter and for checking any constraint violation that may occur. This function calls the Hooke and Jeeves routine to perform the optimization at each value of the penalty parameter.

Penalty Function Method

- Insert data below, and initial step sizes and initial values of the variables in the columns to the right.

- Press function key f 9 to perform the optimization.

Number of variables = 2

Required number of reductions in step size = 14

Maximum number of exploratory moves at each step size = 1000

Initial value of the penalty parameter = 250

Reduction factor for the penalty parameter = 10

variables	initial step	initial values	optimum values	
x_1	10.0	20.0	28.61	Initial value of the function = 125.66
x_2	0.50	1.00	0.1912	
x_3				Constrained minimum = 34.37
x_4				
x_5				
x_6				
x_7				
x_8				
x_9				Make no changes to cells within this border!
x_{10}				
x_{11}				
x_{12}				
x_{13}				
x_{14}				
x_{15}				
x_{16}				
x_{17}				

Fig. 5.6 Spreadsheet 'penalty function method'

5.3.1 'Penalty Function Method'

The interior penalty function method is applied to the well-used problem of the circular tube in compression in Example 2.1 and again in Example 5.4. To show agreement with the earlier analytical optimum, only Euler and local buckling constraints, P_E and P_L in Eqs. (2.1) and (2.2), are included, with no minimum thickness limitation. As in the spreadsheet for the Hooke and Jeeves method, the intention is that only function FN need be modified (or replaced) to define a different problem. This has to be programmed by the user in Visual Basic. The spreadsheet is shown in Fig. 5.6.

An additional function PF for the penalty function method reads the initial variables, initial penalty parameter, reduction factor and certain other parameters from the spreadsheet. It also returns the results of optimization to the spreadsheet. Again, the spreadsheet itself is reserved for input of data and display of results. In addition, function PF takes care of the progressive reduction (for an interior penalty function) of the penalty parameter, each time dividing by the specified reduction factor. It also checks for possible constraint violation. Function PF repeatedly calls function HJ (the Hooke and Jeeves routine) to locate the minimum of the penalty function for successive values of the penalty parameter.

Function FN is modified to calculate values of both the objective function and constraints and, for the interior penalty function method, to assemble the penalty function as in Eq. (5.10) with the current value of the penalty parameter. As set up for the circular tube, constraints are expressed in the form: $P_E/P - 1 \geq 0$ and $P_L/P - 1 \geq 0$. Here, these are simple analytic formulae, but in general, they can be expected to be more complex functions. The objective function is the cross-sectional area A of the tube. Variables are its radius R and thickness t. The necessary geometric, material and other data are included in function FN. With an interior penalty function, it will be recalled that this applies *only* to the feasible region. The function checks, therefore, for negative (or near zero) values of all constraints with each move in the Hooke and Jeeves routine. If a constraint violation occurs ($ICV = -1$), the move is treated simply as a failure, the move is retracted, and the procedure continues. It also checks for infeasible initial values, violating one or more constraints, in which case a message is displayed. Both checks should be removed for an exterior penalty function, since this is valid in the entire design space and works essentially in the infeasible region. In this case, the function FN need do no more than calculate the objective function and constraints and assemble the penalty function.

As previously, function HJ repeatedly calls function FN, now to perform the minimization of the penalty function for the current value of the penalty parameter. The only change to the Hooke and Jeeves routine in function HJ is that ICV for constraint violation is included in certain 'If' statements, so that a move is only regarded as a success if there is both a reduction in the value of the penalty function *and* $ICV = 0$ to indicate no constraint violation.

Table 5.4 Data entry for spreadsheet program 'penalty function method'

Parameters	
Number of variables	Enter the value in cell D8 (maximum 20)
Required number of reductions in step size in Hooke and Jeeves method	Enter the value in cell D11
Maximum number of sequences of exploratory moves at any step size in the Hooke and Jeeves method	Enter the value in cell D15 (see Sect. 4.3.1)
Initial value of the penalty parameter	Enter a positive value in cell D18
Reduction factor for the penalty parameter	Enter a positive value in cell D21
Required number of penalty function iterations	Enter the value in cell D24
Initial step size for variables x_i in the Hooke and Jeeves method	Enter values in order in column G, starting at cell G5 (remaining cells may be left blank)
Variables	
Initial values of variables x_i	Enter values in order in column H, starting at cell H5 (remaining cells may be left blank)
Create the penalty function in function FN (to access function FN, click Visual Basic on the Developer tab to open the Visual Basic Editor)	

The initial penalty parameter, reduction factor and the chosen number of penalty function iterations (i.e. reductions in the penalty parameter) all have to be entered where indicated on the spreadsheet, as well as the required number of step size reductions in the Hooke and Jeeves procedure and certain other data. Initial values of the variables, with corresponding initial step sizes, have to be entered in the appropriate columns. The number of variables is limited to 20, unless further changes are made. There are no specific limits on the number of constraints. For the circular tube, we have only two variables: $X(1) = R$ and $X(2) = t$. The remaining variables should be left blank. A sufficient number of penalty function iterations should be chosen to ensure that the solution has converged to the optimum. The number of reductions in step size determines the precision to which the optimum can be located. Values currently in the spreadsheet might be regarded as typical. If the constraints have been normalized, the initial penalty parameter can be chosen based on an estimated value of the objective function after optimization. Note that for an interior penalty function, the reduction factor must be greater than unity, so that the penalty parameter is progressively reduced. If function FN is modified for an exterior penalty function, the penalty parameter must be less than unity, so that the penalty parameter is increased. Parameters and variables to be entered are listed in Table 5.4.

After entry of the required data in the spreadsheet, optimization is started by pressing function key f 9 (again 'Manual Workbook Calculation' has been selected in Options). Optimum values of the variables are returned in the column 'optimum values' on the spreadsheet, together with the resulting constrained minimum of the function and the number of function evaluations performed. Results for the circular tube are given in Table 5.1. Again, it should be noted that cells I5:I26 in the column headed 'optimum values' contain an array formula for return of the results of optimization to the spreadsheet. No change may be made to cells in this array unless the array formula is first deleted.

5.4 Summary

In practice, most optimization problems, certainly in structural design, will involve some or many constraints. The choice of method depends on the nature of the problem and in particular that of the constraints. For sufficiently smooth, well-behaved problems, methods for constrained optimization generally fall into one of the two categories. We refer to the first as constraint-following methods, in which the active constraints are followed as closely as possible around the design space, or in another way, constraints remain satisfied during the search for an optimum. The generalized reduced gradient method in Solver falls under this category. The second category is by means of a penalty function, in which a penalty term is added to the objective function either to avoid the constraints until close enough to the optimum or to penalize violation of them. These are the methods discussed in this chapter. Methods in either category apply in general to both linear and nonlinear problems. However, if both the constraints and the objective function are linear functions of the design variables, linear programming offers a powerful and effective alternative, although such problems are uncommon in structural design. For discontinuous problems such as those with discrete variables or problems where more local minima exist, the genetic algorithm (discussed in Chap. 8), in which more potential solutions are retained, offers a satisfactory if substantially more time-consuming outcome. Both linear programming and the genetic algorithm are available in Solver. For some discrete variable problems, Solver also offers the useful option to restrict variables to integer values in all three of its methods.

In the gradient projection method, a constraint-following method, a search direction is found on the intersection of the currently active constraints, so that they remain satisfied during the line search. The search direction has to be updated when a previously inactive constraint is encountered or when one is no longer active. Lagrange multipliers can be evaluated to distinguish between constraints that must be retained and those that have to be rejected. The line search is then repeated, and the process continued. The method is more readily applied to inequality constrained problems, but can also be applied to problems with equality constraints. The gradient projection method is based on a linear approximation to the constraints at a current point, so for nonlinear constraints, a correction is necessary when the solution deviates too far from the true constraints. Move limits are usually set to limit this. Use of inverse variables frequently improves the linearity of nonlinear constraints.

In the generalized reduced gradient method, again a constraint-following method and the method in Solver more generally applicable in structural design, a different procedure is followed. Here, surplus variables are added to convert all inequality constraints to equalities, while this does increase the total number of variables. All constraints are then treated in the same way. Variables are selected as dependent variables to satisfy the same number of constraints, and optimization is performed on the remaining independent variables. By suitable partitioning of a matrix containing the derivatives of each constraint, a reduced gradient in the independent variables is found from which a search direction is deduced. Surplus variables have to remain positive or zero throughout the optimization (in effect, this is selection of active constraints). The reduced gradient is re-evaluated when a surplus variable reduces to zero during the line search. Again, the generalized reduced gradient method is based on linearization of nonlinear constraints, with a restoration move to return to the true constraints and move limits to avoid too great a deviation.

The penalty function method, an alternative to the constraint-following methods for constrained optimization, reduces the problem to one of unconstrained optimization, since all constraints are now contained in the penalty term. The unconstrained problem can then be solved by any of the previous methods. By taking a series of values of a penalty parameter, the solution is drawn progressively closer to the required optimum with each unconstrained optimization. Penalty function methods are classed as interior or exterior, depending on whether they approach the optimum from the feasible or the infeasible side. The augmented Lagrangian penalty function is a more advanced form, with the Lagrange multipliers included in the penalty term, and is more rapidly convergent. Penalty function methods can be applied to both equality and inequality constraints, whether linear or nonlinear, and are readily incorporated into an existing unconstrained optimization program. The spreadsheet in this chapter to demonstrate the penalty function method makes use of the Hooke and Jeeves routine in the previous chapter for the unconstrained optimization. This is simply for convenience in the small problem chosen as example. For large problems, more efficient methods would inevitably be used.

Exercises

5.1 Locate the intersection line of the two constraints in Example 5.1 in a three-dimensional design space and mark the minimum point found in the example. Evaluate the Lagrange multipliers to confirm that both constraints are active at this point. Verify the optimum found in the example analytically.

Find two convenient points that satisfy both constraints to draw the line in a design space. Calculate the Lagrange multipliers by Eq. (5.4), making use of the matrix calculation in the example. For an analytical solution use the two constraints to eliminate two variables from the objective function, then differentiate for a minimum.

5.2 Use the gradient projection method to minimize the function:

$$f(\mathbf{x}) = x_1^2 + x_2^2,$$

subject to the constraint:

$$g(\mathbf{x}) = x_1 + 2x_2 \geq 1.$$

Follow the method in Example 5.1. With only one constraint and two variables the matrix calculations can readily be done by hand. Verify the solution analytically.

5.3 Calculate the reduced gradient at the point reached after the initial line search in Example 5.3. Determine a new search direction, and perform the next line search.

Follow the method in Example 5.3 to calculate the reduced gradient, with the same dependent variables, making use of the matrix calculations in the example where possible. Surplus variable x_5 became zero in the initial line search, therefore set the component of the new search direction relating to x_5 to zero if it would cause it to become negative. Compare the result after the new line search with the solution in Example 5.1.

5.4 Repeat Example 5.4 for the circular tube from different initial points and with different series of r values.

Use the spreadsheet 'Penalty Function Method'. Note the results at each step, and plot the path of the optimization in a design space.

5.5 Modify the spreadsheet 'Penalty Function Method' for an exterior penalty function, and repeat Example 5.4.

Use the IF function in Excel to select the maximum of the two terms in the penalty function. Note that a series of increasing r values is now required.

5.6 Modify the spreadsheet again for the augmented Lagrangian penalty function, and repeat Example 5.4.

Note that the r-value is now constant. Iteration is performed with the values of the Lagrange multipliers.

5.7 Change the formulae for the objective function and constraints in the spreadsheet 'Penalty Function Method' to minimize the cost of the container in Exercise 3.4.

Choose a suitable sequence of r values.

5.8 Change the formulae in the spreadsheet to minimize the function in Example 3.4, subject to the same linear constraints.

Plot the optimization path in a design space in any two of the three variables.

5.9 Change the formulae in the spreadsheet 'Penalty Function Method' to optimize the cross section of a hollow steel beam of thin, square section and uniform thickness, under a symmetrically applied shear force of

50×10^3 N and bending moment 10×10^6 N mm. Take into account the maximum equivalent stress at the corners of the section in combined bending and shear, and buckling of the different sides in compression or shear. The maximum allowable stress is 500 N/mm^2, at which any effect of yielding of the material on the buckling stresses can be neglected, and elastic modulus $E = 200 \times 10^3$ N/mm^2.

Variables are the width b of the square and its thickness t. Use the formula in Sect. 3.4.2 *for the equivalent stress. For buckling of the side in compression use the formula* $\sigma_b = 3.62\,E(t/b_e)^2$, *and for buckling of the sides in shear* $\tau_b = 4.83\,E(t/b_e)^2$, *where* b_e *is the effective width (i.e. measured to the mid-thickness of the section). Use the average shear stress in the sides for buckling in shear.*

References

1. Abadie J, Carpentier J (1969) Generalization of the Wolfe reduced gradient method to the case of nonlinear constraints. In: Fletcher R (ed) Optimization. Academic Press, New York, pp 37–47
2. Arora JS (2012) Introduction to optimum design. Academic Press, Waltham MA
3. Dantzig GB, Thapa MN (1997) Linear programming 1: introduction. Springer, Berlin
4. Dantzig GB, Thapa MN (2003) Linear programming 2: theory and extensions. Springer, Berlin
5. Fiacco AV, McCormick GP (1968) Nonlinear programming: sequential unconstrained minimization techniques. Wiley, New York
6. Gabriele GA, Ragsdell KM (1977) The generalized reduced gradient method: a reliable tool for optimal design. ASME J Eng Ind 99:384–400
7. Murty KG (1983) Linear programming. Wiley, New York
8. Nering ED, Tucker AW (1993) Linear programs and related problems. Academic Press, Boston
9. Powell MJD (1978) A fast algorithm for nonlinearly constrained optimization calculations. In: Proceedings of the 1977 Dundee conference on numerical analysis. Lecture notes in mathematics, vol 630. Springer, Berlin, pp 144–157
10. Rosen JB (1960) The gradient projection method for nonlinear programming: part I, linear constraints. SIAM J Appl Math 8:181–217
11. Rosen JB (1961) The gradient projection method for nonlinear programming: part II, nonlinear constraints. SIAM J Appl Math 9:514–532
12. Vanderplaats GN (1984) Numerical optimization techniques for engineering design. McGraw-Hill, New York
13. Zoutendijk G (1960) Methods of feasible directions. Elsevier, Amsterdam

Chapter 6
Optimization of Beams

Abstract Beam optimization involves both the design of the cross section and the distribution of material along its length. For a relatively solid beam, the necessary cross-sectional area under a given bending moment at any section is determined largely by its permitted height and width. At lower load, when the cross section is relatively thin, a beam becomes liable to buckling under compressive and shear stress. A spreadsheet program is made for an I-section beam loaded in bending and shear under stress, buckling and stiffness constraints. For a geometrically similar family of beams, the cross-sectional area can be related to the bending moment by a non-dimensional coefficient depending only on the chosen shape ratios of the cross section. Geometric similarity is adopted in a spreadsheet program for the optimum distribution of material along the length of a beam, with a finite element analysis for the bending moment distribution. Provided some degree of yielding is permitted, the capacity of a beam is not exhausted when the elastic limit is reached at some point. With increasing load, yield hinges are formed at points along the span, leading to the ultimate collapse of the beam. This is the classic problem of limit design.

Beams and trusses in their many different forms must be amongst the most widely encountered of all engineering structures. Truss structures were discussed in earlier chapters, in particular to illustrate some basic principles of structural optimization, also as a convenient model to demonstrate use of Solver. In the present chapter, we shall explore beam optimization in some detail. We can distinguish two different but closely related aspects of beam optimization: first, optimization of the cross section of a beam and second the optimum distribution of material along its length. Since the load-carrying capacity of a beam is not exhausted when some part of it reaches the elastic limit of the material, we have also to consider the redistribution of stress within the cross section due to yielding, and the increase in load that can be carried. These last aspects are introduced later in this chapter.

© Springer International Publishing AG 2017　　　　　　　　　　　　　　　147
A. Rothwell, *Optimization Methods in Structural Design*, Solid Mechanics
and Its Applications 242, DOI 10.1007/978-3-319-55197-5_6

6.1   Beam Cross Section

We consider first the simple rectangular section beam in Fig. 6.1, under a bending moment M about the horizontal axis. By the conventional theory of bending, the maximum stress in the beam is as follows:

$$\sigma_{max} = \frac{Mh}{2I} = \frac{6M}{bh^2},$$

where $I = bh^3/12$ is the second moment of area of the beam, or with cross-sectional area $A = bh$:

$$\sigma_{max} = \frac{6M}{Ah}.$$

If the maximum stress is an allowable stress σ_0 (assumed for the present to be within the elastic limit of the material) we have the following:

$$A = \frac{6M}{\sigma_0 h}.$$

For given M and σ_0, the required cross-sectional area A of the beam depends only on its height h, and it is clear that to minimize A, we require the largest possible h. We must, therefore, impose a limit h_{max} on the height of the beam for any sensible result, as would invariably also be imposed for purely practical reasons. Therefore, we can write as follows:

$$A = \frac{6M}{\sigma_0 h_{max}}. \tag{6.1}$$

The above formula demonstrates the well-known importance of the height of a beam.

To improve the design of the beam, we have to change its shape from the current simple rectangle. It might be tempting to do this in the manner of a fully stressed design, by attempting to make the stress equal to the maximum allowable stress

Fig. 6.1 Rectangular section beam

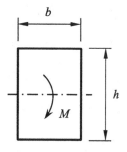

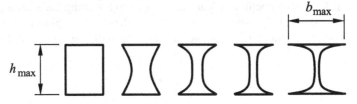

Fig. 6.2 Progressive removal of material from the cross section

over the whole cross section. However, this is fundamentally impossible because, as long as the behaviour remains elastic, the bending stress retains its linear distribution whatever changes in shape are made. Instead, we can remove material from the lesser stressed parts of the section around the neutral axis, where it is also less effective in carrying the bending moment. This is done step by step in Fig. 6.2, purely for illustration and not according to any particular rule. Note that, as for the simple rectangular section, a maximum height is still imposed. It is seen how the familiar shape of an I-section beam begins to emerge.

Since we have up to now considered a beam under only a bending moment, we could in principle continue with removal of material until the web of the beam has entirely disappeared. However, there will usually be a shear force on the beam, and the web is required to carry the largest part of this as well as serving to stabilize the cross section as a whole. The shear stress in the section, combined with the bending stress, limits the minimum thickness of both the web and the flanges. Further, it should be pointed out that in Fig. 6.2, material has been removed from the two sides of the cross section, but this could of course equally well have been done from the middle of the section, ending up with a rectangular, tubular section beam.

At some stage, with continued removal of material, the excess capacity of the beam will have been exhausted. To permit further removal of material from the lesser stressed parts of the beam, this has to be compensated by increase in width of the flanges (referring still to an I-section beam). If no maximum width is specified, with progressively wider and thinner flanges, these may eventually become so thin that buckling becomes the dominant design condition. For a minimum weight beam, we are concerned then with both stress limits and buckling of the cross section. In practice, we are likely to have to specify both a maximum height and a maximum width for the beam, and the design of the cross section will be confined to within these limits.

It will be clear then that for the most efficient shape of cross section the material should be concentrated as closely as possible to the edges of the 'box' defined by these limits, such as in the I-beam or rectangular tube already referred to. For the rest, the cross-sectional shape will be determined by practical considerations of the use of the beam, its connections to other parts of the structure and the manner in which the load is applied. For a chosen shape of beam, its individual dimensions may then be optimized, subject to dimensional and other constraints, to refine the design. Formulae for the minimum volume of a variety of beam sections are given by Rees [3].

Equation (6.1) with coefficient 6 applies strictly to a rectangular section, in pure bending. However, the formula can be generalized by replacing this by a non-dimensional coefficient n_h (where subscript h indicates a height limitation $h_{\max}$):

$$A = n_h \cdot \frac{M}{\sigma_0 h_{\max}}, \tag{6.2}$$

or, if preferred, in non-dimensional form:

$$\frac{A}{h_{\max}^2} = n_h \cdot \frac{1}{\sigma_0} \cdot \frac{M}{h_{\max}^3}. \tag{6.3}$$

Cross-sectional area A is again the area required to limit the maximum stress to the allowable stress σ_0. The quantity $M/h_{\max}^3$ is a useful parameter to represent the severity of the bending moment in a beam, relating this directly to the allowable stress. The value of n_h (given by substituting appropriate values of A and M in the above formulae) depends only on the shape of the cross section and can be used therefore to compare different cross sections. The theoretical minimum value is $n_h = 2$. For the symmetric I-section beam in Fig. 6.3, with $b = h/2$ and $t = b/10$, we obtain $n_h = 3.32$. As already stated, for a solid rectangular section $n_h = 6$.

6.1.1   Thin-Walled Beams

The possibility of buckling in the cross section of a beam was already referred to briefly. At smaller values of $M/h_{\max}^3$ in Eq. (6.3), the section becomes relatively thin, and buckling is likely to be a critical design condition. For the I-section beam discussed earlier this will be buckling of the upper or lower flange, principally in compression. Buckling of the web will be principally in shear. In a flat-sided section, buckling typically becomes a critical design condition at a width/thickness ratio of around 10 or more for a 'side' free on one edge and supported on the other

Fig. 6.3 Symmetric I-section beam

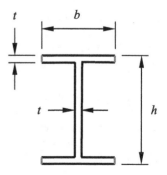

(the flange of an I-beam), or at a ratio of around 25 or more for a 'side' supported on both edges. These values are given only as a guide and should not be taken as design values. The effect of buckling constraints is that, to reduce the width/thickness ratio of the individual parts of the cross section, a specified maximum height and width of the cross section of the beam may not be reached at smaller values of M/h^3_{max}.

The mode of buckling referred to above is local buckling of the cross section. As well this, the possibility of lateral buckling should be referred to. This is buckling out of the plane in which the beam is loaded, coupled with rotation of the cross section, due to insufficient lateral or torsional stiffness. This is a long-wave mode, depending on the length of the beam, its support conditions and the manner of loading. It cannot, therefore be treated simply at the level of the beam cross section. Minimum lateral and torsional stiffness may be imposed, or more simply a minimum width of the cross section. A minimum width is in any case likely to be imposed for practical reasons. A minimum bending stiffness in the plane of loading may also be specified to limit deflection of the beam under load. In general, constraints for optimization of the cross section of a beam include now: stress limitations under combined bending and shear, buckling constraints, dimensional limitations and minimum stiffness.

Figure 6.4 shows in the upper figure how the minimum cross-sectional area A of the I-section beam in Fig. 6.3 varies over a wide range of applied bending moment M when subject to both buckling and stress constraints (note that this is a log–log plot). Corresponding optimum values of height h and width b are plotted in the lower figure. For the purpose of this figure, the beam is in pure bending with no applied shear force. The thickness of the web is arbitrarily chosen, therefore, to be equal to that of the flanges. The bending moment is assumed to be applied in either direction, resulting in a doubly symmetric section. Both a maximum and a minimum width of the cross section are specified, as well as a maximum height. At higher values of bending moment (right hand side of the figure), the optimized beam is relatively thick, and only the stress constraint is active. As expected, here both the height and the width of the beam reach their maximum values h_{max} and b_{max}, and only the thickness varies. At reduced bending moment, buckling of the flanges becomes a critical constraint, together with the previous stress constraint. This leads directly to a reduction in width, $b < b_{max}$, in the optimized design until it reaches the minimum width b_{min}. At that same point, the height of the beam begins to reduce, and these two effects together produce a marked change in slope of the $\log A/\log M$ graph. At still lower bending moment, only the buckling constraint remains active. There is then practically no further change in the height of the beam, and again only the thickness varies. Due to the change in shape of the cross section with bending moment, the value of n_h in Eq. (6.3) varies with M. At $M = 10^8$ Nmm ($M/h^3_{max} = 29.6$) at which $h = h_{max}$ and only the stress constraint is active, we have a cross-sectional area $A = 6273$ mm^2 giving a value of n_h in Eq. (6.3) of $n_h = 4.70$. Figure 6.4 was plotted using the spreadsheet program 'I-section Beam' described in Sect. 6.4.1.

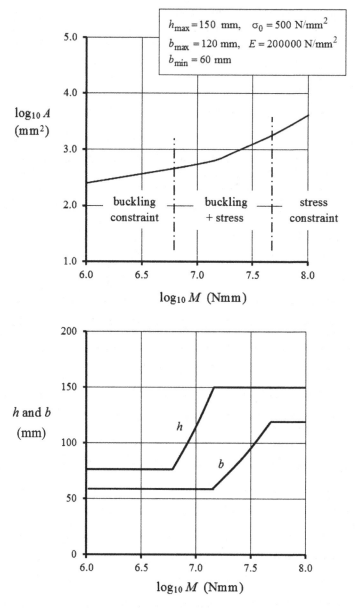

Fig. 6.4 Cross-sectional area A (*upper figure*), height h and width b (*lower figure*) plotted against bending moment M for an optimized I-section beam with equal flange and web thickness

6.1.2 Geometrically Similar Sections

A practical approach to many beam optimization problems is to adopt a geometrically similar family of designs. Geometric similarity means that, for any chosen shape of beam, its cross-sectional dimensions are all directly related to a single, arbitrarily selected dimension of the cross section, for example the height h of the beam. Then, for any change in that dimension, all other dimensions are changed in the same proportion. In other words, in any family of designs, every cross section is a scaled-up or scaled-down replica of the original design. This implies that the cross-sectional area A is proportional to h^2, and the second moment of area I is proportional to h^4. In that case, we can write as follows:

$$I = CA^2, \tag{6.4}$$

where coefficient C can be regarded as the 'shape efficiency' of that family of designs, depending only on the shape ratios of the cross section. These are the fixed ratios between the various dimensions of the cross section, the shape of the beam being defined by these shape ratios. With given shape ratios, the buckling stress remains the same for all members of a geometrically similar family of beams. In fact, many standard ranges of commercially available beam sections roughly correspond to a geometrically similar series of sections. Note that the range of beams plotted in Fig. 6.4, while all I-sections, are not geometrically similar.

For geometrically similar beams, the maximum distance y from the neutral axis to the outer edge of the beam can also be expressed in terms of its cross-sectional area A and a second non-dimensional coefficient C_1 by:

$$y = C_1 A^{1/2}.$$

With the above formula and the previous relation $I = CA^2$, the maximum stress in the beam is:

$$\sigma = M \cdot \frac{C_1 A^{1/2}}{CA^2},$$

which we set equal to an allowable stress σ_0. Therefore:

$$A = n_g \left(\frac{M}{\sigma_0}\right)^{2/3}, \tag{6.5}$$

where coefficient $n_g = (C_1/C)^{2/3}$ depends only on the shape ratios of the cross section. Equation (6.5), for a geometrically similar family of beams, replaces the previous Eq. (6.2). Note that no maximum height is now included in the above formula. Also, the height of an optimum cross section no longer increases without

Table 6.1 Values of C and n_g for different geometrically similar beam sections (height h, width b, thickness t)

□	▣	I	I	T	T
$\dfrac{b}{h} = 1$	$\dfrac{b}{h} = 1$	$\dfrac{b}{h} = 0.5$	$\dfrac{b}{h} = 0.5$	$\dfrac{b}{h} = 0.5$	$\dfrac{b}{h} = 0.5$
	$\dfrac{t}{h} = 0.25$	$\dfrac{t}{h} = 0.2$	$\dfrac{t}{h} = 0.1$	$\dfrac{t}{h} = 0.2$	$\dfrac{t}{h} = 0.1$
$C = 0.0833$	$C = 0.1389$	$C = 0.3542$	$C = 0.7593$	$C = 0.3587$	$C = 0.7221$
$n_g = 3.302$	$n_g = 2.585$	$n_g = 1.840$	$n_g = 1.341$	$n_g = 2.189$	$n_g = 1.756$

limit (as in Sect. 6.1). This is because the width of the beam increases along with its height, restricting its growth.

Values of C and n_g for a number of shapes of beam, together with their shape ratios, are given in Table 6.1. For all sections in the table, h is the height (outside dimension), b is the width and t is the uniform thickness of the cross section. Not surprisingly, there are substantial differences between the more solid and the thinner sections. For doubly symmetric sections, the lower of the allowable stresses in tension and compression should be used for σ_0, if these differ. For sections not symmetric about their neutral axis, such as the T-section in Table 6.1, two different values of coefficient n_g refer to the upper and lower edges of the beam, with the appropriate allowable stresses (the greater value of n_g is given in the table). Equation (6.5) remains applicable to unsymmetrical cross sections, but then with due consideration to the location and orientation of the neutral axis.

Geometric similarity provides a useful relationship between the second moment of area and cross-sectional area of a given shape of beam. Any required change in the shape of the cross section is reflected in a new value of C. The relationship in Eq. (6.4) will be used in Sect. 6.4.2 for the optimum distribution of material over the span of a beam and can, for example, be used to define the properties of the individual stiffeners in a stiffened panel.

6.2 Optimum Spanwise Distribution

Unless practical considerations demand otherwise, for minimum weight, we can expect the cross section of a beam to vary along its length in response to the bending moment in it. This might be by discrete changes in cross section at various points along the beam, or by a more continuous variation. In the previous sections, we explored the design of the cross section of a beam, and in particular relations between the minimum cross-sectional area and bending moment. These relations will be used now for the optimum variation of cross section along the span. For this, we have to make a distinction between a statically determinate beam such as one on two simple-supports or a cantilever beam, and a statically indeterminate beam such as a beam clamped at both ends or one supported at more than two points. In the former case, the bending moment and shear force in the beam are determined entirely by the applied loading, and the optimum spanwise distribution follows directly. In the latter case, the bending moment and shear force depend on the stiffness variation along the span and are therefore directly affected by changes in cross section. For a statically indeterminate beam, this becomes a classic problem in the mathematical calculus of variations, but we shall approach it differently here by a purely numerical method.

6.2.1   Statically Determinate Beams

Just as for the truss structure we studied in the first pages of Chap. 1, we begin with the classic problem of a beam of span L, simply supported at each end with a single load P at mid-span, as in Fig. 6.5. The maximum bending moment, at the point of loading, is $M_{max} = PL/4$. In the first place, we assume that the beam is uniform along its length to obtain a basic formula to relate the volume of the beam to the applied load. If we take the rectangular cross section in Sect. 6.1, with maximum height h_{max} and minimum cross-sectional area A given in Eq. (6.1), we find the following:

$$V = AL = \frac{6M_{max}}{\sigma_0 h_{max}} \cdot L.$$

Substituting for M_{max}:

$$V = \frac{3}{2} \cdot \frac{PL}{\sigma_0} \cdot \frac{L}{h_{max}}.$$

It will be observed that this last formula has much in common with formulae for a pin-jointed truss structure in Chap. 1. However, rather than proceeding further with the rectangular section above, we choose now the more practical option of a family of geometrically similar designs to represent the cross section. It is assumed that these are thick enough, in relation to their other dimensions, to avoid any

Fig. 6.5 Statically determinate beam

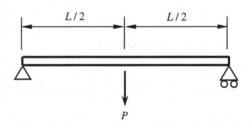

buckling of the beam. With Eq. (6.5) for the cross-sectional area, we obtain an alternative formula for the beam in Fig. 6.5, uniform along its length:

$$V = AL = n_g \left(\frac{M_{\max}}{\sigma_0} \right)^{2/3} \cdot L.$$

Substituting again for $M_{\max}$:

$$V = n_g \left(\frac{PL}{4\sigma_0} \right)^{2/3} \cdot L.$$

In non-dimensional form the above formula becomes as follows:

$$\frac{V}{L^3} = 0.397 \times n_g \left(\frac{1}{\sigma_0} \cdot \frac{P}{L^2} \right)^{2/3}.$$

Here, we have identified a structural index P/L^2, in units of stress, common to all problems of this type with geometrically similar sections. The same structural index was found in Chap. 1 for a truss structure.

To refine the design, we have to allow the cross section of the beam to vary along its length. Since the bending moment is unaffected by the distribution of material along the length of a statically determinate beam, we can directly match the cross-sectional area to the bending moment, in the manner of a fully stressed design. This means, of course, that every beam section is loaded to its maximum capacity, and not that the allowable stress is reached over the entire cross section. The bending moment in the beam in Fig. 6.5 varies linearly from $M_{\max}$ at mid-span to zero at each end. For the rectangular section of given height, this would imply simply a halving of the volume of the beam. For geometrically similar beam sections, the calculation is slightly more complicated:

$$V = 2 \int_{x=0}^{x=L/2} A \, dx = 2 \int_{x=0}^{x=L/2} n_g \left(\frac{M}{\sigma_0} \right)^{2/3} dx = 2 \int_{x=0}^{x=L/2} n_g \left(\frac{P}{2\sigma_0} \right)^{2/3} x^{2/3} dx.$$

Performing the integration:

$$V = 0.238 \, n_g \left(\frac{P}{\sigma_0} \right)^{2/3} L^{5/3}.$$

In non-dimensional form:

$$\frac{V}{L^3} = 0.238 \, n_g \left(\frac{1}{\sigma_0} \cdot \frac{P}{L^2} \right)^{2/3}.$$

Expressed as a strength-to-weight ratio (as in Eq. (1.9) for a truss structure) we have the following:

$$\frac{P}{W} = \frac{4.20}{n_g} \cdot \frac{\sigma_0^{2/3}}{\rho_w} \cdot \left(\frac{P}{L^5}\right)^{1/3},$$

where ρ_w is the specific weight of the material. Coefficients 0.238 and 4.20 in the above formulae refer, of course, specifically to a simply supported beam loaded at mid-span. These can be used to compare with other loading systems and forms of support.

Example 6.1 Derive an expression for the minimum volume of a cantilever beam of length L under a total load P uniformly distributed along the length of the beam.

The bending moment in the beam is as follows:

$$M = \frac{P}{2L} \cdot x^2,$$

(x measured from the free end). With Eq. (6.5) for the cross-sectional area:

$$V = \int_{x=0}^{x=L} A\,dx = \int_{x=0}^{x=L} n_g \left(\frac{M}{\sigma_0}\right)^{2/3} dx.$$

Substituting for M:

$$V = \int_{x=0}^{x=L} n_g \left(\frac{P}{2L\sigma_0}\right)^{2/3} x^{4/3}\,dx.$$

Performing the integration and evaluating the numerical constant, the minimum volume is as follows:

$$V = 0.270\,n_g \left(\frac{P}{\sigma_0}\right)^{2/3} L^{5/3}.$$

In non-dimensional terms this becomes as follows:

$$\frac{V}{L^3} = 0.270\,n_g \left(\frac{1}{\sigma_0} \cdot \frac{P}{L^2}\right)^{2/3}.$$

For comparison, the coefficient 0.270 in the above formula becomes 0.630 for the same cantilever beam if it is uniform along its length.

6.2.2 Statically Indeterminate Beams

As already stated, the bending moment in a statically indeterminate beam is governed by stiffness variation along the span. This is, of course, directly related to the spanwise distribution of material, which we seek to optimize to minimize weight. The simple approach of the previous section, in which the cross section of the beam is matched to the known bending moment at each point along the span, cannot therefore be used to optimize the beam. Instead, we turn to a numerical method and illustrate the optimization of a statically indeterminate beam in a particular example. This is a beam of solid square section, clamped at one end and simply supported at the other, under a single load applied at three-quarter span from the clamped end, as shown in Fig. 6.6. In Table 6.1 we find coefficients $C = 0.08333$ and $n_g = 3.302$ for a solid square section.

Again, we consider first a beam that is uniform along its length. The maximum bending moment is:

$$M_{max} = 0.1582\,PL,$$

which occurs at the point where the load is applied (see [4]). With the minimum cross-sectional area A given by Eq. (6.5), we obtain for the weight W_k of the uniform beam:

$$W_k = n_g \left(\frac{M_{max}}{\sigma_0}\right)^{2/3} \rho L,$$

where subscript k denotes 'weight' expressed in kg and ρ is the density of the material. For comparison with the numerical solution below for an optimized beam, we take the following data: applied load $P = 4800$ N, span $L = 1200$ mm, allowable stress $\sigma_0 = 500$ N/mm^2 and density $\rho = 7850$ kg/m^3. Substituting in the above formulae gives for the weight of the beam:

$$W_k = 4.64 \text{ kg.}$$

If the beam need not be uniform along its length, it can be optimized by the spreadsheet program 'Beam under Lateral Load' in Sect. 6.4.2, with a finite

Fig. 6.6 Statically indeterminate beam

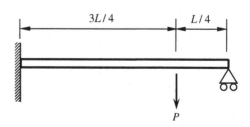

element analysis for the bending moment distribution and again Solver to perform the optimization. Design variables are the second moment of area I of each element of the beam. From Eqs. (6.4) and (6.5), we derive the following expression for the maximum allowable bending moment in an element, in terms of its second moment of area:

$$M_{\text{all}} = \frac{\sigma_0}{(Cn_g^2)^{3/4}} \cdot I^{3/4}. \tag{6.6}$$

The bending moment in each element may not exceed this allowable moment. The spreadsheet is described further in Sect. 6.4.2. Figure 6.7 shows the bending stiffness distribution after optimization, with a resulting weight for the beam:

$$W_k = 2.85 \text{ kg.}$$

(While we refer here to 'stiffness', it is actually second moment of area that is plotted in the figure, on the basis that the elastic modulus is the same for all elements.) The finite element analysis employs conventional beam elements of uniform stiffness, so that the solution is strictly for a 'stepped' beam with 24 uniform-stiffness segments. The optimum stiffness distribution, the solid line in Fig. 6.7, is therefore for this stepped beam. The spreadsheet shows that after optimization the beam has become 'fully stressed' (see the column M_e/M_{all} in the spreadsheet), in the sense that the allowable bending moment is reached at one or other end of each element.

The solution can be further refined as follows. Clearly, the minimum weight already found is an overestimate for a beam with a gradual taper, rather than one

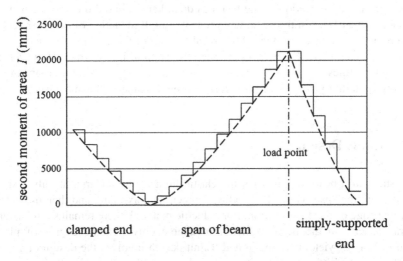

Fig. 6.7 Optimum spanwise stiffness distribution for a beam clamped at one end, simply supported at the other end, loaded at three-quarter span from the clamped end

made in discrete steps. This is because the bending moment in each element is taken to be the greater of the values at either end, and the minimum value of I for the element is based on this greater value. However, we now know the bending moment corresponding to the optimized stiffness distribution from the finite element analysis. Rewriting Eq. (6.6), minimum values of second moment of area I at each node can be recalculated by the following:

$$I = Cn_g^2 \left(\frac{M}{\sigma_0}\right)^{4/3} \tag{6.7}$$

and a new weight obtained. This is done in the spreadsheet under the heading 'Recalculation of second moment of area and minimum weight'. The weight is then further reduced to:

$$W_k = 2.61 \text{ kg.}$$

The resulting stiffness distribution is superimposed on Fig. 6.7 (the continuous curves). It is seen that the bending stiffness reduces to zero between nodes 7 and 8 (slightly beyond quarter span from the clamped end), implying that in effect a hinge is formed at this point. In other words, the originally statically indeterminate beam has developed into a statically determinate one. Part of the beam acts like a simply supported beam to carry the applied load, this being supported in turn by a cantilever beam at the clamped end. This is reminiscent of the statically indeterminate truss in Sect. 1.1, which reduced to an optimum, fully stressed statically determinate one. It can be anticipated that, under a single load case and unless additional requirements apply, reduction of an initially statically indeterminate beam to an optimum, statically determinate one will be a general outcome. It should be noted, however, that optimization of the beam as done here is based only on the bending moment. Shear force in the beam will prevent the full reduction of the cross section to zero at the supposed 'hinge'. Also, recalculation of the stiffness of the beam, as done here, will affect the bending moment distribution in the beam. However, the change in stiffness is relatively small therefore the change in bending moment will be correspondingly small. No further correction is made in the spreadsheet.

6.3   Limit Design

The stress in a beam might reach the elastic limit of the material at only a single point along the span where the bending moment is maximum, and then only at the extreme edge of the cross section. For a ductile material, there remains a substantial extra capacity in the beam, which is available only if some degree of plastic deformation, or yielding, is permitted. Limit design involves the design of a beam on the basis of its ultimate failure, or collapse, rather than simply for a given maximum stress (see [1, 2] and other texts).

6.3.1 Yield Moment

We consider first yielding of the material within the cross section of a beam. This is illustrated in Fig. 6.8. Yielding begins in the outer regions of the cross section and progresses inwards with further increase in bending moment (from left to right in the figure). The last diagram in Fig. 6.8 is the idealized, fully yielded state, which can never in reality be reached. Increase in stress in the inner regions allows these to contribute more to the bending moment on the beam. While the stress distribution is changing with increase in bending moment, the strain distribution remains essentially linear, from zero at the neutral axis to a maximum at the outer edges. The bending moment is obtained by integration over the cross section:

$$M = \int_A \sigma y \, dA,$$

where the stress σ corresponding to the linear strain distribution is obtained from the stress–strain curve of the material, and y is the distance from the neutral axis. However, with even a moderate degree of yielding, this moment is relatively insensitive to the actual shape of the stress–strain curve. This is because regions of lower stress are necessarily close to the neutral axis, where differences in stress have relatively little effect. It is generally sufficient, therefore, to replace the actual stress–strain curve by the idealized one in Fig. 6.9a, or even by that in Fig. 6.9b. We shall continue here with the first of these. This represents only an initial elastic behaviour, with modulus E, followed by yielding at constant stress. Yielding is restricted by specifying a maximum permissible strain ε_{max}, usually well below the strain at failure for the material. The supposed yield stress σ_y is then the stress at ε_{max}, with yielding commencing at a strain ε_e:

$$\varepsilon_e = \sigma_y / E.$$

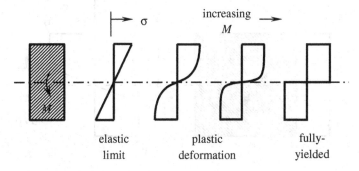

Fig. 6.8 Development of yielding with increase in bending moment

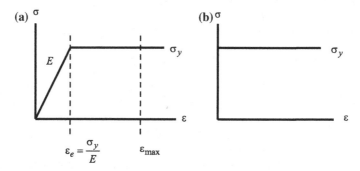

Fig. 6.9 Idealized stress–strain relations

If we take, for example, a maximum strain ε_{max} corresponding to the 0.5% proof stress of the material, taken to be 500 N/mm^2, and $E = 200,000$ N/mm^2, we have the following:

$$\varepsilon_e = \frac{500}{200,000} = 0.0025.$$

The plastic component of strain at the 0.5% proof stress is 0.005, giving the following:

$$\varepsilon_{max} = 0.0025 + 0.005 = 0.0075.$$

For the square section of side a in Sect. 6.2.2, with $\varepsilon_{max}/\varepsilon_e = 3$ yielding begins at a distance $a/6$ from the neutral axis, as shown in Fig. 6.10. Within the elastic region $(0 \leq y \leq a/6)$, the stress at distance y from the neutral axis is as follows:

$$\sigma = \frac{6\sigma_y}{a} \cdot y.$$

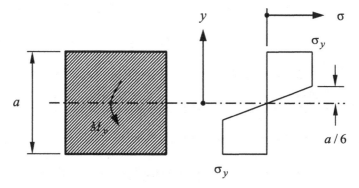

Fig. 6.10 Stress distribution in a square-section beam with $\varepsilon_{max}/\varepsilon_e = 3$

The maximum bending moment, or yield moment M_y, is then as follows:

$$M_y = 2 \int\limits_0^{a/6} \frac{6\sigma_y}{a} y \cdot ay\,dy + 2 \int\limits_{a/6}^{a/2} \sigma_y ay\,dy,$$

which gives:

$$M_y = 0.241 a^3 \sigma_y. \tag{6.8}$$

For comparison, if we chose the more extreme idealization in Fig. 6.9b, the coefficient 0.241 would become 0.25.

From Eq. (6.5), the maximum elastic moment for the square section, based on the idealized stress–strain curve in Fig. 6.9a, therefore with the stress at the elastic limit equal to the yield stress σ_y, is:

$$M_e = \left(\frac{a^2}{n_g}\right)^{3/2} \sigma_y = 0.167 a^3 \sigma_y,$$

with $n_g = 3.302$ for a solid square section from Table 6.1. The ratio $M_y/M_e = 1.44$ shows a substantial increase in capacity over a purely elastic analysis. This ratio is referred to as the 'form factor' for the beam, the name implying that it depends primarily on the shape of the beam and not on the particular stress–strain curve of the material. The above calculations apply, of course, specifically to the chosen square-section beam. For other shapes of beam, the integration above has to be performed over the appropriate cross section. Note that for an unsymmetrical cross section, the neutral axis is no longer the same as the elastic neutral axis. The stress distribution in Fig. 6.10b then has to be adjusted so that, in pure bending, there is no resulting end load on the section.

For a statically determinate beam, the yield moment M_y can be treated as the maximum bending moment at the ultimate load on a beam, always assuming it is of sufficiently ductile material. The weight saving that can be obtained by use of M_y in place of the maximum elastic moment M_e in a particular problem depends, amongst other things, on the maximum strain ε_{max} chosen to avoid significant plastic deformation at the working load.

6.3.2 Limit Load

For a statically indeterminate beam, a further step is now possible. This is to determine the limit load of a beam, or in other words the load at which unlimited further deformation of the beam will, in principle, occur in a so-called collapse mode. When the load on a beam is increased to the extent that the cross section is close to being fully yielded at a point where the bending moment is maximum, the capacity of a statically indeterminate beam is still not exhausted. It is supposed that a 'yield hinge' is then formed at that point, permitting further increase in load by redistribution of the load within the span of the beam. A yield hinge implies that unlimited rotation can occur at a constant moment M_y. The analogy is with a rusty hinge, which can be rotated once its resistance is overcome, while its resistance to rotation does not disappear.

We continue now with the beam in Sect. 6.2.2, with uniform cross section along its length (see Fig. 6.6). The maximum bending moment is at the point of loading, and a yield hinge is first formed at this point. Further increase in load is now taken by the 'cantilever' part of the beam between the clamped end and the yield hinge. This continues until the yield moment is reached at some other point. In the example in Sect. 6.2.2, this is inevitably at the clamped end. This new yield hinge is sufficient to create a 'collapse mechanism' in the beam, as shown in Fig. 6.11a. Note that the figure shows only deformation in the collapse mode, and not the elastic deformation, which can be ignored. The corresponding bending moment diagram is shown in Fig. 6.11b. The limit, or collapse, load can be calculated directly from the collapse mode. Displacement δ under load P and the corresponding angles α_1 and α_2 are shown in Fig. 6.11c. The limit load P is then calculated by equating the work done by the applied load to the work done on the hinges in the collapse mode:

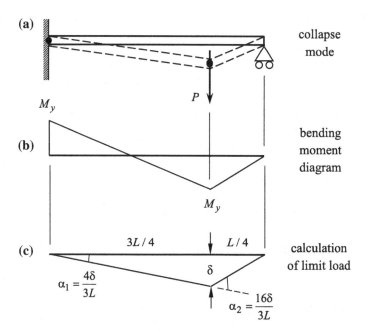

Fig. 6.11 Collapse mode of a clamped/simply supported beam

$$P\delta = M_y\left(\frac{4\delta}{3L} + \frac{16\delta}{3L}\right) = \frac{20}{3}M_y\frac{\delta}{L},$$

which gives:

$$P = 6.67\frac{M_y}{L}.$$

Note that this calculation does not in itself provide any information about the sequence of events leading up to collapse.

If we add a second load to the beam, as in Fig. 6.12, we now have the two possible collapse modes shown. Each of these has to be calculated as above. This gives for mode 1:

$$P_1\delta + P_2\frac{\delta}{2} = M_y\left(\frac{2\delta}{L} + \frac{4\delta}{L}\right), \qquad (6.9)$$

or:

$$P_1 + \frac{P_2}{2} = \frac{6M_y}{L}.$$

For mode 2:

$$P_1\frac{2\delta}{3} + P_2\delta = M_y\left(\frac{4\delta}{3L} + \frac{16\delta}{3L}\right), \qquad (6.10)$$

or:

$$\frac{2P_1}{3} + P_2 = \frac{20M_y}{3L}.$$

Here, the work done by both loads, with their respective displacements, has been included in the formulae for both modes. If loads P_1 and P_2 are increased in proportion, or if one load remains unchanged while the other is increased, the correct limit load is, of course, the lower of the loads calculated by Eqs. (6.9) and (6.10).

To optimize the beam in Fig. 6.12 under given applied loads, we might, for example, choose different beam sections in each of the three segments of the span, as in Fig. 6.13. We can define these by their yield moments M_1, M_2 and M_3. Equations (6.9) and (6.10) then have to be modified to account for the different yield moments. This is continued in Example 6.2 below.

Example 6.2 Find optimum values M_1, M_2, M_3 for the beam in Fig. 6.12, with loads $P_1 = 300$ N, $P_2 = 1200$ N. The beam has a span $L = 1000$ mm, with a solid square cross section. The yield stress of the material $\sigma_y = 500$ N/mm^2.

Fig. 6.12 Alternative
collapse modes

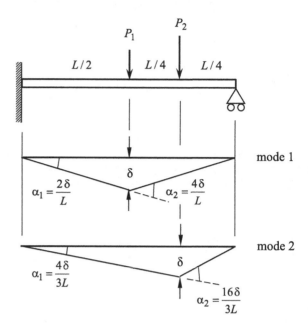

Fig. 6.13 Yield moments in
each segment of the beam

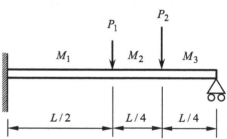

With different values of yield moment immediately either side of the yield hinge
at load P_1 or P_2, we have to take the lesser value of M_1 and M_2, or M_2 and M_3, at
these points in Eqs. (6.9) and (6.10), respectively. Since it is not known beforehand
which will be the lesser value, for optimization this is more easily handled by
treating mode 1 and mode 2 each as two separate modes, as follows.

For mode 1, modifying Eq. (6.9), the limit loads are as follows:

$$P_1 + \frac{P_2}{2} = 2\frac{M_1}{L} + 4\frac{M_1}{L} = 6\frac{M_1}{L} \quad (M_1 \leq M_2),$$

or

$$P_1 + \frac{P_2}{2} = 2\frac{M_1}{L} + 4\frac{M_2}{L} \quad (M_2 \leq M_1).$$

For mode 2, modifying Eq. (6.10) the limit loads are as follows:

$$\frac{2P_1}{3} + P_2 = \frac{4}{3} \cdot \frac{M_1}{L} + \frac{16}{3} \cdot \frac{M_2}{L} \quad (M_2 \leq M_3),$$

or

$$\frac{2P_1}{3} + P_2 = \frac{4}{3} \cdot \frac{M_1}{L} + \frac{16}{3} \cdot \frac{M_3}{L} \quad (M_3 \leq M_2).$$

With Eq. (6.8) for a square section of side a:

$$M_y = 0.241 a^3 \sigma_y,$$

or:

$$A = a^2 = \left(\frac{1}{0.241 \sigma_y}\right)^{2/3} M_y^{2/3} = 0.0410 \, M_y^{2/3},$$

where M_y refers to each of the yield moments M_1, M_2, M_3 of each segment, and A refers to the corresponding cross-sectional areas A_1, A_2, A_3. The volume of the beam to be minimized is then:

$$V = A_1 \frac{L}{2} + A_2 \frac{L}{4} + A_3 \frac{L}{4},$$

or:

$$V = 0.0410 \left(\frac{M_1^{2/3}}{2} + \frac{M_2^{2/3}}{4} + \frac{M_3^{2/3}}{4}\right) L.$$

With loads $P_1 = 300$ N, $P_2 = 1200$ N and $L = 1000$ mm, the above formulae for the limit loads lead to the following constraints:

$$6M_1 \geq 900 \times 10^3 \text{ Nmm},$$

$$2M_1 + 4M_2 \geq 900 \times 10^3 \text{ Nmm},$$

$$\frac{4}{3} M_1 + \frac{16}{3} M_2 \geq 1400 \times 10^3 \text{ Nmm},$$

$$\frac{4}{3} M_1 + \frac{16}{3} M_3 \geq 1400 \times 10^3 \text{ Nmm}.$$

With variables M_1, M_2, M_3, we can set up a spreadsheet to use Solver to find optimum values of the yield moments and the minimum volume of the beam. This gives the following:

$$M_1 = 150 \times 10^3 \text{ Nmm},$$

$$M_2 = M_3 = 225 \times 10^3 \text{ Nmm},$$

$$V = 133.8 \times 10^3 \text{ mm}^3.$$

Corresponding cross-sectional areas can be calculated from the previous formula for A. We find that constraints in both modes are active at the optimum; in other words, the beam now collapses in modes 1 and 2 simultaneously.

If we constrain the solution in Solver so that $M_1 = M_2 = M_3$ for a uniform beam, we find the following:

$$M_1 = M_2 = M_3 = 210 \times 10^3 \text{ Nmm},$$

$$V = 144.9 \times 10^3 \text{ mm}^3.$$

The uniform beam, under the given loading, collapses in mode 2. Comparing with the previous result, the optimized beam shows a modest reduction in volume of material over the uniform beam.

6.4 Spreadsheet Programs

The spreadsheets in this and the following chapters represent problems of more practical interest than those up to now, which were intended primarily to demonstrate optimization methods and setting up an optimization problem. In Sect. 6.4.1, an I-section beam is optimized under stress, buckling and stiffness constraints, with specified maximum and minimum dimensions. Margins of safety in the different failure modes are given.[1] In Sect. 6.4.2, the optimum distribution of material along the span of a beam is found. The bending moment in the beam is obtained by finite element analysis. Loading on the beam can be specified, as well as conditions of support.

---

[1]Margin of safety is defined as: $\left(\dfrac{\text{actual strength}}{\text{required strength}}\right) - 1$, and is required therefore to have a value greater than or equal to zero.

Fig. 6.14 I-section beam
with unequal flanges

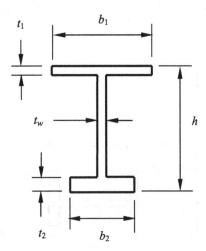

6.4.1   'I-Section Beam'

The spreadsheet uses Solver to optimize the I-section beam in Fig. 6.14, under bending moment about the horizontal axis and shear force in the plane of the web. The cross-sectional area of the beam is minimized subject to material stress limits and buckling of the flanges and web. A minimum bending stiffness may be specified in addition to required limits on cross-sectional dimensions. The spreadsheet is shown in Fig. 6.15.

6.4.1.1   Modelling

The bending and shear stress at critical locations in the cross section are calculated by conventional bending theory under a maximum positive and negative bending moment with corresponding shear forces (positive bending moment causes tension in the upper flange). The critical locations are at the junction of each flange and the web, where the maximum bending stress is combined with the shear stress built up along the flange and at the neutral axis where the shear stress is maximum and the bending stress is zero. The shear stress at the junction of the flange and web is estimated by calculating the shear stress in both the flange and the web at this point (necessary because of difference in thickness). These are treated as separate conditions for optimization. For combined bending and shear stress, the equivalent (von Mises) stress is calculated by:

$$\sigma_{eq} = \sqrt{\sigma^2 + 3\tau^2},$$

where σ is the tensile or compressive bending stress and τ the shear stress. The equivalent stress is related to the allowable stress σ_t in tension or σ_c in

I-section Beam

Parameters

$M^{+} =$ 5.00E+06 Nmm
$M^{-} =$ 2.00E+06 Nmm
$Q^{+} =$ 1.00E+04 N
$Q^{-} =$ 4.00E+03 N
$\sigma_{t} =$ 400 N/mm²
$\sigma_{c} =$ 300 N/mm²
$\tau_{all} =$ 225 N/mm²
$\sigma_{2} =$ 275 N/mm²
$E =$ 72800 N/mm²
$m =$ 15
$h_{max} =$ 500 mm
$b_{1\,max} =$ 200 mm
$b_{1\,min} =$ 25 mm
$b_{2\,max} =$ 100 mm
$b_{2\,min} =$ 25 mm

Variables

$h =$ 100.0 mm
$b_{1} =$ 50.0 mm
$b_{2} =$ 50.0 mm
$t_{w} =$ 5.00 mm
$t_{1} =$ 5.00 mm
$t_{2} =$ 5.00 mm

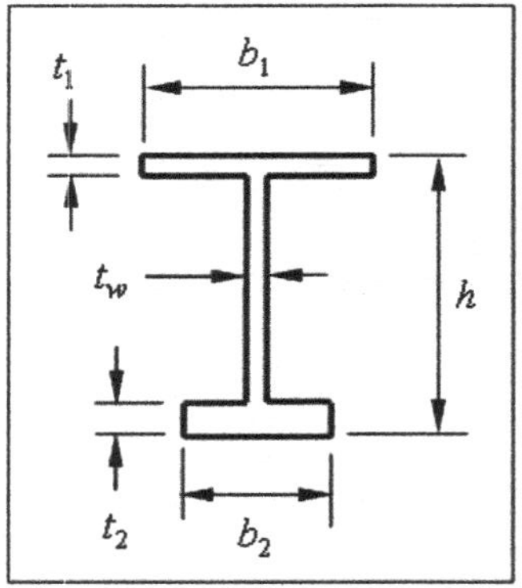

Cross-sectional area = 950.0 mm²

$y =$ 0.00 mm
$I =$ 1.43E+06 mm⁴
$EI =$ 1.04E+11 Nmm²

Margins of safety

	upper flange	lower flange	web top	web bottom	neutral axis
max. tensile stress	1.28	4.71	1.26	4.66	
max. comp. stress	3.28	0.71	3.24	0.70	

Fig. 6.15 Spreadsheet 'I-section Beam'

compression, as appropriate. The von Mises criterion is discussed in Chap. 7. The maximum shear stress, on the neutral axis, is related directly to the allowable shear stress τ_{all}.

For the buckling stress σ_b of the flanges in compression, the following formula is used:

$$\sigma_b = 0.385E\left(\frac{t}{b}\right)^2,$$

where t is the thickness of the upper or lower flange (t_1 or t_2, respectively), and b is its *half*-width ($b = b_1/2$ or $b = b_2/2$). Buckling in compression is based on the compressive stress in the middle plane of each flange. For the buckling stress τ_b of the web in shear:

$$\tau_b = 4.83E\left(\frac{t_w}{h_e}\right)^2,$$

where t_w is the thickness of the web and h_e is an effective height of the web measured between the middle planes of the flanges. Buckling in shear is based on the average shear stress in the web. Both above formulae assume simple support at the junction of the flange and web, ignoring any mutual restraint at this point, and are likely therefore to give a conservative estimate of the buckling stresses. To allow for reduction in modulus with the approach of yielding, the tangent modulus E_t is used for buckling of the flanges and the secant modulus $E_s(=\sigma/\varepsilon)$ for buckling of the web in shear. These are based on the Ramberg–Osgood formula used earlier in Sect. 3.4.1. In shear, the secant modulus is calculated at an equivalent compressive stress $\sigma = \sqrt{3} \cdot \tau$. Buckling formulae, and use of reduced moduli, are discussed further in Chap. 7.

6.4.1.2 Optimization

Design variables are the height h of the beam, the flange widths b_1 and b_2, the web thickness t_w, and the flange thicknesses t_1 and t_2. A maximum height of the beam has to be specified, also maximum and minimum flange widths and thicknesses. A minimum web thickness is specified to prevent this becoming vanishingly small when the shear force is zero or very small. None of the above may be left blank. The minimum flange width should be chosen to avoid the possibility of lateral buckling of the beam. Other limits on dimensions may be added in the Solver dialog box as required.

Constraints are expressed in terms of margins of safety (greater than or equal to zero). These can be summarized as follows:

- maximum combined (von Mises) stress in tension and compression in the upper flange at the junction of the flange and web

- maximum combined (von Mises) stress in tension and compression in the lower flange at the junction of the flange and web
- maximum combined (von Mises) stress in tension and compression at the top of the web
- maximum combined (von Mises) stress in tension and compression at the bottom of the web
- maximum shear stress in the web at the neutral axis of the beam
- buckling in compression of the upper and lower flanges of the beam
- buckling in shear of the web of the beam
- minimum bending stiffness.

With both positive and negative bending moments, this is in total 13 constraints, in addition to dimensional constraints.

Maximum positive and negative bending moments with corresponding shear forces, material data and minimum bending stiffness (if required, otherwise this may be zero) have to be entered under 'Parameters' in the spreadsheet, as well as maximum and minimum dimensions. Negative bending moment and shear force are entered as *positive* numbers. Suitable initial dimensions of the cross section have to be entered under 'Variables'. Data entry is summarized in Table 6.2. Solver can then be run with the GRG Nonlinear method to minimize the cross-sectional area of the beam, subject to the given margins of safety and dimensional constraints.

At higher loads, when the beam is relatively thick, we find that the height of the beam and the width of the flanges expand to their maximum values, and the maximum allowable stress is reached at one or more points in the cross section. At lower loads, buckling intervenes and the height of the beam and flange widths are reduced. When a minimum bending stiffness is specified, this may override both the stress and buckling limits in the beam.

Table 6.2 Data entry for spreadsheet program 'I-section Beam'

Parameters	
Max. bending moments M^+, M^- and *corresponding* shear forces Q^+, Q^-	Enter values in cells C6:C9 all as *positive* numbers (positive bending moment causes tension in the upper flange)
Allowable stresses in tension, compression and shear σ_t, σ_c, τ_{all}	Enter values in cells C10:C12 all as *positive* numbers
0.2% proof stress σ_2, elastic modulus E, Ramberg–Osgood index m	Enter values in cells C13:C15
Max. and min. dimensions h_{max}, $b_{1\,max}$, $b_{1\,min}$, $b_{2\,max}$, $b_{2\,min}$, $t_{w\,min}$, $t_{1\,min}$, $t_{2\,min}$	Enter values in cells C16:C23 (none of these may be zero or left blank)
Min. flexural stiffness EI_{min}	Enter value in cell C24 (may be zero or left blank if not required)
Variables	
Height h, upper flange width b_1, lower flange width b_2, web thickness t_w, upper flange thickness t_1, lower flange thickness t_2	Enter initial values in cells F6:F11

Additional constraints may be added, for example to restrict the beam to a symmetrical cross section, or to a family of geometrically similar sections. Alternative formulae for reduction in modulus with yielding are readily substituted if required. The spreadsheet might also be adapted for other shapes of beam. Attention is again drawn to the proper choice of initial values of the design variables. Values too far removed from the expected optimum may result in Solver being unable to find a solution (see the last paragraph of Sect. 2.3.2). If necessary, suitable initial values can be chosen by setting the margins of safety given in the spreadsheet to acceptable levels.

6.4.2 'Beam Under Lateral Load'

The spreadsheet uses Solver coupled with a finite element analysis to optimize the bending stiffness distribution along the length of a beam under lateral applied load. Distribution of load on the beam and its type of support can be defined as required. Suitable relations between cross-sectional area, allowable bending moment and second moment of area are defined. The beam may be optimized for minimum weight under the applied loading, subject to a specified allowable bending stress, or for maximum stiffness. With the finite element analysis, the spreadsheet illustrates use of Solver for a problem significantly larger than in previous chapters and might serve as a model for other problems of similar type. The spreadsheet is shown in Fig. 6.16.

6.4.2.1 Modelling

A finite element model with 24 equal-length elements along the length of the beam is built in the function Stiffness to analyze the bending moment distribution. The second moment of area is defined for each element. Lateral loads may be applied at any of the nodes. Chosen support of the beam, either simply supported or clamped, can be applied at any node by introducing appropriate constraints on deflection and rotation at those points. At least two conditions of support must be defined to eliminate rigid body movement of the beam. Initial values of second moment of area I have to entered in the spreadsheet, together with applied loads P, the length L of the beam, elastic modulus E, allowable bending stress σ_{all}, density ρ, and values of m, n, p, q to be explained shortly. With these values, the spreadsheet calculates the deflection v, rotation θ and bending moment M at each node, and the weight W of the beam. The required data input for the spreadsheet, including the method for defining supports, is listed in Table 6.3.

Standard, uniform-stiffness beam elements are used for the finite element analysis. Values of bending stiffness EI for each element are scaled by dividing by the current largest value of EI, and the element length is reduced to unity before being entered into the function Stiffness, to avoid extreme values of the determinant of the

Beam under Lateral Load

Parameters

length L =	1.20E+03	mm	coeff. p =	3.4610
modulus E =	2.00E+05	N/mm²	coeff. q =	1.0746
allow. stress σ_{all} =	5.00E+02	N/mm²	index m =	0.5
density ρ =	7.85E+03	kg/m³	index n =	0.75

*Make no changes
to cells within
the red borders!*

		constraints		load
node	x/L	v	θ	P
1	0.00	1	1	0.00E+00
2	0.04	0	0	0.00E+00
3	0.08	0	0	0.00E+00
4	0.13	0	0	0.00E+00
5	0.17	0	0	0.00E+00
6	0.21	0	0	0.00E+00
7	0.25	0	0	0.00E+00
8	0.29	0	0	0.00E+00
9	0.33	0	0	0.00E+00
10	0.38	0	0	0.00E+00
11	0.42	0	0	0.00E+00

Variables

element	I
1	1.00E+04
2	1.00E+04
3	1.00E+04
4	1.00E+04
5	1.00E+04
6	1.00E+04
7	1.00E+04
8	1.00E+04
9	1.00E+04
10	1.00E+04
11	1.00E+04

Fig. 6.16 Spreadsheet 'Beam under Lateral Load'

Table 6.3 Data entry for spreadsheet program 'Beam under Lateral Load'

Parameters	
Length of beam L, modulus E, allowable bending stress σ_{all}, density ρ	Enter values in cells C7:C10
Coefficients p, q, indices m, n	Enter values in cells F7:F10
Supports	Enter '1' in cells C16:C40 to constrain vertical deflection v at a node, enter '1' in cells D16:D40 to constrain rotation θ at a node, otherwise enter '0' N.B. at least 2 constraints are required
Applied loads P (positive downwards)	Enter values of applied load at a node in cells E16: E40, otherwise enter '0'
Variables	
Second moment of area I of elements	Enter initial values in cells H16:H40 N.B. it is recommended to choose values so that the maximum value of M_e/M_{all} in cells L49:L72 is typically around 1

stiffness matrix. Standard Excel functions are used in the spreadsheet to invert the stiffness matrix set up by the function Stiffness, and for multiplication of the inverted stiffness matrix by the load vector. The stiffness matrix, after application of constraints, and the inverted stiffness matrix are contained on sheet 2 of the workbook.

Before we can proceed to optimize the beam, we have to define suitable relations between the second moment of area I of the cross section, its cross-sectional area A and the allowable bending moment M_{all}. From Eqs. (6.4) and (6.5) we see that, for geometrically similar sections:

$$A \propto I^{1/2},$$

$$M_{all} \propto A^{3/2} \sigma_{all} \propto I^{3/4} \sigma_{all},$$

where σ_{all} is the allowable bending stress. For the rectangular section of given height in Sect. (6.1), also for any thin-walled section in which the overall dimensions are fixed and only the thickness of all parts increases or decreases in the same proportion, we have the simpler relations:

$$A \propto I,$$

$$M_{all} \propto A \sigma_{all} \propto I \sigma_{all}.$$

We generalize the above by writing:

$$A = pI^m,$$

$$M_{all} = qI^n \sigma_{all}.$$

Relating coefficients p and q to those in Sect. 6.1.2 for geometrically similar sections:

$$p = \frac{1}{C^{1/2}} \quad \text{and} \quad q = \left(\frac{1}{Cn_g^2}\right)^{3/4},$$

with $m = 1/2$ and $n = 3/4$ as above. Note that, for geometrically similar sections, p and q depend only on the *shape* of the cross section, and not on the actual dimensions. For other than geometrically similar sections, the above formulae may be adopted as approximate rule for scaling the cross section, in which case p and q are no longer dimensionless coefficients. By performing the necessary calculations for two or more beam sections in the range of interest, coefficients p and q, and indices m and n in the formulae can be determined on a 'best fit' basis.

6.4.2.2 Optimization

The finite element analysis determines the bending moment distribution in the beam for the required loading, type of support and stiffness distribution. For given elastic modulus, the bending stiffness of the beam can be represented simply by its second moment of area I. Design variables are then the second moment of area of each element. To proceed with optimization of the beam, initial values of I should be chosen so that the largest calculated bending moment M_e is of similar magnitude to the allowable bending moment M_{all} (see column M_e/M_{all} in the spreadsheet). Unless otherwise required, initial I values may be chosen the same for all nodes. To avoid possible numerical failure (note the fractional values of the indices m and n above), a minimum value of I equal to 0.0001 times its maximum value is specified for optimization. The only constraints, apart from minimum I above, are that the bending moment M_e in each element must not exceed the allowable bending moment M_{all}, itself of course dependent on the second moment of area. Bending moment M_e is taken to be the greatest numerical value of the moments at either end of the element. Other practical limits on I can be added in the Solver dialog box if required. After optimization, the initial values of I for each element, the deflection v, rotation θ and bending moment M at each node, and the weight W of the beam are replaced by the optimized ones.

The spreadsheet and Solver dialog box are initially set up for the beam in Sect. 6.2.2, which is of solid square section, clamped at one end and simply supported at the other, under a single load applied at three-quarter span from the clamped end. Appropriate values of m, n, p, q for a square-section beam have been entered in the spreadsheet. In this example, after optimization, we obtain the stiffness distribution shown in Fig. 6.7. This is the true optimum stiffness for a 'stepped' beam consisting of uniform-stiffness segments, as in the present model. However, the solution can be further refined as follows. The minimum weight already found is an *overestimate* for a beam that is actually tapered, rather than one made in discrete steps. This is because, as already stated, the bending moment in each element is taken to be the *greatest* value at either end. With the bending moment corresponding to the optimum stiffness distribution now known, minimum values of I at each node can be recalculated by:

$$I = \left(\frac{M}{q\sigma_{\text{all}}}\right)^{1/n}.$$

With $A = pI^m$, a new, reduced weight can then be calculated. This is done in the spreadsheet under the heading 'Recalculation of second moment of area and minimum weight'. Recalculation of the stiffness of the beam, as done here, will then affect the bending moment distribution in the beam. Provided the change in stiffness distribution (as opposed to its magnitude) is relatively small, the corresponding change in bending moment will also be small. This is not taken into account in the spreadsheet.

The spreadsheet can be modified if required to increase the number of elements. The cells allocated to arrays in the spreadsheet will then have to be redefined, as well as some detail changes in the function Stiffness. It can also be modified for elements of different lengths. This will again require changes in the function Stiffness. Note that any change to an array requires any array formula in the spreadsheet referring to that array first to be deleted, then re-entered with the necessary changes. Arrays are indicated in the spreadsheet with a red border. For further information on the use of array formulae refer to the Appendix.

Finally, a comment on use of the finite element method in this spreadsheet is appropriate. For small problems, such as here, optimization based on finite element analysis using Solver presents little difficulty, although it will be noticed that the time taken for the computer to perform the optimization is considerably longer than in any of the previous spreadsheets. Measurement of changes in constraint values resulting from changes in each design variable, such as for constraint linearization, is fundamental to an efficient numerical procedure to satisfy constraints while searching for an optimum. In a general purpose routine such as Solver, this is done by finite difference, implying that the finite element analysis has to be repeated many times. In a large problem, with many variables and constraints, this would rapidly lead to unacceptable computer times and become highly impractical. Special methods exist in finite element analysis to deduce the so-called constraint sensitivity, by means of which changes in constraint values with change in each design variable are obtained at the same time as the actual analysis, making economical use of a single finite element analysis at each iteration. This is discussed further in Chap. 9.

6.5   Summary

Beam optimization has two distinct but closely related aspects: optimization of the cross section and optimization of the distribution of material along the length of the beam. When only stress limits apply, optimization of the cross section places the material as far as possible from the neutral axis of the beam. For a realistic solution, it is necessary then to define both a maximum height and a maximum width for the cross section. By progressive removal of material from the cross section, a typical I-section beam emerges, with flanges in which the combined bending and shear stress does not exceed the maximum allowable and a thickness of web sufficient for the shear force on the beam. However, at smaller values of a parameter $M/h_{\max}^3$, the cross section may become thin enough that buckling, rather than maximum allowable stress, becomes the dominant design condition. The transition from a stress-critical to a buckling-critical design under a combination of bending moment and shear force can be explored with the spreadsheet 'I-section Beam'. A different approach to the design of the cross section is by choice of a geometrically similar series of sections, that is, a cross section of given form in which all dimensions are scaled up or scaled down in proportion to increase or decrease the size of the cross section. This provides a fixed relation between second moment of area and cross-sectional area, useful for example in optimization of the cross-sectional area along the length of a beam. In practice, many ranges of commercially available beams correspond quite closely to a series of geometrically similar sections.

The bending moment in a statically indeterminate beam depends on the stiffness distribution along its length, whereas for a statically determinate beam it does not. For this reason, for optimization of the cross section along the length of a beam, we have to distinguish between the two cases. For a statically determinate beam, it is possible to directly match the necessary cross-sectional area to the bending moment point by point along its length to optimize the beam. We do this for a family of geometrically similar sections, subject to a maximum allowable stress. In contrast, for a statically indeterminate beam, we adopt a numerical approach in which the bending moment in the beam is obtained by finite element analysis. With a given relation between the allowable bending moment in each element and its second moment of area, and a further relation between cross-sectional area and second moment of area, Solver is used to optimize the stiffness distribution. This is done in the spreadsheet 'Beam under Lateral Load'. By reduction in stiffness to zero at some point or points along the beam during optimization, it is found that the originally statically indeterminate beam is in fact reduced to a statically determinate one.

The strength of a beam made of a ductile material is not exhausted when the stress at the outer edge of the cross section, at one or more points along its length, reaches the elastic limit of the material. Provided some limited degree of plastic deformation is permitted, redistribution of stress within the cross section enables a greater bending moment to be reached without excessive deformation. With a suitable definition of maximum strain, this is termed as the yield moment of the section. For a statically determinate beam, the yield moment can, in appropriate circumstances, be used directly for optimization of the beam. For a statically indeterminate beam, we can go a step further. It is supposed that after the yield moment is reached, a hinge is in effect formed at that point. This is a hinge at which rotation is then possible, while the yield moment is maintained during rotation. When a sufficient number of yield hinges have developed, we have a collapse mode for the beam, from which the limit load, or the maximum load the beam can sustain, can be directly calculated. When there are more possible collapse modes, the one with the lowest load is the correct one. For optimization, if we allow a different cross section in different parts of the span, then the collapse load depends on the yield moment in each part. Each collapse mode leads to an additional constraint for a numerical optimization using Solver, in which the variables are the yield moments of each part. Limit load is the ultimate or maximum load on a structure, providing some margin over the working load at which there is typically no significant permanent deformation due to yielding of the material.

Exercises

6.1 Calculate the coefficient n_h in Eq. (6.2) for an I-section beam with a maximum height and width of 100 mm, under a bending moment of 50×10^6 Nmm. The allowable stress of the material is 500 N/mm^2. The beam has equal thickness web and flanges.
Calculate the required thickness so that the maximum stress is equal to the allowable stress. Calculate the cross-sectional area and substitute A, M, σ_0 and h_{max} in Eq. (6.2). Compare the value of n_h with that of a solid rectangular section.

6.2 Calculate the coefficients C and n_g for a solid circular section bar, and for a hollow circular section with inner diameter equal to one-half of the outer diameter.
To calculate coefficients C and n_g for the solid and hollow circular bars, first choose an arbitrary diameter and calculate A, I and coefficient C. Then for some chosen allowable stress σ_0, calculate the corresponding maximum bending moment M and coefficient n_g. Compare the values of n_g with those in Table 6.1.

6.3 Derive a formula for the minimum volume V of a beam, simply supported at each end, with geometrically similar cross section under loads $P/2$ at one-third span and two-thirds span.
Calculate the bending moment distribution, and use Eq. (6.5) for the minimum cross-sectional area at any section. Compare the volume with that of a beam loaded by a load P at mid-span in Sect. 6.2.1.

6.4 Derive a formula for the yield moment M_y of a hollow, square-section beam of side a and thickness $a/4$, with yield stress σ_y. Take $\varepsilon_{max}/\varepsilon_e = 3$.

Follow the method in Sect. 6.3.1. *There are now three integrals to evaluate for* M_y. *Compare the result with the maximum elastic moment for the same beam.*

6.5 Repeat Example 6.2 for a beam clamped at both ends, under the same loading. *The two possible collapse modes now have three yield hinges. Modify the formulae in Example 6.2 for the additional yield hinge. With these as constraints, and variables the yield moments* M_1, M_2, M_3, *set up a spreadsheet to use Solver to minimize the volume of the beam. Compare the volume with that of a uniform beam, clamped at both ends.*

6.6 Use the spreadsheet 'I-section Beam' to show the effect of a required minimum bending stiffness EI_{min} on the minimum cross-sectional area of the beam, with minimum flange widths $b_{1min} = b_{2min} = 25$ mm, under the loading and with the material properties given on the spreadsheet.
Take a range of EI_{min} *from* 20×10^9 *to* 500×10^9 Nmm^2. *Use the values on the spreadsheet for other maximum and minimum dimensions. Observe the margins of safety and the optimized dimensions of the cross section at different values of* EI_{min}.

6.7 Use the spreadsheet 'I-section Beam' to show the effect of a required minimum flange width on the minimum cross-sectional area of the beam, under the loading and with the material properties given on the spreadsheet.
Take a range of minimum flange widths $b_{1\,min} = b_{2\,min}$ *from* 20 *to* 50 mm. *Set* $EI_{min} = 0$ *(to avoid influencing the results). Use the values on the spreadsheet for other maximum and minimum dimensions. Plot the cross-sectional dimensions against minimum flange width, indicating on the plot the ranges over which different constraints are critical.*

6.8 Use the spreadsheet 'Beam under Lateral Load' to find the optimum stiffness distribution for a beam of length 1200 mm, clamped at both ends, with:

    (a) a load of 4800 N applied at mid-span,
    (b) a load of 4800 N applied at quarter span,
    (c) loads of 3600 and 1200 N applied at quarter span and three-quarter span.

    Compare the stiffness distribution and the minimum weight of the beam in each case.
    Use the parameters initially present in the spreadsheet. Modify the constraints for a beam clamped at both ends. Place the load(s) at the appropriate node(s) in each case. Choose an initial second moment of area $I = 10{,}000$ mm^4 *for all elements. Use Solver to minimize the weight W. Compare the weight of the stepped beam with that of the tapered beam after recalculation. Plot the bending stiffness to locate points at which it reduces to zero. Note how the beam is reduced to a statically determinate one in each case.*

6.9 Use the spreadsheet 'Beam under Lateral Load' to optimize a beam of length 1200 mm, simply supported at each end and at a third support at mid-span, carrying a load of 4800 N uniformly distributed along the span.

Apply loads of 200 N *at all unsupported nodes to represent the uniformly distributed load. Modify the constraints as necessary to represent the proper conditions of support. Choose an initial second moment of area* $I = 2000$ mm^4 *for all elements. Use Solver to minimize the weight W.*

6.10   Repeat exercise 6.9 by modelling only one-half of the symmetric beam. Examine any difference in the two results.

Reduce the length of the beam to 600 mm *for the half-beam, with loads reduced to* 100 N *at each node to represent the same uniformly distributed load. Change the support at the node on the plane of symmetry to displacement v and rotation* θ *both constrained, to represent the condition of the symmetric beam at mid-span. Choose an initial second moment of area* $I = 2000$ mm^4 *for all elements. The weight of the whole beam is now twice that given in the spreadsheet. Compare the results with those in exercise 6.9, now with in effect twice as many elements for the whole beam, to indicate the accuracy of the finite element model (note that we are using a uniform-stiffness beam element here to model the tapered beam).*

References

1. Hoff NJ (1956) The analysis of structures. Wiley, New York
2. Neal BG (1956) The plastic methods of structural analysis. Springer, New York
3. Rees DWA (2009) Mechanics of optimal structural design: minimum weight structures. Wiley, New York
4. Young WC, Budynas RG (2001) Roark's formulas for stress and strain. McGraw-Hill, New York

Chapter 7
Reinforced Shell Structures

Abstract Formulae are given for the bending and shear stress in a long shell structure such as an aircraft wing, together with formulae for the different modes of buckling. These are used in the optimization of some reinforced shell structures. Alternative methods for modelling the reinforcement are discussed. At higher stresses, yielding of the material causes a redistribution of stress in the cross section and reduction in buckling stress. The Ramberg–Osgood formula is used to define the stress–strain curve of the material and to derive formulae for the tangent and secant moduli. Efficiency formulae are developed for stiffened panels in compression and shear. A spreadsheet program for a panel with integral, unflanged stiffeners under compression and shear optimizes the cross section, subject to minimum stiffener spacing and plate thickness, with reduced moduli for buckling. The same program is incorporated in a spreadsheet for the optimization of a rectangular box beam under bending and shear, with variable rib spacing, subject to material stress limits and buckling in different modes. A third spreadsheet program optimizes the cross section of a circular fuselage section, with variable skin thickness and stringer dimensions around the cross section.

Reinforced shell structures are typical of practically all aerospace structures and countless other applications where low weight is at a premium. Such structures, built up from thin sheet material, carry load essentially by membrane stress, in other words by direct and shear stress confined to the plane of the sheet material. This is a direct result of the low bending stiffness of the thin sheet, whereby 'thin' is implied relative to the other dimensions of the structure. In the special case of a pressure vessel, whether of circular, spherical or other shape, pressure is taken by tensile stress in the shell, and in principle, no further reinforcement is required. However, in a structure such as a box beam designed to carry loads applied along its length, additional reinforcement is generally necessary to resist buckling of the thin sheet under the compressive and shear stresses that arise, as well as for input of locally applied loads on the structure. We shall restrict attention in this chapter to the design of a box beam, such as forms the main load carrying structure of an aircraft wing or fuselage and is the basis of many other forms of lightweight structure.

© Springer International Publishing AG 2017
A. Rothwell, *Optimization Methods in Structural Design*, Solid Mechanics and Its Applications 242, DOI 10.1007/978-3-319-55197-5_7

A typical box beam is shown in Fig. 7.1, assumed here to carry a vertical load causing a bending moment about the horizontal axis. It consists of upper and lower panels with stiffeners which break up the thin sheet into smaller panels to resist buckling. These also take part with the sheet in carrying bending stress. The side panels, or 'shear webs', carry the main shear force on the beam, and these commonly also require stiffeners to resist buckling. Intermediate shear webs may be introduced for a better distribution of shear stress. Twisting moment on the beam, for example due to eccentrically applied load, is taken by shear stress around the whole cross section. For this, the closed section of a beam box is strongly preferred, since an open section (any section containing no closed cells) has much lower torsional stiffness. In addition, transverse members, referred to as 'ribs, frames or bulkheads', are required at intervals along the length of the beam to provide support for the stiffeners and for input of major loads. Which term is used will depend on the context, just as the terms 'sheet' and 'plate' are used variously throughout the text.

With regard to locally applied loads on a box beam, such as the pressure distribution on the upper and lower surfaces of an aircraft wing, it is useful to consider how these loads are transmitted into the structure as a whole. In spite of its low bending stiffness, such loads have to be carried by local bending of the sheet into the adjacent stiffeners, often assisted by the tensile stress that develops in the sheet with increasing load when deformation of the sheet is resisted by the surrounding structure. The stiffeners then transmit the resulting load from the sheet, again in bending, into the ribs. Other major loads may be applied directly to appropriately placed ribs. These distribute the load on them by shear stress at their edges into the shear webs and into the upper and lower panels. This shear stress contributes directly to the shear force in the shear webs, while the shear stress in the upper and lower panels develops the tensile or compressive stress in the sheet and stiffeners resulting from the bending moment on the beam.

In the present chapter formulae for the stress analysis of a box beam under a combination of applied shear force, bending moment and twisting moment are reviewed, with particular attention to the treatment of discrete stiffeners. Appropriate formulae are given for the buckling of stiffened panels in a box beam. These are used in the first place to deduce a theoretical efficiency formula for a stiffened panel. Later in the chapter, a spreadsheet program for the optimization of a

Fig. 7.1 Rectangular box beam

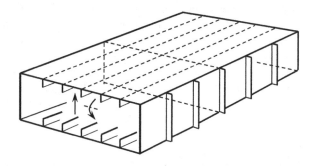

stiffened panel is introduced. This, in addition to buckling constraints and stress limits, allows dimensional and other limits to be included and reduced moduli for yielding. The same spreadsheet is incorporated into a spreadsheet program for the optimization of a rectangular box beam, making use of the bending and shear stress analysis in the following sections. A further spreadsheet for the optimization of a circular fuselage section uses an alternative means of modelling the cross section for optimization of the stiffened panels at different locations around the fuselage.

7.1 Bending Stress

Due to different sheet thicknesses, various types of reinforcement and often an irregular shape of cross section, it cannot necessarily be assumed that the cross section of a box structure is symmetric about any axis. Formulae given here are, therefore, for any shape of cross section, under general biaxial loading.[1] The axis system and definition of bending moments about each axis are shown in Fig. 7.2. Axes x, y are through the centre of gravity G of the section, with any convenient orientation. Bending moments M_x and M_y are assumed positive when they produce tensile stress in the first (positive x, y) quadrant. Shear forces Q_x and Q_y are defined as positive in the positive x, y directions. The longitudinal z-axis is defined in agreement with the right-hand rule (i.e. out of the paper in Fig. 7.2).

In accordance with the conventional theory of bending, we shall assume a bilinear strain distribution in the cross section:

$$\varepsilon = C_1 y + C_2 x, \tag{7.1}$$

Fig. 7.2 Definition of bending moments

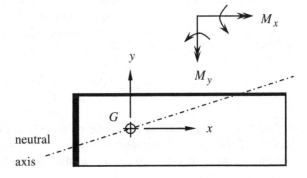

where C_1, C_2 are actually the curvatures about each axis, but are treated here simply as unknown constants. With an elastic modulus E, the corresponding bending stress is:

$$\sigma = E\,\varepsilon. \tag{7.2}$$

By integration over the cross section A of the beam, we have bending moments:

$$
\begin{aligned}
M_x &= \int_A \sigma y\,dA = EC_1 \int_A y^2\,dA + EC_2 \int_A xy\,dA, \\
M_y &= \int_A \sigma x\,dA = EC_1 \int_A xy\,dA + EC_2 \int_A x^2\,dA,
\end{aligned} \tag{7.3}
$$

or:

$$
\begin{aligned}
M_x &= C_1 EI_x + C_2 EI_{xy}, \\
M_y &= C_1 EI_{xy} + C_2 EI_y,
\end{aligned} \tag{7.4}
$$

where

$$I_x = \int_A y^2\,dA, \quad I_y = \int_A x^2\,dA, \quad I_{xy} = \int_A xy\,dA$$

are the second moments of area of the section. The product second moment I_{xy} is a measure of the lack of symmetry in the section. Taking axes x, y through the centre of gravity ensures that the bending stress causes no resultant end load on the section.

With known applied bending moments M_x and M_y, Eqs. (7.4) can be solved for the coefficients C_1 and C_2, which are substituted in Eq. (7.1) for the strain ε at any point, and the bending stress σ is then obtained from Eq. (7.2). This gives finally the general formula:

$$\sigma = \frac{(M_x I_y - M_y I_{xy})\,y \;+\; (M_y I_x - M_x I_{xy})\,x}{I_x I_y - I_{xy}^2}. \tag{7.5}$$

The line $\sigma = 0$ (through the centre of gravity) defines the neutral axis of the section. If the section is symmetric about either axis ($I_{xy} = 0$) then under, say, bending moment M_x Eq. (7.5) reduces to the familiar form:

$$\sigma = \frac{M_x}{I_x}\,y.$$

7.1.1 *Effect of Yielding*

If yielding of the material occurs at higher stresses in some parts of the cross section, a reduced modulus has to be used in those parts. While many common metallic materials have a clearly defined yield point below which there is little reduction in modulus, for most high-grade alloys yielding takes place progressively with the increase in stress, with a smooth stress–strain curve such as in Fig. 7.3. The Ramberg–Osgood formula [15] is commonly used for a close approximation to the actual stress–strain curve for such materials. This formula was already referred to in Chap. 3, in the context of a reduced modulus for buckling. Note that, other than in Sect. 6.3 of the previous chapter, to avoid a severe loss of buckling strength in a shell structure, we are mostly concerned with reduction in modulus at a stress at which a relatively small degree of plastic deformation has taken place. The effect of yielding is therefore treated differently.

The Ramberg–Osgood formula for the strain ε at stress σ is:

$$\varepsilon = \frac{\sigma}{E} + \varepsilon_R \left(\frac{\sigma}{\sigma_R} \right)^m,$$

where ε_R is the plastic component of strain at a reference stress σ_R, and E is the initial elastic modulus. For example, if σ_R is chosen to be the 0.2% proof stress σ_2 (the stress at which the total strain less the elastic, or recoverable, strain component is 0.2%), then $\varepsilon_R = 0.002$. The index m defines the 'sharpness' of the stress strain curve and is regarded as a material property, generally taking values in a wide range between 5 and 50. Figure 7.4 illustrates stress–strain curves for different values of m. The secant modulus E_s, or simply the ratio of stress to strain after yielding, is obtained directly from the above formula:

$$E_s = \frac{\sigma}{\varepsilon} = \frac{E}{1 + \frac{\varepsilon_R E}{\sigma} \left(\frac{\sigma}{\sigma_R} \right)^m}, \qquad (7.6)$$

Fig. 7.3 Definition of the secant and tangent modulus

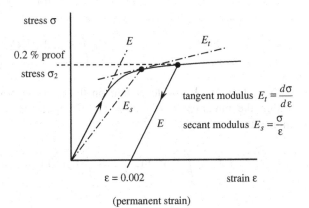

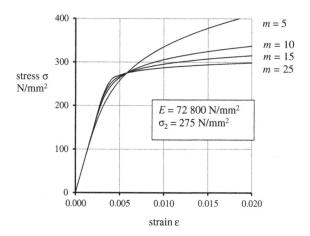

Fig. 7.4 Stress–strain curves by the Ramberg–Osgood formula

while the tangent modulus is obtained by differentiation:

$$E_t = \frac{d\sigma}{d\varepsilon} = \frac{E}{1 + \frac{\varepsilon_R E m}{\sigma_R}\left(\frac{\sigma}{\sigma_R}\right)^{m-1}} . \tag{7.7}$$

The secant and tangent moduli are indicated on the stress–strain curve in Fig. 7.3. The secant modulus is used below for the bending stress distribution after the onset of yielding. Both moduli are used in the buckling formulae in Sect. 7.3.

Returning to Eq. (7.2) for the bending stress in the beam, the elastic modulus E now has to be replaced by the secant modulus E_s:

$$\sigma = E_s \varepsilon.$$

The strain distribution is assumed to remain bilinear, as in Eq. (7.1). With secant modulus E_s, Eqs. (7.3) for the bending moments become:

$$M_x = \int_A \sigma y \, dA = C_1 \int_A E_s y^2 \, dA + C_2 \int_A E_s xy \, dA,$$

$$M_y = \int_A \sigma x \, dA = C_1 \int_A E_s xy \, dA + C_2 \int_A E_s x^2 \, dA. \tag{7.8}$$

Note that, E_s, being dependent on the stress at any point, has to be retained within the integrals. It is convenient in equations to define 'weighted' second moments of area $\bar{I}_x$, $\bar{I}_y$ and $\bar{I}_{xy}$ by:

$$\bar{I}_x = \int_A E_s y^2 \, dA, \quad \bar{I}_y = \int_A E_s x^2 \, dA, \quad \bar{I}_{xy} = \int_A E_s xy \, dA.$$

To ensure no end load on the section as a result of bending, axes x and y must now be taken through an effective centre of gravity, given by:

$$\bar{x} = \frac{\int_A E_s x \, dA}{\int_A E_s \, dA}, \quad \bar{y} = \frac{\int_A E_s y \, dA}{\int_A E_s \, dA},$$

where x and y refer to any convenient reference axes, and the area of each part of the cross section has again been weighted by the value of the secant modulus in that part. Equations (7.8) can now be written more concisely:

$$M_x = C_1 \bar{I}_x + C_2 \bar{I}_{xy},$$
$$M_y = C_1 \bar{I}_{xy} + C_2 \bar{I}_y.$$

These can be solved for coefficients C_1 and C_2, which are then substituted in Eq. (7.1) for the strain ε. Multiplying the strain by the secant modulus at any point in the cross section gives finally a modified form of Eq. (7.5):

$$\sigma = E_s \frac{(M_x \bar{I}_y - M_y \bar{I}_{xy})y + (M_y \bar{I}_x - M_x \bar{I}_{xy})x}{\bar{I}_x \bar{I}_y - \bar{I}_{xy}^2} \tag{7.9}$$

for the stress at that point, where again x and y refer to the effective centre of gravity.

Since the secant modulus depends on the stress at any point in the cross section, the weighted second moments of area and effective centre of gravity in the above formula cannot, of course, be calculated beforehand. An iterative calculation is therefore required to ensure that the secant modulus used in the calculation of $\bar{I}_x$, $\bar{I}_y$ and $\bar{I}_{xy}$ is consistent with the stress obtained by Eq. (7.9).

7.1.2 Modelling of Discrete Stiffeners

As already discussed, the low bending stiffness of the sheet in a thin shell structure makes it necessary to reinforce the structure to resist buckling. Stiffeners placed along the length of a box beam also contribute to its bending stiffness and take their share of the bending stress. The cross-sectional area of individual stiffeners has, therefore, to be included in the second moments of area of the section to calculate the stress distribution. To simplify the calculation, stiffeners may be modelled in one of two ways. If they are uniformly spaced across the panel, their area can be included with the sheet to give an equivalent (or 'smeared') thickness:

$$\bar{t} = t + \frac{A_s}{b},$$

where A_s is the actual stiffener cross-sectional area, b is the stiffener spacing, and t is the sheet thickness. In this simple approach, the conventional engineering formulae can readily be used to calculate second moments of area.

Alternatively, the cross-sectional area of the sheet on either side of a stiffener can be included with the stiffener itself, resulting in an effective stiffener area:

$$\bar{A}_s = A_s + bt.$$

If there are a large number of stiffeners, several adjacent stiffeners and the appropriate area of sheet may be lumped into a single substitute stiffener at the centre of gravity of the group, as illustrated in Fig. 7.5 for a group of three stiffeners. In this way, the cross section is represented by a series of discrete areas, which simplifies the numerical calculation of second moments of area for more complicated shapes of cross section and, with stress defined at the same discrete points, also simplifies the use of the secant modulus. This modelling is used in the following example and in the shear stress calculation in the next section.

Example 7.1 Calculate the bending stress distribution in the rectangular box beam shown in Fig. 7.6, with a width of 1500 mm and a height of 500 mm. The upper panel has a thickness of 4 mm, with nine stiffeners of area 150 mm^2 at a spacing of 150 mm. The lower panel has a thickness of 2 mm, with nine stiffeners of area 75 mm^2, also at a spacing of 150 mm. The shear webs have a thickness of 8 mm on the left-hand side and 4 mm on the other side. The beam carries a vertical upward shear force causing a negative bending moment of 500 kNm at the section shown.

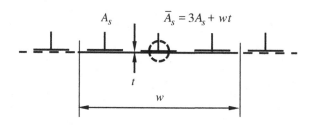

Fig. 7.5 Lumping of stiffeners and adjacent sheet into a single substitute stiffener

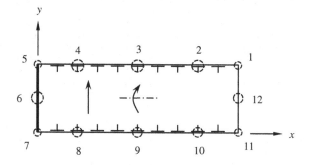

Fig. 7.6 Rectangular box beam in Example 7.1

Note that the beam is unsymmetric as a consequence of the different thicknesses and stiffener areas. To calculate the bending stress distribution, the upper and lower panels are each represented by three uniformly spaced, substitute stiffeners. Additional substitute stiffeners represent the shear webs. Stiffeners are numbered as shown in Fig. 7.5, with axes x, y now referred to a convenient point in the cross section. It is assumed in this example that the stresses are low enough that no yielding occurs.

The area $\bar{A}_s$ of substitute stiffeners 2, 3 and 4 is:

$$\bar{A}_s = 3 \times 150 + 3 \times 150 \times 4 = 2250 \text{ mm}^2.$$

Similarly, the area of substitute stiffeners 8, 9 and 10 is:

$$\bar{A}_s = 1125 \text{ mm}^2.$$

Each shear web can be represented by substitute stiffeners of area:

$$\bar{A}_s = \frac{A_w}{6}$$

at the two corners, where A_w is the cross-sectional area of the web, and a third substitutes stiffener at mid-height of area:

$$\bar{A}_s = \frac{2}{3} A_w.$$

This results in a second moment of area equal to that of the actual web, while the third stiffener corrects for the loss of cross-sectional area that would otherwise occur. Noting that the main function of the box beam in this example is to carry vertical load, the reasoning behind this modelling is to obtain an accurate value of I_x with minimum error in I_y and I_{xy}. The area of substitute stiffener 1 is then:

$$\bar{A}_s = \frac{500 \times 4}{6} + 75 \times 4 = 633 \text{ mm}^2.$$

The second term in the calculation above is to account for a width 75 mm of the upper panel at each corner which (as a result of the specific placing of the substitute stiffeners) has not already accounted been for. The areas of substitute stiffeners 5, 7, and 11 are calculated in a similar way, with values:

$$\bar{A}_s = 967, \ 817, \ 483 \text{ mm}^2,$$

respectively. The area of substitute stiffener 6 is:

$$\bar{A}_s = \frac{2}{3} \times 500 \times 8 = 2667 \text{ mm}^2,$$

and similarly for stiffener 12:

$$\bar{A}_s = 1333 \text{ mm}^2.$$

The areas found above, together with the location x, y of each substitute stiffener, are entered in Table 7.1, from which the position of the centre of gravity of the box beam $\bar{x}$, $\bar{y}$ and second moments of area I_x, I_y and I_{xy} are then calculated (note that areas $\bar{A}_s$ are denoted in the table simply by A). Finally, the bending stress distribution is calculated in the last column of Table 7.1 by Eq. (7.5), in this case:

$$\sigma = \frac{M_x}{I_x I_y - I_{xy}^2} \left[I_y(y - \bar{y}) - I_{xy}(x - \bar{x}) \right].$$

Comparing the stresses at the four corners of the box, it is seen that a skewed stress distribution is obtained, as expected for an unsymmetric cross section. The maximum tensile stress is at the lower right corner and maximum compressive stress at the upper left corner. The stresses calculated at the different substitute stiffeners can, if required, be interpolated between stiffeners for the stress at some specific point in the cross section, or the number of substitute stiffeners can be increased for a more detailed stress distribution.

This example is continued in Example 7.2, in which the shear stress distribution around the cross section is calculated.   ∎

7.2 Shear Stress

The shear stress distribution in a box beam is derived directly from the bending stress. For this, we have first to consider equilibrium of the sheet and stiffener, as shown in Fig. 7.7. It is assumed that the area of the sheet has been incorporated into a substitute stiffener, as described in the previous section, so that the sheet in effect carries only shear stress. There is a discrete increment $\Delta\tau$ in shear stress on passing from one side to the other side of the stiffener, related to the increment in direct stress $\Delta\sigma$ in the stiffener over a small distance Δz. For equilibrium in the direction of the stiffener:

$$\Delta\tau \cdot t \Delta z + \Delta\sigma \cdot \bar{A}_s = 0,$$

where $\bar{A}_s$ is the area of the substitute stiffener. To be independent of the sheet thickness t, it is convenient to refer instead to 'shear flow':

$$q = \tau t$$

rather than shear stress in the above equation, so that the corresponding increment in shear flow Δq is:

Table 7.1 Calculation of second moments of area and bending stress distribution in Example 7.1

Substitute stiffener	A, mm²	x, mm	y, mm	Ax, mm³ × 10⁶	Ay, mm³ × 10⁶	Ax², mm⁴ × 10⁶	Ay², mm⁴ × 10⁶	Axy, mm⁴ × 10⁶	σ, N/mm²
1	633	1500	500	0.950	0.317	1425	158	475	−120
2	2250	1200	500	2.700	1.125	3240	563	1350	−123
3	2250	750	500	1.688	1.125	1266	563	844	−128
4	2250	300	500	0.675	1.125	203	563	338	−132
5	967	0	500	0.000	0.483	0	242	0	−135
6	2667	0	250	0.000	0.667	0	167	0	28
7	817	0	0	0.000	0.000	0	0	0	192
8	1125	300	0	0.338	0.000	101	0	0	195
9	1125	750	0	0.844	0.000	633	0	0	200
10	1125	1200	0	1.350	0.000	1620	0	0	205
11	483	1500	0	0.725	0.000	1087	0	0	208
12	1333	1500	250	2.000	0.333	3000	83	500	44
Σ	17,025			11.269	5.175	12,575	2338	3506	

$\bar{x} = \frac{\Sigma Ax}{\Sigma A} = 662 \text{ mm}$ $I_x = \Sigma Ay^2 - A\bar{y}^2 = 764 \times 10^6 \text{ mm}^4$

$\bar{y} = \frac{\Sigma Ay}{\Sigma A} = 304 \text{ mm}$ $I_y = \Sigma Ax^2 - A\bar{x}^2 = 5116 \times 10^6 \text{ mm}^4$

(position of the $I_{xy} = \Sigma Axy - A\bar{x}\bar{y} = 81 \times 10^6 \text{ mm}^4$

centre of gravity)

$\bar{x}, \bar{y}$ (referred to the centre of gravity)

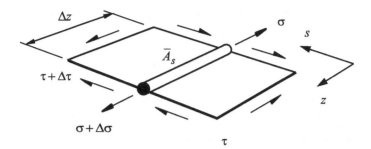

Fig. 7.7 Equilibrium of direct and shear stresses at a stiffener (coordinate s is around the cross section)

$$\Delta q \cdot \Delta z + \Delta \sigma \cdot \bar{A}_s = 0,$$

or:

$$\Delta q = -\bar{A}_s \cdot \frac{\mathrm{d}\sigma}{\mathrm{d}z}. \tag{7.10}$$

With the use of substitute stiffeners, the shear flow can be treated as constant between adjacent stiffeners.

By conventional bending theory, the bending moment M and shear force Q on a beam are related by:

$$\frac{\mathrm{d}M}{\mathrm{d}z} = Q,$$

so we can write:

$$\frac{\mathrm{d}\sigma}{\mathrm{d}z} = \frac{\mathrm{d}\sigma}{\mathrm{d}M} \cdot \frac{\mathrm{d}M}{\mathrm{d}z} = \frac{\sigma}{M} \cdot Q, \tag{7.11}$$

provided that the beam remains linear elastic. From Eq. (7.10), the increment in shear flow at each stiffener then becomes:

$$\Delta q = -\bar{A}_s \frac{Q}{M} \sigma. \tag{7.12}$$

As alternative to Eq. (7.10) we could write:

$$\Delta q = -\frac{\Delta P}{\Delta z}, \tag{7.13}$$

where P is the force in the stiffener (assumed tensile positive). This form may be preferable when there is a change in cross section of the beam between two adjacent

Fig. 7.8 Calculation of shear flow

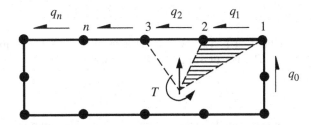

sections, or if some yielding of the material occurs (when Eq. (7.12) is no longer applicable).

With Eq. (7.12), the shear flow can be calculated step-by-step around the cross section, as in Fig. 7.8. Starting with an unknown shear flow q_0, the shear flow becomes:

$$q_1 = q_0 - \left[\bar{A}_s \frac{Q}{M} \sigma \right]_1,$$

$$q_2 = q_1 - \left[\bar{A}_s \frac{Q}{M} \sigma \right]_2$$

and so on, or in general:

$$q_n = q_0 - \sum_{i=1}^{n} \left[\bar{A}_s \frac{Q}{M} \sigma \right]_i, \tag{7.14}$$

where q_n is understood to mean the shear flow between stiffeners n and $n + 1$. Note that the second term above may be positive or negative, depending on the location around the cross section.

To determine the unknown q_0, we have to consider the resulting twisting moment on the section. The net twisting moment of the shear flows q_n about some arbitrary point must be equal to the applied twisting moment about that same point, that is, the moment of the shear force about the chosen point together with any externally applied twisting moment. As shown in Fig. 7.9, the moment of the shear flow on a length Δs of the cross section is:

Fig. 7.9 Twisting moment due to shear flow about an arbitrary point

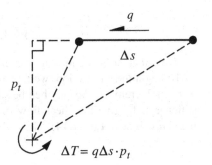

$$\Delta T = q \Delta s \cdot p_t. \tag{7.15}$$

Note that $p_t \Delta s$ is twice the area of the triangle formed by Δs and the chosen point. The net twisting moment is then:

$$T = \sum q_n \Delta s \cdot p_t, \tag{7.16}$$

summed around the whole cross section. The unknown q_0 is found by substituting the calculated values of q_n from Eq. (7.14), each including q_0, in the above formula and equating T to the required applied twisting moment. The procedure for this is illustrated in Example 7.2.

As we have seen, calculation of shear flow is based entirely on equilibrium between bending stress and shear flow. The shear flow distribution is then automatically in equilibrium with the applied shear force. This also implies that the shear flow calculation still remains valid if yielding occurs, provided that Eq. (7.13) is used instead of Eq. (7.12). With discrete modelling of stiffeners it is, of course, necessary that the same model is used for both calculation of second moments of area and the shear flow calculation. Calculation of q_0 ensures equilibrium with the resulting twisting moment. If the structure is subject only to a twisting moment T (no shear force), then Eq. (7.12) shows $\Delta q = 0$, that is, constant shear flow q_0 around the whole section, and from Eq. (7.16):

$$q_0 = \frac{T}{2A_0}, \tag{7.17}$$

where A_0 is the enclosed area of the cross section.

Example 7.2 Calculate the shear stress distribution in the unsymmetric box beam in Example 7.1 by the method described above. A vertical upward shear force of 100 kN acts on the beam through a point 300 mm from the left-hand side, as indicated in Fig. 7.6. The bending stress under a negative bending moment of 500 kNm has already been calculated in Example 7.1.

The bending stress σ given in the last column of Table 7.1 is copied to Table 7.2 for the present example. With the same substitute stiffeners as in the previous example, increments in shear flow Δq at each stiffener are calculated by Eq. (7.12) in the fourth column of Table 7.2:

$$\Delta q = -\bar{A}_s \frac{Q}{M} \cdot \sigma = \frac{100 \times 10^3}{500 \times 10^6} \bar{A}_s \cdot \sigma = 0.2 \times 10^{-3} \bar{A}_s \cdot \sigma \text{ N/mm}.$$

(Note that M is negative in this example.)

To begin the calculation, we take the shear flow in segment 12–1 to be zero (i.e. $q_0 = 0$). Increments Δq can then be summed around the cross section, starting from stiffener 1, to give the shear flow q_n in the sheet between each pair of stiffeners n and $n+1$, as in Eq. (7.14). These shear flows now have to be corrected by adding

Table 7.2 Calculation of the shear stress distribution in Example 7.2

Subs. stiff.	σ, N/mm²	A, mm²	Δq, N/mm	Segment	q_n, N/mm	Δs, mm	p_t, mm	ΔT, Nmm $\times 10^6$	q_{total}, N/mm	τ, N/mm²
1	−120	633	−15.2							
				1–2	−15	300	250	−1.1	58	15
2	−123	2250	−55.4							
				2–3	−71	450	250	−7.9	3	1
3	−128	2250	−57.6							
				3–4	−128	450	250	−14.4	−55	−14
4	−132	2250	−59.4							
				4–5	−188	300	250	−14.1	−114	−29
5	−135	967	−26.1							
				5–6	−214	250	300	−16.0	−140	−18
6	28	2667	14.9							
				6–7	−199	250	300	−14.9	−125	−16
7	192	817	31.4							
				7–8	−167	300	250	−12.6	−94	−47
8	195	1125	43.9							
				8–9	−123	450	250	−13.9	−50	−25
9	200	1125	45.0							
				9–10	−78	450	250	−8.8	−5	−3
10	205	1125	46.1							
				10–11	−32	300	250	−2.4	41	20
11	208	483	20.1							
				11–12	−12	250	1200	−3.7	61	15
12	44	1333	11.7							
				12–1	0	250	1200	−0.2	73	18
								$\Sigma \Delta T = -110 \times 10^6$ Nmm		

$$q_0 = -\frac{\Sigma \Delta T}{2A_0} = \frac{110 \times 10^6}{2 \times 0.75 \times 10^6} = 73 \text{ N/mm}$$

the shear flow q_0 necessary to ensure that the twisting moment of the shear flows around the whole cross section is equal to the twisting moment of the applied shear force about some chosen reference point.

In this example (since there is no externally applied twisting moment), it is convenient to measure the perpendicular distance p_t to the point at which the shear force acts, about which the twisting moment is then zero. Values of p_t are shown in Table 7.2, together with the distance Δs between adjacent stiffeners, from which increments in twisting moment ΔT due to the shear flow in each segment are calculated by Eq. (7.15) in column 9.

With an enclosed area of the box beam:

$$A_0 = 1500 \times 500 = 0.75 \times 10^6 \text{ mm}^2$$

the resulting twisting moment $\sum \Delta T$ is eliminated by the shear flow q_0 calculated in the table. This is then added to the previously calculated shear flow q_n to give the final shear flow q. Corresponding shear stresses are given in the last column of Table 7.2.

It is readily verified that the calculated shear flow is in fact in equilibrium with the applied shear force. Multiplying the shear flow in the vertical segments 5–6, 6–7, 11–12 and 12–1 by the width 250 mm of each of the four segments gives a resulting vertical shear force:

$$\frac{(140 + 125 + 61 + 73) \times 250}{1000} = 99.75 \text{ kN.}$$

(The signs of the first two terms within brackets have been changed so that all shear flows are directed upwards.) The small difference between this and the actually applied shear force (100 kN) is due to rounding of the bending stresses taken from Table 7.1. It can similarly be verified that the horizontal resultant of the shear flows is zero, as required. ∎

7.2.1 Torsional Stiffness

The shear strain around the cross section of a box beam (or other closed, thin-walled section) loaded in torsion, or under shear force not through the flexural centre of the section, causes it to twist along its axis. Shear strain can be expressed in terms of the rate of twist θ of the beam (angle of twist per unit length) and the so-called warping displacement w in the axial z direction, as shown in Fig. 7.10. Twisting of the beam causes a displacement $p_{t_c} \theta \, \delta z$ in the circumferential s direction, where p_{t_c} is the perpendicular distance from the tangent at a point in the cross section to the centre of rotation C (inset in Fig. 7.10). Warping causes a displacement $\frac{dw}{ds} \cdot \delta s$ in the axial direction. The net shear strain is then:

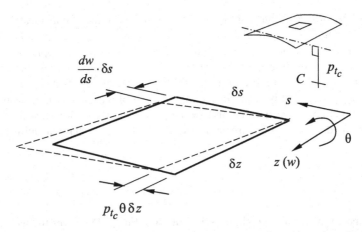

Fig. 7.10 Shear strain in the sheet in terms of rate of twist θ and warping displacement w (coordinate s is around the circumference of the cross section)

$$\gamma = p_{t_c}\,\theta + \frac{dw}{ds} = \frac{q}{Gt}.$$

The last term above simply relates the shear strain to the shear flow, where G is the shear modulus of the material and t the local sheet thickness.

Integrating the shear strain γ around the entire cross section:

$$\theta \oint p_{t_c}\,ds + \oint \frac{dw}{ds}\,ds = \frac{1}{G}\oint \frac{q}{t}\,ds.$$

The second integral on the left-hand side must be zero for no discontinuity in axial displacement w on completing the integration around the closed section, leaving:

$$\theta \oint p_{t_c}ds = \frac{1}{G}\oint \frac{q}{t}\,ds.$$

Since:

$$\oint p_{t_c}\,ds = 2A_0$$

(where A_0 is again the enclosed area) we have for the rate of twist θ caused by the shear flow in the beam:

$$\theta = \frac{1}{2A_0 G}\oint \frac{q}{t}\,ds. \qquad (7.18)$$

It will be observed that it is not necessary to know the actual location of the centre of rotation C to calculate the rate of twist by the above formula. Also, it should be noted that Eqs. (7.18) and (7.19) that follows apply *only* to a closed-section beam (for an open section, such as a channel section, warping displacement plays a much more prominent role in determining the twist of a beam).

Equation (7.18) can be used for the rate of twist of a closed-section beam under any combination of shear force and torsion, once the shear flow distribution has been obtained. Under pure torsion T, from Eq. (7.17), we have a uniform shear flow:

$$q = \frac{T}{2A_0}$$

around the entire section which, on substitution in Eq. (7.18), gives the well-known formula for torsional stiffness:

$$k = \frac{T}{\theta} = 4A_0^2 G \oint \frac{1}{t} \mathrm{d}s. \tag{7.19}$$

Minimum torsional stiffness is one of the constraints applied in the spreadsheet programs later in this chapter.

7.2.2 von Mises Criterion

Unless it is loaded purely in bending, a box beam is subject to a combination of axial stress and shear stress around the whole cross section. The von Mises criterion is generally used to define an equivalent stress to predict failure under combined stress. In fact, the von Mises criterion is the condition for *yielding* of a ductile material under any combination of stress. It is fundamentally for a three-dimensional state of stress and in its usual form is expressed by:

$$(\sigma_1 - \sigma_2)^2 + (\sigma_2 - \sigma_3)^2 + (\sigma_3 - \sigma_1)^2 = 2\,\sigma_y^2,$$

where σ_1, σ_2 and σ_3 are the three principal stress components, and σ_y is the yield stress in simple tension. This formula represents a combination of stresses to cause distortion of the material, after removal of the uniform, or 'hydrostatic', stress components that cause only change in volume. Under two-dimensional plane stress, the criterion reduces to:

$$\sigma_1^2 - \sigma_1 \sigma_2 + \sigma_2^2 = \sigma_y^2.$$

As stated above, the von Mises criterion is in principle for yielding, but the same criterion is widely used to predict *failure* when σ_y is replaced an equivalent stress:

$$\sigma_{eq} = \sqrt{\sigma_1^2 - \sigma_1\,\sigma_2 + \sigma_2^2}.$$

The equivalent stress σ_{eq} is taken to be equal to the ultimate tensile stress of the material at failure under combined stress. It is convenient to substitute the usual formula for principal stress for σ_1 and σ_2 above, when we obtain:

$$\sigma_{eq} = \sqrt{\sigma_x^2 - \sigma_x\sigma_y + \sigma_y^2 + 3\tau_{xy}^2},$$

where σ_x, σ_y and τ_{xy} are the conventional stress components. If any of the stresses in the above formulae is compressive, it must take a negative sign, thereby increasing the value of the equivalent stress and predicting earlier failure. For a combination of only axial stress σ and shear stress τ, the formula reduces further to:

$$\sigma_{eq} = \sqrt{\sigma^2 + 3\tau^2}.$$

Only this last equation above is in fact of interest to us here. Later in this chapter, we shall refer to equivalent stress when defining constraints for optimization. The same formula was used for the optimization of an I-section beam in the previous chapter. The theoretical derivation of the von Mises criterion can be found in many texts on strength of materials and theory of plasticity.

7.3 Buckling Formulae

Much of the cross section of the box beam structure in the previous sections is subject to a combination of compressive stress and shear stress, and if it is made of relatively thin sheet material requires reinforcement to resist buckling. As shown in Fig. 7.1, this reinforcement consists of stiffeners in the lengthwise direction on the upper and lower panels, supported by ribs at intervals along the length of the box, and commonly transverse stiffeners on the shear webs. Stiffeners are sufficiently closely spaced to break up the sheet into smaller panels, with the necessary increase in buckling stress. The purpose here is to review methods and formulae appropriate to the buckling of this type of structure. These are used, together with the material allowable stresses, to determine stress limits for the optimization of a typical stiffened panel and reinforced shell structures later in this chapter.

7.3.1 Buckling in Compression

A stiffened panel in compression buckles into a series of waves along its length involving simultaneous deformation of both the sheet and the stiffeners. The

Fig. 7.11 Buckling of a
stiffened panel in compression

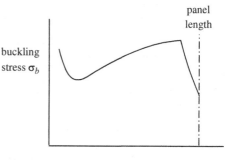

buckling mode can best be characterized by its wavelength. Figure 7.11 shows a typical plot of buckling stress against wavelength (in the figure, λ is actually the half-wavelength). Other plots may show more local minima, mainly at shorter half-wavelengths, depending on the type of stiffener. With unsymmetric stiffeners, the sharp discontinuity at longer half-wavelength generally disappears. However, we can broadly identify two principal buckling modes: short wave and long wave. In short-wave buckling, referred to as local buckling, the half-wavelength is related to the stiffener spacing. In long-wave buckling, referred to as flexural buckling, the half-wavelength is typically equal to the length of the panel, that is, the distance over which the panel is supported.

A stiffened panel can be treated as an assembly of long, thin plates, relatively stiff in their own plane but flexible out-of-plane. The lines of intersection of these plates can be regarded, at least in the first place, as nodal lines. These are, points in the cross section where there is no displacement in the buckling mode, while rotation takes place about those points. Figure 7.12 illustrates the corresponding buckling deformation for three different stiffener types. For a wide panel, the buckling mode is usually repetitive across the width of the panel and in a long panel is in principle sinusoidal along its length. The modes shown in Fig. 7.12 are the classical short-wave or local buckling. Other modes of buckling can be regarded as due to breakdown of one or more of these nodal lines. In the case of the flanged stiffener in Fig. 7.13a, if the flange is not sufficiently wide it may not possess enough in-plane stiffness to support the node at the top. This gives rise to a more general form of local buckling, with twisting of the stiffener as shown in the figure, and reduction in buckling stress. If the stiffener has insufficient height, as in Fig. 7.13b, both nodes may be unsupported, and buckling is then with lateral displacement of the whole cross section. Note that in this case, the cross section is not distorted. Support is then provided only at the ends of the panel, with buckling taking place in a long-wave, flexural buckling mode, commonly accompanied by some twisting of the stiffener.

Fig. 7.12 Local buckling
modes

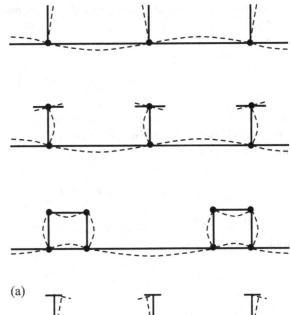

Fig. 7.13 Breakdown of
nodes

(a)

(b)

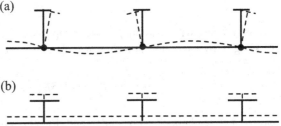

A detailed numerical analysis is required to produce the buckling stress curve in
Fig. 7.11. The program in ESDU Data Item 98016 [10][2] can conveniently be used
for this. However, for an initial design of stiffened panel, serving as starting point
for further optimization, other methods may be employed with appropriate formulae
to predict the buckling stress. These are discussed below.

7.3.1.1   Local Buckling

Due to the relatively close stiffener spacing, the sheet between stiffeners as well as the
plates making up the stiffeners themselves can be regarded as long plates, that is, long
enough that the buckling mode is unaffected by conditions at their ends. Standard
formulae are available for the buckling of long, flat plates with some specific edge

---

[2]ESDU Data Items are listed by number, with the title, relevant section number in the Structures
series and date of issue in References at the end of this chapter.

conditions. For a plate simply-supported (free to rotate) on both long edges, the buckling mode is sinusoidal along its length, with a half-wavelength equal to the width of the plate, and a single half-wave across its width. The buckling stress is:

$$\sigma_b = \frac{\pi^2}{3\left(1 - v^2\right)} E \left(\frac{t}{b}\right)^2,$$

where b is the width of the plate, t is its thickness, E is the modulus of the material, and v is Poisson's ratio. If Poisson's ratio is taken to be 0.3, the formula reduces to the more familiar form:

$$\sigma_b = 3.62E\left(\frac{t}{b}\right)^2.$$

If the long edges are clamped (no rotation), buckling is at a smaller wavelength with increased buckling stress:

$$\sigma_b = 6.31E\left(\frac{t}{b}\right)^2.$$

For a plate with one long edge simply-supported and the other edge free, buckling is by exception at a *long* wavelength with buckling stress:

$$\sigma_b = 0.385E\left(\frac{t}{b}\right)^2,$$

and for a plate clamped on one edge and the other edge free, the buckling stress is:

$$\sigma_b = 1.13E\left(\frac{t}{b}\right)^2.$$

Poisson's ratio v is again taken to be 0.3 in all the above formulae. For a plate under combined compressive stress σ and shear stress τ, a widely used criterion for buckling is:

$$\frac{\sigma}{\sigma_b} + \left(\frac{\tau}{\tau_b}\right)^2 = 1, \tag{7.20}$$

where in this formula, σ takes a *positive* sign for compression, and σ_b, τ_b are the buckling stresses in pure compression and shear, respectively. Under combined tensile and shear stress, the corresponding formula is:

$$\frac{1}{2} \cdot \frac{\sigma}{\sigma_b} + \frac{\tau}{\tau_b} = 1, \tag{7.21}$$

where σ now takes a *negative* sign for tension (thereby delaying buckling). Buckling in shear is discussed in the next section. The derivation of the above formulae can be found in Timoshenko and Gere [18], Brush and Almroth [1], Vinson [19] and in many other texts.

For an approximate solution to the local buckling stress of a stiffened panel, the individual plates making up the cross section could all be treated as simply-supported at their edges, taking as buckling stress the lowest value found from the above formulae. However, significant interaction takes place between the different parts of the cross section, due to their different buckling stresses and preferred wavelengths. Nevertheless, being simply an assembly of flat plates, the general formula for local buckling taking account of the interaction between adjacent plates has the same form as for the single plates above:

$$\sigma_L = KE\left(\frac{t}{b}\right)^2, \qquad (7.22)$$

where we use σ_L to denote local buckling, and b is now the stiffener spacing and t is the sheet thickness. A numerical analysis is necessary for an accurate determination of the buckling coefficient K, which depends only on the shape of the panel. Extensive graphical data for the local buckling coefficient of various types of panel is provided in a number of Data Items in the ESDU 'Structures' series, together with computer programs to calculate values for other types. Local buckling coefficients for the panel with integral, unflanged stiffeners in Fig. 7.14 are plotted in Fig. 7.15. At smaller h/b ratios, buckling of the sheet is resisted by the stiffeners, increasing the local buckling stress above that of a simply-supported plate, eventually approaching the buckling stress of a clamped plate with thicker stiffeners. At larger h/b ratios, buckling of the stiffeners is resisted by the sheet, reducing then the local buckling stress. The discontinuity in some of the curves in Fig. 7.15 is due to change in wavelength at that point. As for a simply-supported plate, the half-wavelength of the local buckling mode is mostly in the neighbourhood of the stiffener spacing, showing the strong influence of the sheet. To correct for yielding at higher stresses in local buckling, the elastic modulus E can be replaced by the secant modulus E_s, or by a function of E_s and the tangent modulus E_t giving a somewhat greater reduction in buckling stress than by use of the secant modulus alone. Unlike for a simple column, the reduced modulus is governed by the biaxial stress distribution in the sheet in the buckling mode, giving rise to different reduced moduli depending on the nature of the buckling mode [see 13]. The spreadsheet

Fig. 7.14 Panel with integral, unflanged stiffeners

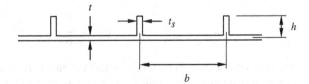

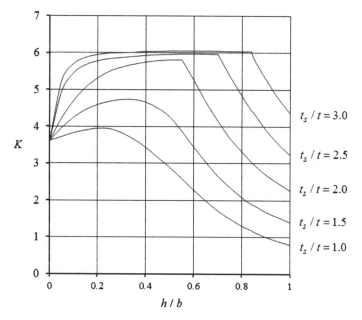

Fig. 7.15 Local buckling coefficient K for a panel with integral, unflanged stiffeners (based on data in ESDU Data Item 70003, with permission from IHS ESDU)

program for a panel with integral, unflanged stiffeners in Sect. 7.4.1 uses the more accurate reduced modulus for local buckling in ESDU Data Item 70003 [4].

7.3.1.2 Flexural Buckling

The mode of buckling in Fig. 7.13b shows lateral displacement but no deformation of the cross section of the panel, and is therefore not substantially different to that of a simple column. Long-wave, or flexural, buckling deformation takes place in bending, typically in a single half-wave between the two ends, or adjacent supports. Euler's formula can be used to predict the buckling stress. Principal difference is that the relatively high in-plane stiffness of the sheet prevents in-plane displacement, so that displacement in the buckling mode is necessarily confined to a direction perpendicular to the sheet. The formula for the flexural buckling stress σ_F is then:

$$\sigma_F = \frac{\pi^2 EI}{AL^2},\qquad(7.23)$$

where I is the second moment of area of the panel about a neutral axis parallel to the sheet, A is its cross-sectional area, and L is the length of the panel (i.e. distance between supports). With even a relatively small number of stiffeners, the buckling behaviour in the middle of the panel is little affected by support at the edges, so for a repetitive pattern of stiffeners I and A are conveniently taken to refer to a single stiffener and its corresponding width of sheet. For a continuous panel, supported at intervals along its length, adjacent panels are likely to deform in opposite directions in each bay, and it is commonly assumed that the panel is effectively simply-supported at each support. Otherwise, an additional factor should be included in the above formula to account for the degree of restraint at each support. For flexural buckling, the elastic modulus E is normally replaced by the tangent modulus E_t to correct for yielding at higher stresses. Justification for use of the tangent modulus lies in the presence of some initial deviation from perfectly straight over the length of the panel. Significant deformation, progressively matching the buckling mode with increasing load, takes place before the theoretical buckling load has been reached. With the stress still *increasing* over the whole cross section, the tangent modulus governs the continued deformation of the panel until failure.

Example 7.3 Calculate the flexural and local buckling stresses of the panel with integral, unflanged stiffeners in Fig. 7.14. The stiffener spacing $b = 150$ mm, the stiffener height $h = 30$ mm, the plate thickness $t = 4$ mm, and the stiffener thickness $t_s = 4$ mm. The panel is taken to be simply-supported over a length $L = 375$ mm. It is made of aluminium alloy with an elastic modulus $E = 72{,}800$ N/mm^2, 0.2% proof stress $\sigma_2 = 275$ N/mm^2, and index $m = 15$.

It is assumed that the panel is relatively wide, so that for flexural buckling it is sufficient to consider a single stiffener together with a width 150 mm of plate. Note that the height of the stiffener in Fig. 7.14 is measured to the mid-plane of the plate. The cross-sectional area A of the combined stiffener and plate is calculated to be 712 mm^2, its neutral axis is located at a distance 2.52 mm from the middle-plane of the plate, and its second moment of area I about the neutral axis is 32,280 mm^4.

Calculate first the buckling stresses neglecting any reduction in modulus due to yielding. From Eq. (7.23), the flexural buckling stress is then:

$$\sigma_F = \frac{\pi^2 EI}{AL^2} = \frac{\pi^2 \times 72\,800 \times 32\,280}{712 \times 375^2} = 231.6 \text{ N/mm}^2.$$

For local buckling, we use Fig. 7.15 for the buckling coefficient K. The ratio of stiffener height to stiffener spacing $h/b = 0.2$, and the ratio of stiffener thickness to plate thickness $t_s/t = 1.0$, giving $K = 3.95$. From Eq. (7.22), the local buckling stress is:

$$\sigma_L = KE \left(\frac{t}{b}\right)^2 = 3.95 \times 72{,}800 \times \left(\frac{4}{150}\right)^2 = 204.5 \text{ N/mm}^2.$$

If instead of this more accurate value of K we use the elementary values $K = 3.62$ (for the plate) and $K = 0.385$ (for the stiffener), we obtain buckling stresses of 187.4 N/mm^2 and 498.3 N/mm^2 for the plate and stiffener, respectively. Taking the lower of these, the local buckling stress $\sigma_L = 187.4$ N/mm^2, a difference (in this example) of 8%.

The buckling stresses found above will now be corrected for the effect of yielding of the material. For flexural buckling, we replace E in Eq. (7.23) by the tangent modulus E_t. Equation (7.23) can be rewritten as:

$$\frac{\sigma_F}{E_t} = \frac{\pi^2 I}{AL^2} = \frac{\pi^2 \times 32\,280}{712 \times 375^2} = 3.182 \times 10^{-3}.$$

We require a value of σ_F which, when substituted in the formula for E_t in Eq. (7.7), gives a ratio σ_F/E_t that satisfies the above condition. This can be done by some trial and error, or by use of Goal Seek in Excel. The reduced flexural buckling stress is 205.0 N/mm^2, a reduction of 11% on the original value of 231.6 N/ mm^2. While this stress is well below the proof stress of 275 N/mm^2, it is seen that there is a significant reduction in buckling stress.

For local buckling, we replace E in Eq. (7.22) by the secant modulus E_s in this example. Equation (7.22) can be rewritten as:

$$\frac{\sigma_L}{E_s} = K\left(\frac{t}{b}\right)^2 = 3.95 \times \left(\frac{4}{150}\right)^2 = 2.809 \times 10^{-3}.$$

Proceeding as above, with the formula for E_s in Eq. (7.6), we find a ratio σ_L/E_s to satisfy the above condition. This gives a reduced local buckling stress of 202.7 N/mm^2, which scarcely differs from the original value of 204.5 N/mm^2. ∎

7.3.2 Buckling in Shear

A flat plate loaded in shear buckles into a series of skew waves along its length, with a half-wavelength related to the width of the plate. For a long plate of width b, thickness t and elastic modulus E, with simply-supported edges, the shear buckling stress is:

$$\tau_b = 4.83\,E\,\left(\frac{t}{b}\right)^2, \tag{7.24}$$

and if the edges are clamped:

$$\tau_b = 8.11\,E\,\left(\frac{t}{b}\right)^2.$$

The thickness of a shear web of a typical box beam may be such that no stiffeners are required, and the web can be treated simply as a long plate. In that case, either of the two above formulae for the shear buckling stress may be used, depending on the nature of the attachment at the upper and lower edges.

For thinner webs requiring stiffeners to improve the buckling stress, these are normally placed across the width of the web, as in Fig. 7.1. These stiffeners are unable to take any part of the shear force on the web, but break up the web into smaller panels with the required increase in shear buckling stress. Due to the small half-wavelength of the buckling mode, stiffeners should be closely spaced to be effective, in any case at a spacing not greater than the width of the web. With increasing bending stiffness of the stiffeners, there is a progressive increase in buckling coefficient, until a critical stiffness is reached at which they become fully effective. Buckling is then essentially confined to the space between stiffeners. The bending stiffness of the stiffeners $E_{st}I$ is expressed in a parameter:

$$\mu = \left(\frac{E_{st}Id}{Eh^2t^3}\right)^{1/2}, \tag{7.25}$$

where E_{st} is the elastic modulus of the stiffeners, and E and t refer to the web itself. For a stiffened web, we use d for the stiffener spacing and h for the height of the web. Critical values μ_c of the above parameter for some values of ratio h/d are given in Table 7.3. This is based on data in ESDU Data Item 02.03.02 [11], which also gives shear buckling coefficients for stiffeners having less than the critical bending stiffness and for stiffeners with torsional stiffness. In practice, calculation of the bending stiffness of the stiffener should be for the stiffener itself together with an appropriate width of web, about the neutral axis of the combination. If it is assumed that stiffeners with greater than this critical stiffness provide simple-support for the web (e.g. if they are open sections of low torsional stiffness, such as angle-section stiffeners) and the web is also simply-supported on its upper and lower edges, the shear buckling stress for the square or rectangular panels between stiffeners is given by:

$$\tau_b = KE\left(\frac{t}{d}\right)^2,$$

where

$$K = 4.83 + 3.61\left(\frac{d}{h}\right)^2, \quad 0 \le \frac{d}{h} \le 1.$$

Table 7.3 Critical values μ_c in Eq. (7.25) for stiffeners with no torsional stiffness (based on graphical data in ESDU Data Item 02.03.02)

h/d	Web simply-supported		Web clamped	
	μ_c	K	μ_c	K
1.0	0.68	8.50	0.43	11.40
1.5	1.13	6.50	0.54	7.30
2.0	1.78	5.90	1.05	6.30

This is a conventional formula for buckling in shear of a simply-supported, rectangular plate. In fact, the parabolic formula above is based on $K = 4.83$ for a long plate and $K = 8.44$ for a square plate, and slightly underestimates K for intermediate values of d/h due to change in the number of half-waves along the plate. The corresponding formula for a web clamped on its upper and lower edges is:

$$K = 8.11 + 3.22\left(\frac{d}{h}\right)^2, \quad 0 \le \frac{d}{h} \le 1.$$

Accurate values of K for rectangular plates, also for other edge conditions, are widely available in the literature and in ESDU Data Item 71005 [5]. The secant modulus may be used to correct for yielding of the web, or alternatively the function of secant modulus and tangent modulus given in Sect. 7.4.1, at an equivalent stress $\sigma = \tau/\sqrt{3}$.

7.3.3 Efficiency Formula for a Compression Panel

In Sect. 7.3.1, we identified the two principal modes of buckling of a stiffened panel in compression as a short-wave mode, or local buckling, and a long-wave mode, or flexural buckling. Based on these two modes, we can develop an efficiency formula for a stiffened panel in compression, in a similar way to the efficiency formula for a circular tube in Chap. 2. However, unlike the circular tube with simply a diameter and thickness, we need more dimensions to define the geometry of a stiffened panel. Therefore, we identify two principal dimensions of the cross section, of different order of magnitude, choosing for these the thickness t of the sheet and the stiffener spacing b. All other dimensions are expressed as shape ratios, in terms of the appropriate principal dimension. For the panel with integral, unflanged stiffeners in Fig. 7.14, we have shape ratios t_s/t and h/b, and similar shape ratios for other types of panel.

The general formula for the local buckling stress of a stiffened panel in Eq. (7.22) is:

$$\sigma_L = KE\left(\frac{t}{b}\right)^2,$$

which is already expressed in terms of the principal dimensions t and b. The buckling coefficient K depends only on the shape ratios of the particular panel. The local buckling coefficient for a panel with integral, unflanged stiffeners is plotted as function of h/b for various t_s/t in Fig. 7.15.

For flexural buckling, we can express the buckling stress in Eq. (7.23) as:

$$\sigma_F = K_F E\left(\frac{b}{L}\right)^2,$$

where the flexural buckling coefficient K_F also depends only on the shape ratios of the panel. Again for a panel with integral, unflanged stiffeners, with cross-sectional area:

$$A = ht_s + bt$$

the second moment of area about the neutral axis is:

$$I = \frac{h^3 t_s}{3} - \frac{h^4 t_s^2}{4(bt + ht_s)}$$

(both per stiffener with its associated sheet). With the above formulae we obtain, in more compact form:

$$K_F = \frac{\pi^2 H^3 T(4 + HT)}{12(1 + HT)^2},$$

where $H = h/b$ and $T = t_s/t$. It is assumed here that the thickness of the panel is small compared with its other dimensions. As stated earlier, a factor may be applied to coefficient K_F to allow for restraint at the ends of the panel, or an effective length L may be used.

The compressive load on the panel can be defined as a *loading intensity* p (load per unit width) rather than as a load on the panel as a whole, so that for a relatively wide panel its actual width need not enter into the problem. For the same reason, it is also convenient to refer to the equivalent thickness $\bar{t}$ of the panel:

$$\bar{t} = t + \frac{A_s}{b} = Ct,$$

rather than to its actual cross-sectional area. Again, coefficient C depends only on the shape ratios of the panel.

Fig. 7.16 Design space for a stiffened panel of given shape subject to local and flexural buckling

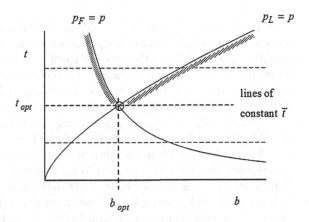

The optimization problem is now:

$$\text{minimise } \bar{t}$$

$$\text{subject to constraints :}$$

$$p_L \geq p,$$

$$p_F \geq p,$$

where $p_L = \sigma_L \cdot \bar{t}$ and $p_F = \sigma_F \cdot \bar{t}$. This is illustrated in the design space in Fig. 7.16. This is representative of any shape of panel, but the actual constraint lines and equivalent thickness depend, of course, on the particular panel and chosen shape ratios. It is seen that, as for the circular tube, the optimum is at the point of simultaneous buckling in the flexural and local modes. With this as optimality criterion, we can write for the stress at this point:

$$\sigma = \frac{p}{Ct} = K_F E \left(\frac{b}{L}\right)^2 = KE \left(\frac{t}{b}\right)^2. \tag{7.26}$$

By eliminating b and t from the three simultaneous equations above, we find:

$$\sigma = \left(\frac{K_F K}{C^2}\right)^{1/4} \left(\frac{pE}{L}\right)^{1/2},$$

or

$$\sigma = \eta \, E^{1/2} \left(\frac{p}{L}\right)^{1/2}, \tag{7.27}$$

where

$$\eta = \left(K_F K / C^2\right)^{1/4}$$

is the efficiency of the panel. Here, we have identified a new structural index p/L. Since K_F, K and C all depend on shape ratios h/b and t_s/t for the panel in Fig. 7.14, and other shape ratios for other types of panel, we can perform a numerical search for the optimum values of these ratios for maximum efficiency. With the maximum stress in an optimized panel now known from Eq. (7.27), together with the optimum shape ratios, the corresponding dimensions b and t are readily calculated from Eq. (7.26). For a panel with integral, unflanged stiffeners, the maximum efficiency $\eta = 0.81$ [2], and for a panel with individual, plain Z-section stiffeners $\eta = 0.95$ [12]. More values are found for other shapes of panel [16].

It is found that the optimum ratio h/b for most types of panel is not far removed from unity, meaning that we have a relatively small stiffener spacing. Practical considerations will commonly demand a greater stiffener spacing. While the unconstrained optimum is not difficult to find, to apply constraints on individual dimensions is a more extensive numerical task. The spreadsheet program in Sect. 7.4.1 optimizes a panel with integral, unflanged stiffeners, subject also to dimensional, material and other constraints, and calculates the efficiency. It will be observed that no reduced modulus has been included in the efficiency formula. This is because we define here the 'geometric' efficiency of a panel. However, when reduced moduli are included in the buckling stress calculation, as in the spreadsheet in Sect. 7.4.1, reduction in buckling stress due to yielding causes a corresponding reduction in the apparent or 'achieved' efficiency calculated in the spreadsheet.

Up to now, it has been assumed that the panel is free of imperfections. In particular, we are concerned here with some local waviness of the sheet, which can have a significant effect on the performance of an optimized panel based on simultaneous buckling modes. Any initial waviness grows in amplitude with the approach of local buckling. This results in loss of stiffness of the panel as a whole and therefore reduction in flexural buckling load, depending of course on the magnitude of the initial imperfection. Common practice is to design a panel so that flexural buckling occurs *before* local buckling, to reduce the significance of imperfections. In other words, we introduce an additional constraint in the optimization that σ_F/σ_L is less than some required value, typically in the range of 0.85–0.95. By optimization with this additional constraint, the shape of the panel is adjusted to minimize the resulting loss of efficiency. A suitable value of this constraint may be chosen in the spreadsheet in Sect. 7.4.1.

Example 7.4 Calculate the efficiency of the panel in Example 7.3.

For the efficiency of the cross-sectional shape without reduction in buckling stress by yielding, we take the elastic buckling calculation in Example 7.3. Since the local buckling stress ($\sigma_L = 204.5$ N/mm^2) is less than the flexural buckling stress ($\sigma_F = 231.6$ N/mm^2), we base the loading intensity for the efficiency calculation on the lower of these two. With cross-sectional area $A = 712$ mm^2 (for a single stiffener with plate) and stiffener spacing $b = 150$ mm, the equivalent thickness is:

$$\bar{t} = \frac{A}{b} = \frac{712}{150} = 4.75 \text{ mm},$$

and the loading intensity is:

$$p = \sigma \bar{t} = 204.5 \times 4.75 = 971 \text{ N/mm}.$$

Substituting in the efficiency formula in Eq. (7.27), with $E = 72{,}800$ N/mm^2 and $L = 375$ mm, we have:

$$204.5 = \eta \left(\frac{971 \times 72{,}800}{375} \right)^{1/2},$$

giving an efficiency:

$$\eta = 0.471.$$

As might be expected, the efficiency of this panel, which does not have simultaneous buckling modes and has a ratio $h/b = 0.2$ far removed from the optimum (at around $h/b = 0.7$), is much lower than the theoretical maximum efficiency $\eta = 0.81$ for this type of panel.   ∎

7.3.4 Shear Web Efficiency

We consider now a shear web with transverse stiffeners, such as forms one of the side panels of the typical box beam in Fig. 7.1. Formulae for the buckling of both stiffened and unstiffened shear webs are given in Sect. 7.3.2. To derive an efficiency formula for a stiffened shear web, a different approach has to be taken to that for a compression panel in Sect. 7.3.3. This is due to the stiffness requirements for the stiffeners, which do not themselves take any part of the shear force on the web but break it up into smaller panels, thereby improving its buckling stress. We shall assume that the stiffeners belong to a geometrically similar family of stiffeners, such that all dimensions (including the thickness) remain strictly in proportion. This is the assumption also made for a beam in Sect. 6.1.2, where geometric similarity is discussed in more detail. The stiffeners can then be represented by the simple formula:

$$I = CA^2,$$

where I is the second moment of area and A is the cross-sectional area of each stiffener. Coefficient C depends only on the chosen type and shape of stiffener and is readily calculated for a particular stiffener. The value of I (but not A) should include some allowance for an effective area of the web to which the stiffener is attached, if it is placed on one side of the web only.

The buckling stress of a stiffened web depends on the parameter μ in Eq. (7.25):

$$\mu = \left(\frac{Id}{h^2 t^3}\right)^{1/2}, \qquad\qquad (7.28)$$

where d is again the stiffener spacing, h is the height of the web, and t is the web thickness (assuming now that the web and stiffeners are of the same material). As described in Sect. 7.3.2, with increasing bending stiffness of the stiffeners, there is a progressive increase in buckling coefficient until a critical value $\mu = \mu_c$ is reached at which the stiffeners become fully effective. Commonly, the optimum occurs at or close to this critical value of stiffness, and in the first place, this will be assumed here. Values of μ_c for a limited number of ratios of h/d, with corresponding buckling coefficient K, are given in Table 7.3. With $I = CA^2$, the cross-sectional area of the stiffeners can be expressed as an equivalent stiffener thickness:

$$t' = \frac{A}{d} = \frac{1}{d}\sqrt{\frac{I}{C}}. \tag{7.29}$$

From Eq. (7.28), we have:

$$I = \mu^2 \frac{h^2 t^3}{d},$$

and substituting for I in Eq. (7.29), we obtain finally for the equivalent thickness of the stiffeners:

$$t' = \frac{A}{d} = \frac{\mu}{C^{1/2}} \cdot h \left(\frac{t}{d}\right)^{3/2}. \tag{7.30}$$

The corresponding plate thickness t of the web is obtained from its buckling stress. Under a shear force Q the shear flow in the web is $q = Q/h$, and with buckling stress:

$$\tau_b = \frac{q}{t} = KE \left(\frac{t}{d}\right)^2$$

we can solve for t in the above equation to give:

$$t = \left(\frac{q\,d^2}{KE}\right)^{1/3} = \frac{r^{2/3}}{K^{1/3}} \cdot \left(\frac{qh^2}{E}\right)^{1/3}, \tag{7.31}$$

where $r = d/h$. We define now a total equivalent thickness of the stiffened web:

$$\bar{t} = t + t',$$

with t from Eq. (7.31) above and t' from Eq. (7.30). If, as already suggested, we assume that the optimum occurs at the critical value μ_c, we can directly substitute $\mu = \mu_c$ and the corresponding value of K in the formula for $\bar{t}$ to find a minimum.

The expressions for t and t' are of different form and do not lead to a definition of efficiency in a manner similar to that of a compression panel. Therefore, we choose to base the formula for efficiency on that of a long, unstiffened web of height h with buckling stress:

$$\tau_b = KE \left(\frac{t}{h}\right)^2 = \frac{q}{t},$$

from which we find:

$$t = \frac{1}{K^{1/3}} \left(\frac{qh^2}{E}\right)^{1/3}$$

and

$$\tau = K^{1/3} E^{1/3} \left(\frac{q}{h}\right)^{2/3}.$$

This is the efficiency formula for an unstiffened web, with structural index in shear q/h. The value of K depends on the type of support on the long edges of the web. We preserve a similar form of efficiency formula for a stiffened web:

$$\tau' = \eta_s E^{1/3} \left(\frac{q}{h}\right)^{2/3}, \tag{7.32}$$

where η_s is an efficiency coefficient in shear and τ' is an *equivalent* shear stress:

$$\tau' = \frac{q}{t}.$$

By maximizing the equivalent shear stress, we are in fact maximizing the strength-to-weight ratio of the stiffened web. For a particular design, we may calculate t and t' from Eqs. (7.31) and (7.30) to obtain the equivalent shear stress τ', and substitute in Eq. (7.32) for the efficiency of the web. When we do this, we find that the efficiency does retain some weak degree of dependence on the structural index q/h, such that the index of 2/3 in the efficiency formula will be slightly reduced. However, there is no single value of the index that will cover all cases. For a given value of q/h, efficiency η_s can be used to compare webs with different d/h ratios and different types of stiffener.

More extensive data for the relation between buckling coefficient K and stiffener parameter μ are given in ESDU 02.03.02. This data is used in the author's paper [17] to explore the whole range of μ values in an appropriate design space. It is confirmed that, except at unusually small stiffener spacing, the optimum does in fact occur at or close to the critical value μ_c. The index on q/h in the efficiency formula is found to be quite close to the value 2/3 adopted above. There is seen to be a continuous improvement in efficiency with reducing stiffener spacing, and this is the most significant factor affecting efficiency. It is found that the stiffener coefficient C has in fact a relatively small effect on the efficiency of the web. Empirical formulae are produced for efficiency in terms of a parameter:

$$\varphi = \frac{1}{C^{1/2} E^{1/6}} \left(\frac{q}{h}\right)^{1/6},$$

also for stiffeners with some torsional stiffness. The index 1/6 on q/h confirms the weak dependence of efficiency, as defined above, on the structural index.

Example 7.5 A stiffened shear web of height 250 mm is loaded under a shear force of 50 kN. The equal-angle stiffeners (on one side only) have a coefficient $C = 1.0$. The elastic modulus is 72,800 N/mm^2. Calculate the efficiency of the shear web at ratios of web height to stiffener spacing $h/d = 1, 1.5$ and 2.

We shall assume that the optimum occurs when the stiffeners satisfy the criterion $\mu = \mu_c$, at which they become fully effective. Calculate first the efficiency for ratio $h/d = 1$. From Table 7.3, for a web simply-supported along its length:

$$\mu_c = 0.68, \quad K = 8.50.$$

The shear flow at shear force $Q = 50$ kN is:

$$q = \frac{Q}{h} = \frac{50,000}{250} = 200 \text{ N/mm}.$$

From Eq. (7.31), the optimum plate thickness is:

$$t = \frac{r^{2/3}}{K^{1/3}} \cdot \left(\frac{qh^2}{E}\right)^{1/3} = \frac{1}{8.50^{1/3}} \cdot \left(\frac{200 \times 250^2}{72,800}\right)^{1/3} = 2.723 \text{ mm}.$$

Shear stress in the web is:

$$\tau = \frac{Q}{ht} = \frac{50,000}{250 \times 2.723} = 73.45 \text{ N/mm}^2.$$

At this level of stress, it is assumed that any effect of yielding can be neglected. From Eq. (7.30), the equivalent thickness of the stiffeners is:

$$t' = \frac{\mu}{C^{1/2}} \cdot h \left(\frac{t}{d}\right)^{3/2} = 0.68 \times 250 \times \left(\frac{2.723}{250}\right)^{3/2} = 0.193 \text{ mm}.$$

Optimum cross-sectional area of each stiffener is:

$$A = dt' = 250 \times 0.193 = 48.3 \text{ mm}^2.$$

Total equivalent thickness is:

$$\bar{t} = t + t' = 2.723 + 0.193 = 2.916 \text{ mm}.$$

Equivalent shear stress is:

$$\tau' = \frac{q}{\bar{t}} = \frac{200}{2.916} = 68.59 \text{ N/mm}^2.$$

To calculate the efficiency:

$$\tau' = \eta_s \, E^{1/3} \left(\frac{q}{h}\right)^{2/3}.$$

Substituting in the above formula:

$$68.59 = \eta_s \times 72{,}800^{1/3} \times \left(\frac{200}{250}\right)^{2/3},$$

giving $\eta_s = 1.906$.

The calculation is repeated for $h/d = 1.5$ and 2, taking appropriate values of μ_c and K from Table 7.3. The following results are obtained:

h/d	t, mm	A, mm^2	η_s
1	2.72	48.3	1.91
1.5	2.27	61.3	2.21
2	1.94	76.0	2.48

The example illustrates the strong influence of stiffener spacing on the efficiency of a stiffened shear web.   ■

7.3.5 Post-buckled Shear Webs

We considered shear web efficiency in the previous section only from the point of view of initial buckling. However, for a conventional stiffened web, this rarely causes failure of the web as a whole. In many cases, buckling may even be difficult to detect experimentally, at least until it has become sufficiently well developed. After initial buckling, a shear web may still possess considerable remaining strength. Shear webs were one of the earliest applications of post-buckled design, notably in aircraft structures.

The post-buckling behaviour of a stiffened shear web, with transverse stiffeners as in Fig. 7.1, is the result of an alternative load carrying mechanism in which the buckled web behaves like a stretched membrane, in the so-called diagonal tension, while the stiffeners and upper and lower edge members acquire compensating compressive stresses. This diagonal tension roughly follows the lines of folds in the initial shear buckling mode. Failure may occur by failure of the web itself under combined shear and tensile stress, or by buckling of the stiffeners, generally in 'forced crippling', which is a form of buckling of the stiffeners induced by buckling of the web. The practical design of diagonal tension webs is comprehensively discussed by Kuhn [14]. Based largely on this, ESDU Data Item 77014 [8] provides much graphical data, with a computer program for the analysis of diagonal tension webs in Data Item 02005 [3].

Fig. 7.17 Comparison of the efficiency of various shear webs (Figs. 7.17 and 7.19 are taken from the author's Ph.D. thesis, Cranfield University, 1979)

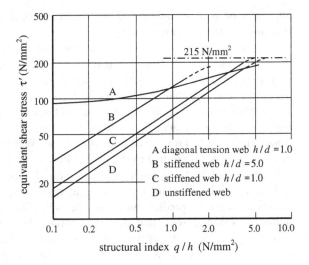

A diagonal tension web $h/d = 1.0$
B stiffened web $h/d = 5.0$
C stiffened web $h/d = 1.0$
D unstiffened web

Fig. 7.18 Shape of stiffener for the shear webs in Figs. 7.17 and 7.19

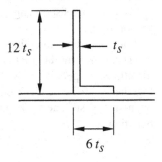

The optimum design of a diagonal tension web is invariably a balanced design, that is, one with simultaneous failure of the web and stiffeners. A comparison between an unstiffened web, conventional stiffened webs and a diagonal tension web over a range of structural index q/h is shown in Fig. 7.17. The equivalent shear stress $\tau' = q/\bar{t}$ as in the previous section. For the diagonal tension web, the ratio of web height to stiffener spacing $h/d = 1.0$. The shape of stiffener for both the stiffened webs and the diagonal tension web is shown in Fig. 7.18. The material has $E = 72,000$ N/mm^2 and 0.5% proof stress $= 400$ N/mm^2. A cut-off for the equivalent stress is shown at $\tau' = 215$ N/mm^2. The lines for the unstiffened and stiffened webs in Fig. 7.17 are based on Eq. (7.32), with appropriate values of η_s (note the log-log plot in the figure). Data from Kuhn are used for the diagonal tension web. Unlike conventional stiffened webs, the efficiency of a diagonal tension web (as measured by the equivalent shear stress that can be reached) is largely independent of stiffener spacing, at least for the single-sided stiffeners adopted here. As shown in the figure, diagonal tension webs are also characterized by their relative insensitivity to the structural index. The substantial improvement in efficiency of a diagonal tension web over conventional stiffened webs at low structural index is apparent in Fig. 7.17.

Except at higher values of structural index, for which material limitations curtail the improvement in efficiency, a balanced design of diagonal tension web leads to a very thin web, substantial stiffeners and a high post-buckling ratio β (ratio of maximum load to initial buckling load). While a high post-buckling ratio may not be acceptable in practice, due to the repeated buckling that this implies, much of the improvement in efficiency—as compared with an unbuckled design—can be achieved at much lower post-buckling ratios. Webs designed to fail at a low post-buckling ratios fail in a mode quite unlike the usual diagonal tension web. This is in a diagonal buckling mode involving both the web and the stiffeners over the web as a whole, for which experimental results support the use of orthotropic buckling theory to predict failure of the web. This is on the assumption of a uniformly stiffened web, that is, one in which the stiffeners are supposed to be 'smeared' over the width of the web. This may be justified, even with more widely spaced stiffeners, by the fact that at lower post-buckling ratios the level of diagonal tension in the web is still relatively low. On the other hand, at higher post-buckling ratios and increased diagonal tension, failure of individual stiffeners becomes predominant.

For orthotropic buckling, the stiffness of the web is expressed in terms of *average* flexural rigidities D_1 and D_2 for bending in the longitudinal and transverse directions, respectively. A third term D_3, which represents the torsional rigidity together with some Poisson's ratio terms, is neglected for webs with simple, open-section stiffeners. With this simplification, the shear force per unit width N_{xy} at buckling is:

Fig. 7.19 Improvement in efficiency of a shear web in the early post-buckled range

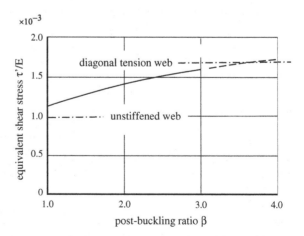

$$N_{xy} = K \frac{(D_1 D_2^3)^{1/4}}{h^2},$$

where the orthotropic buckling coefficient $K = 32.4$ for a long web with simply-supported edges.

The maximum equivalent shear stress for a shear web over a lower range of the post-buckling ratio β is plotted in Fig. 7.19. This is based on the orthotropic buckling formula above, with the web thickness and the cross-sectional area of the stiffeners optimized at each value of β. The ratio of web height to stiffener spacing $h/d = 1.0$. The shape of stiffener is again as shown in Fig. 7.18. The material is the same as in Fig. 7.17. Accurate values of K for initial buckling of the web, according to the actual μ value for the stiffener, are obtained from ESDU Data Item 02.03.02. Beyond $\beta = 3$, the use of the orthotropic buckling formula may be unreliable because of the likelihood of stiffener failure under compressive load due to increasing diagonal tension. Figure 7.19 also indicates the maximum equivalent shear stress that can be achieved for a diagonal tension web with the same h/d ratio and type of stiffener, using data from Kuhn. It is seen that a large part of the improvement in efficiency of a post-buckled shear web can be achieved at a relatively low post-buckling ratio.

7.4   Spreadsheet Programs

The three spreadsheets in this chapter all relate to the optimization of thin shell structures subject to buckling as well as stress and other constraints. The spreadsheet in Sect. 7.4.1 performs the optimization of a panel with discrete longitudinal stiffeners, loaded in compression and shear. Existing graphical data for the buckling of a stiffened panel is first tabulated and then interpolated to make it available for optimization. Stiffened panels form the basic building block of the structures in the following two sections. The spreadsheet in Sect. 7.4.1 is incorporated directly into the spreadsheet in Sect. 7.4.2 for optimization of a rectangular box beam in bending and shear, also with buckling of the shear webs forming the side walls of the box and variable rib spacing. The spreadsheet in Sect. 7.4.3 optimizes a circular section aircraft fuselage under vertical bending moment and shear force, with stiffened panels that can vary around the circumference according to local stress levels. An alternative method is used to represent graphical buckling data for optimization.

7.4.1   'Stiffened Panel'

The spreadsheet uses Solver to optimize the panel with integral, unflanged stiffeners in Fig. 7.14, loaded in axial compression and shear. The mass of the panel is

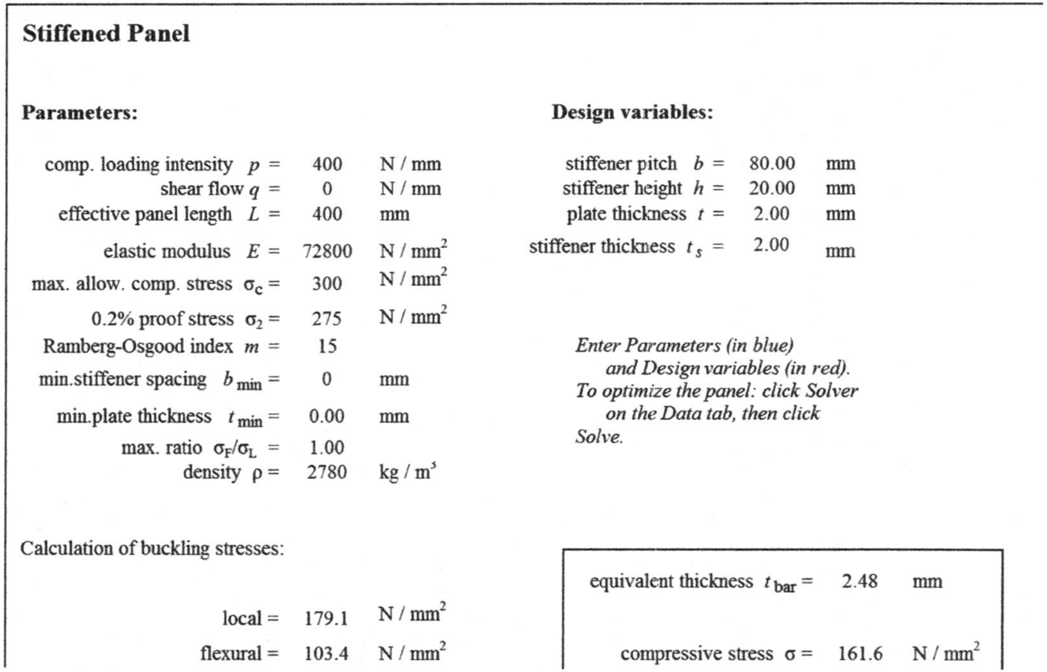

Stiffened Panel

Parameters:

comp. loading intensity	p =	400	N / mm
shear flow	q =	0	N / mm
effective panel length	L =	400	mm
elastic modulus	E =	72800	N / mm²
max. allow. comp. stress	σ_c =	300	N / mm²
0.2% proof stress	σ_2 =	275	N / mm²
Ramberg–Osgood index	m =	15	
min.stiffener spacing	b_{min} =	0	mm
min.plate thickness	t_{min} =	0.00	mm
max. ratio	σ_F/σ_L =	1.00	
density	ρ =	2780	kg / m³

Design variables:

stiffener pitch	b =	80.00	mm
stiffener height	h =	20.00	mm
plate thickness	t =	2.00	mm
stiffener thickness	t_s =	2.00	mm

*Enter Parameters (in blue)
and Design variables (in red).
To optimize the panel: click Solver
on the Data tab, then click
Solve.*

Calculation of buckling stresses:

local =	179.1	N / mm²
flexural =	103.4	N / mm²

equivalent thickness	t_{bar} =	2.48	mm
compressive stress	σ =	161.6	N / mm²

Fig. 7.20 Spreadsheet 'Stiffened Panel'

minimized subject to local and flexural buckling, and material stress limits. Reduced moduli are used to correct for yielding of the material. A minimum stiffener spacing and minimum plate thickness may be specified. Other constraints may be added if required. The efficiency of the optimized panel is calculated. The spreadsheet is shown in Fig. 7.20.

7.4.1.1   Modelling

For local buckling of the panel, use is made of graphical data in ESDU Data Item 70003, in which the buckling coefficient K is plotted against ratio h/b for different values of t_s/t (see Fig. 7.15). ESDU 70003 also provides an appropriate correction for yielding. To make this graphical data available for optimization, a polynomial function in h/b and t_s/t might be fitted to the data to represent the buckling coefficient. This can be done using Solver to minimize the mean square of the error between the polynomial function and the available data at an appropriate number of points. This has the advantage of providing a smooth, continuous function for optimization. However, a simple polynomial cannot properly reproduce the discontinuity in the curves beyond $h/b = 0.5$ at larger values of t_s/t. Unless otherwise constrained, this is the region in which the optimum is generally found. Alternatively, values of K from the graphical data in ESDU 70003 can be tabulated, and the buckling coefficient for required values of h/b and t_s/t obtained by interpolation. This method can generally approach the actual data more closely, especially at the discontinuity referred to above, but does of course provide a discontinuous representation of the original curves as a whole. In particular, this is so for interpolation between the limited number of t_s/t values in the original data. The second approach is the one adopted in the present spreadsheet.

A bilinear interpolation is performed, with as base point the row and column in the table with the largest values of h/b and t_s/t below the required values. These are found in the Visual Basic functions RW and CM. Values of K are read at this point, and at the adjacent row and column below and to the right of the base point, as in Fig. 7.21. Interpolation in the table for K at the required values of h/b and t_s/t is then performed in the Visual Basic function K with the formulae:

Fig. 7.21 Bilinear interpolation for the local buckling coefficient K

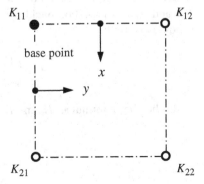

$$K = K_{11}(1 - x)(1 - y) + K_{21}x(1 - y) + K_{12}(1 - x)y + K_{22}xy,$$

$$x = \frac{H - H_{11}}{H_{21} - H_{11}}, \quad y = \frac{T - T_{11}}{T_{12} - T_{11}},$$

$$H = \frac{h}{b}, \quad T = \frac{t_s}{t}.$$

With this value of buckling coefficient K, the local buckling stress σ_L is calculated by Eq. (7.22). When the compressive stress in the panel is accompanied by a shear stress, Eq. (7.24) is first used for the buckling stress τ_b in pure shear, treating the plate as effectively simply-supported at the stiffeners. For the combination of compression and shear, an approximate solution is then obtained by means of the criterion in Eq. (7.20), with σ_b replaced by the local buckling stress σ_L of the panel. The spreadsheet is not intended for a panel in pure shear, as different conditions would then apply to the design of the stiffeners.

The flexural buckling stress σ_F is calculated by Eq. (7.23). It is assumed that the panel is uniform and sufficiently wide that it is unaffected by conditions at the sides of the panel. A continuous panel, supported at regular intervals along its length, is generally treated as though it is simply-supported at each support. The length L of the panel is then the distance between supports. For other conditions at the supports, or at the ends of the panel, a suitable effective length may be used.

Reduced moduli are used to correct for yielding of the material at higher stresses. For local buckling, the same reduced modulus is used as in ESDU 70003. This is:

$$E_{\text{red}} = E_s \cdot \left(\frac{1 - v_e^2}{1 - v^2}\right) \cdot \left[0.5 + 0.25\left(1 + 3\frac{E_t}{E_s}\right)^{1/2}\right]^{1/2}.$$

This formula is specifically for the present type of panel and takes account of the relatively greater role played by the stiffeners in the local buckling of a panel with plain, unflanged stiffeners. In the above formula, v_e is the elastic value of Poisson's ratio, taken to be $v_e = 0.3$, and v is the value of Poisson's ratio after yielding, given by:

$$v = v_p - \left(\frac{E_s}{E}\right) \cdot (v_p - v_e),$$

where v_p is the fully plastic Poisson's ratio $v_p = 0.5$ (see ESDU Data Items 76016 [7] and 83044 [9]). For buckling of the plate in shear, the following reduced modulus is used:

$$E_{\text{red}} = E_s \cdot \left(\frac{1 - v_e^2}{1 - v^2}\right) \cdot \left[0.83 + 0.17\frac{E_t}{E_s}\right].$$

In the above formulae, E_s and E_t are evaluated at an equivalent stress:

$$\sigma_{eq} = \sqrt{\sigma^2 + 3\tau^2}.$$

The tangent modulus is used for flexural buckling. Formulae for the secant modulus E_s and the tangent modulus E_t are given in Eqs. (7.6) and (7.7).

7.4.1.2 Optimization

Design variables are the stiffener spacing b, stiffener height h, plate thickness t, and stiffener thickness t_s. Note that h is measured to the mid-plane of the plate, as in ESDU 70003. A minimum stiffener spacing and minimum plate thickness may be specified. To remain within the valid range of data for local buckling in ESDU 70003, the following limits are imposed on ratios h/b and t_s/t:

$$0 \le h/b \le 1, \quad 0.75 \le \frac{t_s}{t} \le 3.0.$$

Constraints are defined as follows:

(1) local buckling of the panel is expressed in terms of the criterion:

$$\frac{\sigma}{\sigma_L} + \left(\frac{\tau}{\tau_b}\right)^2 \le 1.$$

(2) for flexural buckling:

$$\sigma \le \sigma_F.$$

(3) a maximum ratio of flexural to local buckling stress σ_F/σ_L may be specified to reduce the effect of local imperfections ($\sigma_F/\sigma_L < 1$), otherwise $\sigma_F/\sigma_L = 1$.
(4) for material failure the combined compressive and shear stress is expressed as an equivalent stress:

$$\sigma_{eq} = \sqrt{\sigma^2 + 3\tau^2} \le \sigma_c,$$

where σ_c is the specified allowable stress in compression.
(5) limits on ratios h/b and t_s/t, as referred to above.

In the above formulae, the compressive stress $\sigma = p/\bar{t}$, where $\bar{t}$ is the equivalent thickness of the panel, and the shear stress $\tau = q/t$. Other constraints refer to the maximum or minimum of cross-sectional dimensions. Further constraints may be added in Solver if required. Appropriate reduced moduli are used for local buckling, flexural buckling and buckling in shear, as previously described. It should be noted that these reduced moduli are evaluated at the stress σ and τ actually present in the panel, and do not therefore give the proper values for the buckling stress if the

Table 7.4 Data entry for spreadsheet program 'Stiffened Panel'

Parameters	
Compressive loading intensity p	Enter the value in cell D6 as a *positive* number (may not be zero)
Shear flow q	Enter the value in cell D7 as a *positive* number or zero
Effective length of panel L	Enter the value in cell D8 (actual length if simply-supported at the ends)
Elastic modulus E, allowable compressive stress σ_c, 0.2% proof stress σ_2, Ramberg–Osgood index m	Enter values in cells D9:D12 all as *positive* numbers
Min. stiffener spacing b_{min}, min. plate thickness t_{min}	Enter values in cells D13:D14
Max. ratio flexural/local buckling stress σ_F/σ_L	Enter a value in cell D15 (must be 1.0 or less)
Density ρ	Enter the value in cell D16
Design variables	
Stiffener spacing b, stiffener height h, plate thickness t, stiffener thickness t_s	Enter initial values in cells I6:I9 (stiffener height h measured to the mid-plane of the plate). Ensure that $0 \leq h/b \leq 1.0$, $0.75 \leq t_s/t \leq 3.0$

corresponding constraint is not active. This is taking advantage of the optimization procedure, which implicitly performs the iteration necessary to calculate the tangent and secant modulus while optimizing the panel.

Parameters and design variables to be entered in the spreadsheet are listed in Table 7.4. After optimization initial values of the design variables are replaced by their optimum values, the compressive and shear stress in the optimized panel and its mass are given. The efficiency of the panel is calculated only if at least one of the buckling constraints [1, 2] above are active at the optimum. To maintain the definition of efficiency, the initial *elastic* modulus E is used in Eq. (7.27). However, any reduction in stress due to yielding is included in the calculation of efficiency, reducing therefore its value. It will be noticed that there is a tendency for optimization to stop at the particular values of h/b and t_s/t in the table of local buckling coefficients. This is, of course, due to the discontinuous representation of the local buckling coefficient and the shallow vertices formed at these values by bilinear interpolation. At a sufficiently low compressive loading intensity, both to avoid yielding and to obtain a panel with a small thickness compared with its other cross-sectional dimensions, with shear flow $q = 0$, $\sigma_F/\sigma_L = 1$ and no active dimensional constraints we obtain a theoretical maximum efficiency $\eta = 0.812$ at $h/b = 0.70$ and $t_s/t = 2.50$.

Note that incorrect results will be displayed in the spreadsheet if initial values of the design variables are entered outside the valid range of data: $0 \leq h/b \leq 1$, $0.75 \leq t_s/t \leq 3.0$. Current values of h/b and t_s/t can be seen in the spreadsheet.

Fig. 7.22 Rectangular box
beam

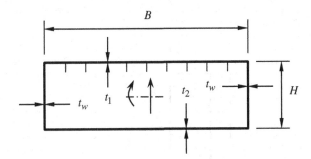

7.4.2   'Rectangular Box Beam'

The spreadsheet uses Solver to optimize the rectangular box beam in Fig. 7.22, loaded under a bending moment about the horizontal axis and a vertical shear force through the mid-point of the box, at a given cross section. The spreadsheet is shown in Fig. 7.23. The bending moment is defined as causing compression in the upper panel, tension in the lower panel (otherwise reverse the loading and regard 'upper panel' as 'lower panel' in the spreadsheet). The mass per metre of the box, including ribs at variable spacing, is minimized subject to buckling of the upper panel, buckling of the side walls or shear webs, and material stress limits under combined stress in the upper and lower panels and shear webs. The spreadsheet in Sect. 7.4.1, for optimization of a panel with integral, unflanged stiffeners, is incorporated as sheet 2 of the present spreadsheet.

7.4.2.1   Modelling

We take advantage of the rectangular shape of the box structure to represent the upper panel, in compression, simply by its equivalent or smeared thickness, since under the given loading the bending stress is uniform over the whole width and is readily calculated. Calculation of the buckling stresses of the upper panel is performed in sheet 2. The lower panel is in practice likely also to be a stiffened panel but, being in tension, we need here to refer only to its equivalent thickness. The shear webs are taken to be unstiffened, otherwise we could again refer to an equivalent thickness. Ribs, at intervals along the box to stabilize the cross section, are defined simply by a substitute thickness. This is the thickness of a plain, flat plate filling the whole cross section, of the same weight as the actual rib. The mass of the ribs is included in the total mass per metre of the box structure. The vertical shear force acts through the mid-point of the box which, with a structure symmetric about the vertical axis, results in a symmetric stress distribution.

The stress distribution at the given cross section is calculated by conventional beam theory. It is assumed that there is no abrupt change in the box structure at the given cross section. The maximum combined (von Mises) bending and shear stress

Rectangular Box Beam

Parameters

$M =$	600	kNm
$Q =$	120	kN
$B =$	1500	mm
$H =$	500	mm
$t_{rib} =$	2.00	mm
$b_{min} =$	120.0	mm
$t_{1\,min} =$	1.00	mm
$t_{2\,min} =$	1.00	mm
$t_{w\,min} =$	1.00	mm
$E =$	72800	N/mm^2
$\sigma_2 =$	275	N/mm^2
$m =$	15	
$\sigma_c =$	300	N/mm^2
$\sigma_t =$	400	N/mm^2
$(\sigma_F/\sigma_L)_{max} =$	1.00	
$\rho =$	2780	kg/m^3

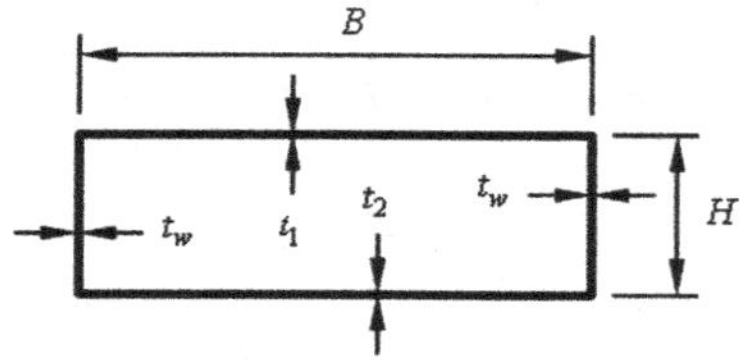

Design variables

upper panel:			
(see figure	$b =$	120.0	mm
on sheet 2)	$h =$	30.0	mm
	$t_1 =$	3.00	mm
	$t_s =$	3.00	mm
lower panel:	$t_2 =$	3.00	mm
shear webs:	$t_w =$	6.00	mm
rib spacing:	$L =$	600.0	mm

*Enter Parameters (in blue)
and Design variables (in red).
To optimize the box beam: click Solver
on the Data tab, then click Solve.*

Upper panel: see Sheet 2
$h / b =$ 0.25
$t_s / t_1 =$ 1.00

mass per metre =   **51.62**   kg/m   (including ribs)

Fig. 7.23 Spreadsheet 'Rectangular Box Beam'

occurs in the corners of the box, either in the web or in the upper or lower panel, depending on the relative thicknesses. The maximum shear stress is at the neutral axis of the box. von Mises stresses and margins of safety for material failure in the web and panels are calculated at these points (for the web only the minimum margin of safety is shown in the table in the spreadsheet).

As already said, the spreadsheet in Sect. 7.4.1 for a stiffened panel is incorporated as sheet 2 of the present spreadsheet to calculate the flexural and local buckling stresses of the upper panel in compression. The necessary data for this is read from sheet 1, with the buckling stresses returned to sheet 1. For flexural buckling, it is assumed that the panel is simply-supported at the ribs. The margin of safety for flexural buckling is included in the table in sheet 1. Local buckling of the upper panel is conservatively based on the compressive and shear stress at the corner of the box and is shown in the spreadsheet as a value of the buckling criterion for combined stress. Buckling of the webs in shear, taken to be simply-supported at their upper and lower edges as in Eq. (7.24), is based on the average shear stress in the webs. Appropriate reduced moduli are used to correct the buckling stresses for the effect of yielding of the material. Again, it should be noted that reduced moduli are based on actual stresses and do not therefore give the proper value of buckling stresses when the corresponding margin of safety is greater than zero.

7.4.2.2 Optimization

Please note that no data may be entered in sheet 2, as this will overwrite the reading of data from sheet 1. Initial values of design variables and parameters have to be entered in sheet 1. For the same reason, optimization must be performed in sheet 1, not in sheet 2.

Design variables are those referring to the stiffened panel that forms the upper panel of the box: stiffener spacing b, stiffener height h, plate thickness t_1, stiffener thickness t_s and additional variables: the (equivalent) thickness t_2 of the lower panel, the thickness t_w of each web and the rib spacing L. Note that the substitute rib thickness t_{rib} has to be specified and is not variable. Design variables and parameters, including the applied bending moment and shear force, to be entered are listed in Table 7.5.

Constraints referring to the upper panel are as described in Sect. 7.4.1. The same limitations on ratios h/b and t_s/t_1 apply. As before, a maximum ratio of flexural to local buckling stress ($\sigma_F/\sigma_L \le 1$) may be chosen, to reduce the effect of initial imperfections. Additional constraints refer to buckling of the shear webs, and maximum combined (von Mises) stresses at critical points in the section. Other constraints are for a minimum stiffener spacing, minimum plate thickness of the upper panel and a minimum web thickness.

After optimization, the initial design variables are replaced by their optimized values, and the resulting mass of the box beam and margins of safety are given. With variable rib spacing, an optimum can be reached with regard to flexural

Table 7.5 Data entry for spreadsheet program 'Rectangular Box Beam'

Parameters	
Bending moment M and shear force Q	Enter values in cells C5:C6 as *positive* numbers ($M \neq 0$)
Width B and height H of box beam	Enter values in cells C7:C8
Rib thickness t_{rib}	Enter the value in cell C9
Minimum stiffener spacing $b_{\min}$, minimum plate thickness $t_{1\min}$, minimum thickness t_2, minimum web thickness t_w	Enter values in cells C10:C13
Elastic modulus E, 0.2% proof stress σ_2, Ramberg–Osgood index m, max. allow. compressive stress σ_c	Enter values in cells C14:C17 all as *positive* numbers
Maximum allowable tensile stress σ_t	Enter the value in cell C18
Maximum ratio flexural/local buckling stress σ_F / σ_L	Enter a value in cell C19 (must be 1.0 or less)
Density ρ	Enter the value in cell C20
Design variables	
Stiffener spacing b (upper panel), stiffener height h (upper panel), thickness of upper panel t_1 stiffener thickness t_s, thickness of lower panel t_2, web thickness t_w, rib spacing L	Enter initial values in cells G5:G11 (stiffener height h is measured to the mid-plane of the plate, thickness t_1 is the plate thickness of the stiffened panel, and both shear webs are of the same thickness t_w). Ensure that $0 \leq h/b \leq 1.0$, $0.75 \leq t_s/t \leq 3.0$

buckling of the panel and the mass per metre of the ribs. If large changes in rib spacing occur, it may be considered appropriate to change the specified (substitute) rib thickness to a more representative value for the problem in hand, and to repeat the optimization.

The spreadsheet can be extended to more loading cases, with bending moment and shear force applied in both directions. In that case, both the upper and the lower panels have to be designed as compression panels. The spreadsheet may also be extended to include a twisting moment (in effect changing the point through which the shear force is applied), and bending moments applied about both axes. More extensive changes to the spreadsheet are then required, as the stress distribution will no longer be symmetric.

Attention is drawn to the final paragraph of Sect. 7.4.1 with regard to optimization of the upper panel.

7.4.3   'Circular Fuselage Section'

The spreadsheet uses Solver to optimize the cross section of an aircraft fuselage of circular form, loaded in bending and shear. The shell structure of the fuselage consists of a thin skin reinforced by Z-section stringers, supported by ring frames at

Fig. 7.24 Fuselage cross section

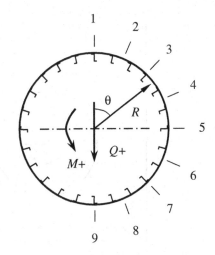

Fig. 7.25 Circular fuselage section: detail of the stringer-skin panels

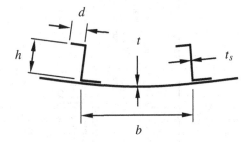

intervals along its length. The cross section is shown in Fig. 7.24, with detail of the stringer-skin panels in Fig. 7.25. The skin thickness and the stringer dimensions vary around the cross section. The cross-sectional area of the frames and the frame spacing have to be specified. Two loading cases are defined. These are a positive bending moment and a negative bending moment about the horizontal axis at a given section, with corresponding vertical shear forces acting through the centre of the fuselage. The mass per unit length of the structure is minimized, subject to buckling of the stringer-skin panels and material stress limits under combined compressive and shear stress. The spreadsheet is shown in Fig. 7.26.

7.4.3.1 Modelling

The stress distribution in the cross section is calculated by conventional beam theory. Under the given loading, the structure will be symmetric about the vertical axis. Only one-half of the section need be considered, therefore. The cross section is represented by the area of skin and stringers concentrated into substitute members at nine uniformly spaced points around the half-section, as indicated on the

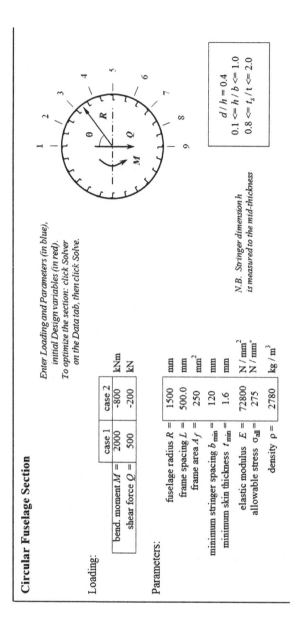

Fig. 7.26 Spreadsheet 'Circular Fuselage Section'

spreadsheet. The calculation of second moment of area is based on this simplified model (values in the table 'Fuselage cross section' refer to the half-section). The stringer-skin panels are taken to be uniform within each segment associated with a particular substitute member. It is not necessary at this stage to consider a discrete stringer spacing, since this will change during optimization and can be adjusted at a later stage. The bending stress σ is calculated at each of the nine points. The shear stress τ is calculated as discussed in Sect. 7.2 and represents an 'average' shear stress at each point. These stresses are given in the tables 'load case 1' and 'load case 2' on the spreadsheet. Equivalent von Mises stresses σ_{eq} are calculated for combined tension or compression and shear stresses in the same tables.

The flexural buckling stress σ_F of the stringer-skin panels is calculated by Eq. (7.23), with effective length equal to the frame spacing. The required properties of the individual stringers are calculated in the table 'stringer cross section'. Note that the stringers have a fixed ratio $d/h = 0.4$. For local buckling of the panels, buckling coefficients from ESDU Data Item 71014 [6] are used. This data (otherwise than in the spreadsheet for a stiffened panel and a rectangular box beam in the previous sections) is represented in a fourth degree polynomial in the ratios $H = h/b$ and $T = t_s/t$ (refer to the Visual Basic function KLZ in the spreadsheet for the coefficients of the polynomial). This gives an error of less than one per cent over practically the whole data. The range of data taken from ESDU 71014 is limited to $0.1 \leq h/b \leq 1$ (the lower limit being to avoid unrealistic values of buckling coefficient) and $0.8 \leq t_s/t \leq 2$. The latter is to avoid impractically thin stringers, but also has the effect of removing curves for small values of t_s/t with discontinuities that cannot properly be represented by the polynomial function. The local buckling stress σ_L by Eq. (7.22) at the different points around the cross section is given in the tables for the two load cases. For the shear buckling stress τ_b of the skin panels, a shear buckling coefficient $K = 4.83$ is used in Eq. (7.24), since the individual skin panels can be assumed to be relatively long. Equations (7.20) or (7.21) are then used to calculate the value of the buckling criterion under combined compression or tension and shear. No correction for yielding is made in any of the buckling formulae. This is on the assumption that, for the anticipated loading on this type of structure, stresses will be well below the elastic limit of the material. A single maximum allowable stress is specified to ensure that the material remains within or close to the elastic limit.

7.4.3.2 Optimization

Design variables are the skin thickness t, stringer spacing b, stringer thickness t_s and stringer height h at each of the nine points in the half-section. Suitable initial values have to be entered in the table. Parameters to be entered, including the two load cases, are listed in Table 7.6. Positive bending moment causes tension in the upper part of the cross section. Constraints refer to the flexural and local buckling stresses, and the equivalent von Mises stresses at each point. Normalized constraint values

Table 7.6 Data entry for spreadsheet program 'Circular Fuselage Section'

Parameters	
Positive and negative bending moments M and corresponding shear forces Q in loading cases 1 and 2	Enter values in cells D8:E9 (positive bending moment causes tension in the upper fuselage)
Fuselage radius R, frame spacing L, frame area A_f	Enter values in cells E12:E14
Minimum stringer spacing b_{min}, minimum skin thickness t_{min}	Enter values in cells E15:E16
Elastic modulus E, allowable stress σ_{all}, density ρ	Enter values in cells E17:E19 all as positive numbers
Design variables	
Stringer spacing b, skin thickness t, stringer height h, stringer thickness t_s	Enter initial values in cells D26:G34 at each location in the half-section (fuselage is assumed symmetric about the vertical axis, and stringer height h is measured to the mid-thickness of the flanges)

for flexural buckling and allowable stress are calculated in tables 'loading case 1' and 'loading case 2'. For local buckling, the constraint values refer to the buckling criterion for combined compression and shear. All values in these three columns have to be positive or zero. Additional constraints are the above limits on h/b and t_s/t. A minimum stringer spacing and a minimum skin thickness also have to be specified (the latter may in any case be required to meet minimum requirements for a pressure cabin). A constraint on the maximum ratio of flexural to local buckling stress (σ_F/σ_L), to reduce imperfection sensitivity, cannot immediately be applied in this spreadsheet, due to the limit on the number of constraints in Solver. To include such constraints would require removal of other, inactive constraints. Depending on the magnitude of the applied loading, constraints on allowable stress may be found far from critical, and safe therefore to remove.

After optimization, the initial design variables are replaced by their optimized values, and the minimum mass per unit length of the fuselage shell structure is given. Margins of safety in the two loading cases are given in a separate table. If necessary, a different stringer spacing might then be chosen to account for actual panel sizes and other practical requirements, and the optimization repeated with chosen stringer spacing as additional constraints. If the frame spacing may be varied, this can be added to the list of variables in Solver to find an optimum. The spreadsheet can be extended to include bending moments about the vertical axis and corresponding shear forces, and a twisting moment. Corrections for yielding can, if required, be made in a similar manner to those in the spreadsheet in Sect. 7.4.2.

7.5 Summary

Formulae are reviewed for the stress distribution in a long, reinforced shell structure, in particular for the box beam that has been the main concern of this chapter. Except near major discontinuities in the structure or in the loading on it, the bending stress at any section can be calculated accurately enough by conventional bending theory. The structure may be unsymmetric either due to its cross-sectional shape or due to the distribution of material in the cross section. In either case, second moments of area about both axes and the product second moment of area have to be calculated, all with reference to the centre of gravity of the section. The product second moment is a measure of the lack of symmetry in the section. A general formula is developed for the bending stress in terms of the second moments.

Above the elastic limit, reduction in stiffness due to yielding is taken into account by means of the secant modulus. The strain distribution in the cross section is assumed to remain linear, but after yielding the stress distribution is no longer linear. The secant modulus is included in the calculation of effective second moments of area and centre of gravity position, as well as in the formula for bending stress. Since the secant modulus depends on the stress at any point in the cross section, this results in an iterative calculation, in fact dealt with in successive iterations of the optimization process itself. For materials with a smooth stress–strain curve, the Ramberg–Osgood formula can be used for calculation of the secant modulus.

A typical box structure will be reinforced by longitudinal stiffeners supported by transverse members spaced along its length. To simplify the model, stiffeners can be represented together with the sheet as an equivalent or 'smeared' thickness; alternatively, the sheet and stiffeners can be represented by substitute stiffeners around the cross section. In the latter case, the second moments of area are based on the substitute stiffeners, and the bending stress is also calculated at those locations. With the use of substitute stiffeners, the shear stress calculation is simplified on the basis that the shear flow is constant between the stiffeners. In a single-cell structure, the shear flow is calculated step by step around the cross section from the bending stress in each substitute stiffener. A constant value of shear flow then has to be added to the result to ensure that the resulting moment is in equilibrium with the twisting moment on the section. This may include an applied twisting moment, but is also the moment of the applied shear force about some reference point. In a multi-cell section, the shear flow cannot be obtained purely by statics, but must ensure the same rate of twist in all cells.

A box beam or similar shell structure, if it is relatively thin, is liable to buckling in combined compression and shear in the individual panels between stiffeners, as well as buckling of the panels as a whole between the transverse reinforcing members (termed ribs, frames or bulkheads). The former is referred to as local buckling, the latter as flexural buckling. An exact analysis of buckling stresses can be highly complex because of various interactions that occur, and is further hindered by unavoidable even if very small imperfections in the structure. Some basic formulae are given for the buckling stress of thin plates in compression and shear, and under combined stress. These formulae can be applied to the individual elements of a stiffened panel, but interactions between adjacent parts can cause a significant change in buckling stress. A large amount of data is available in the literature for a more accurate calculation of the buckling stress of stiffened panels, also with appropriate effective moduli after yielding, depending on the characteristics of the buckling mode. This is used in the spreadsheet programs in this chapter for a stiffened panel in compression, a rectangular box beam and a circular fuselage section.

An efficiency formula for a compression panel with integral, unflanged stiffeners is derived from the formulae for local and flexural buckling. Similar formulae can be derived for other shapes of stiffener. For more complicated problems involving material limitations and limits on stiffener spacing, sheet thickness or other dimensions, a numerical optimization such as in the spreadsheet programs mentioned above becomes necessary. A correction can also be applied to reduce imperfection sensitivity, that is, growth of initial imperfections reducing the buckling load. A different approach has to be followed for the efficiency of a stiffened shear web. This is due to the stiffness requirements for the stiffeners, which before buckling do not themselves take any part of the shear force on the web but increase its buckling resistance. An efficiency formula for the equivalent shear stress is based on that of an unstiffened web, although some weak dependence on the structural index then remains. It is found that, except at small stiffener spacing, the optimum occurs at or close to a critical value μ_c at which the stiffeners become fully effective. After initial buckling, a state of diagonal tension develops in a stiffened shear web. At lower values of structural index, this can result in a substantial improvement in efficiency, but at the same time leads to a high post-buckling ratio. However, it is shown that much of this gain in efficiency can also be achieved at much lower post-buckling ratios.

Exercises

Unless otherwise indicated, in exercises 7.3–7.10 use the loading and other data already entered in the original spreadsheet.

7.1   Make a hand calculation of the flexural and local buckling stresses of the panel with integral, unflanged stiffeners in Fig. 7.14, loaded in axial compression. The stiffener spacing is 120 mm, the stiffener height is 30 mm, and the thickness of both plate and stiffeners is 3.0 mm. The effective length of the panel is 800 mm, and the elastic modulus of the material is 72,800 N/mm^2. Calculate the efficiency of the panel, based on the lesser of the two buckling stresses.

Note that the stiffener height is measured to the mid-plane of the plate. Take the local buckling coefficient $K = 3.95$ in Fig. 7.15. Assume no reduction in modulus due to yielding. Calculate the efficiency from Eq. (7.27).

7.2 A stiffened shear web of height $h = 500$ mm and stiffener spacing $d = 250$ mm has stiffeners designed to satisfy the criterion $\mu = \mu_c$ (taken to be the optimum). The stiffeners have a coefficient $C = 1.0$. The elastic modulus $E = 72{,}800$ N/mm^2 (neglect any effect of yielding). Calculate the equivalent shear stress τ' in the web under a shear force $Q = 60, 120, 240$ kN. Use the three results to deduce an efficiency formula for a stiffened shear web of this type in the form:

$$\tau' = \alpha E^{1-n}(q/h)^n$$

Take μ_c and K from Table 7.3, with the long edges of the web simply-supported. Use Eqs. (7.31) and (7.30) to calculate t and t', and calculate the equivalent shear stress at each load. Make a $\log(\tau')$-$\log(q/h)$ plot to verify the above relation. Deduce values of α and n from the log-log plot, or numerically.

7.3 Use the spreadsheet 'Stiffened Panel' to optimize a panel with minimum stiffener spacing $b_{min} = 100$ mm and minimum plate thickness $t_{min} = 1.0$ mm, and with these dimensional constraints removed. Determine also the maximum efficiency of the panel.

Set b_{min} and t_{min} to the required values and optimize the panel. Compare the dimensions of the panel and its efficiency with those of a panel with these limits removed. For maximum efficiency, we should avoid any reduction in modulus. Set b_{min} and t_{min} to zero, and reduce the loading intensity p until $E_t = E_s = E$ after optimization. Try different initial design variables to ensure that a true optimum has been found, and not a local minimum.

7.4 Use the spreadsheet 'Stiffened Panel' to verify the hand calculation in Exercise 7.1.

Enter the dimensions and effective length from Exercise 7.1 (no optimization required). Compare the second moment of area and buckling stresses with the calculated values. For the correct efficiency, the compressive stress in the panel has to match the lesser of the flexural and local buckling stresses. Try different values of loading intensity p until it is close enough (alternatively use Goal Seek in Excel). Verify that there is negligible reduction in modulus due to yielding.

7.5 Use the spreadsheet 'Stiffened Panel' to show the sensitivity of the flexural and local buckling stresses to stiffener spacing, stiffener height, plate thickness and stiffener thickness for the panel in Exercise 7.1.

Enter the dimensions and effective length from Exercise 7.1. Make a one per cent increase in each dimension in turn, and note the increase or decrease in flexural and local buckling stress. The data can be used to select the best dimension to change to improve one buckling stress with the least effect on the other, or to obtain constraint gradient data.

7.6 Use the spreadsheet 'Rectangular Box Beam' to optimize the beam for a range of rib spacing from 100 to 1000 mm, under the bending moment, shear force and other data given in the spreadsheet.
Remove rib spacing from the list of variables in Solver before optimizing the beam for different rib spacing. Try different initial design variables to verify convergence. Make a plot of minimum mass per metre against rib spacing.

7.7 Optimize the beam in the spreadsheet 'Rectangular Box Beam' under a range of bending moment M from 100 to 1000 kNm with corresponding shear force $Q = \frac{M}{5}$ (in kN), to show the relation between mass per metre of the beam and applied loading.
Set the minimum cross-sectional dimensions to zero to have no effect on the optimized beam. Ensure that rib spacing is included in the list of variables before optimizing the beam under different loading. Make a plot of mass per metre against bending moment. Note the relatively small increase in mass per metre with increase in loading.

7.8 Use the spreadsheet 'Circular Fuselage Section' to show the effect of different frame spacing on the minimum mass per metre of the fuselage structure, both with minimum skin thickness $t_{min} = 1.6$ mm and minimum stringer spacing $b_{min} = 120$ mm and with no minimum skin thickness and stringer spacing.
Repeat the optimization for different frame spacing over a range from 200 to 1000 mm. Note the change in the mass of the frames and the mass of the fuselage shell with different frame spacing. Add frame spacing to the list of variables in Solver for the optimum frame spacing.

7.9 Calculate the efficiency of the stringer-skin panel in the spreadsheet 'Circular Fuselage Section', at the point of maximum compressive stress, after optimization with minimum skin thickness $t_{min} = 1.6$ mm, minimum stringer spacing $b_{min} = 120$ mm and frame spacing $L = 500$ mm. Compare the efficiency with that of the same panel with no constraint on skin thickness and stringer spacing.
Deduce the loading intensity p from the compressive stress and optimized dimensions of the panel, and substitute in Eq. (7.27) for the efficiency. For the maximum efficiency, set the constraints on minimum skin thickness and stringer spacing to zero. Repeat the optimization and recalculate the efficiency. This is the maximum efficiency for this type of panel with $d/h = 0.4$. Observe the difference in dimensions of the panel with and without dimensional constraints.

7.10 Add a constraint on the ratio of flexural to local buckling stress: $\sigma_F/\sigma_L \leq 0.85$ to the spreadsheet 'Circular Fuselage Section', to reduce the imperfection sensitivity of the panels. Take $b_{min} = 120$ mm and $t_{min} = 1.6$ mm. Compare the optimized mass per metre with $\sigma_F/\sigma_L \leq 0.85$ with the optimized mass per metre with no added constraint.
There are substantial margins of safety on allowable stress in both load cases. To avoid the limit on the number of constraints in Solver, these constraints may be removed and replaced by constraints on σ_F/σ_L at each location. Note the small increase in mass as a result of the constraint on σ_F/σ_L.

References

1. Brush DO, Almroth BO (1975) Buckling of bars, plates and shells. McGraw-Hill, New York
2. Catchpole EJ (1954) The optimum design of compression surfaces having un-flanged integral stiffeners. J R Aeronaut Soc 58:765–768
3. ESDU 02005 (2003) Flat panels in shear—post-buckling analysis. Engineering Sciences Data, Structures series, section 9. IHS, London
4. ESDU 70003 (1976) Local buckling of compression panels with unflanged integral stiffeners. Engineering Sciences Data, Structures series, section 21. IHS, London
5. ESDU 71005 (1995) Buckling of flat plates in shear. Engineering Sciences Data, Structures series, section 9. IHS, London
6. ESDU 71014 (1976) Local buckling of compression panels with flanged stringers. Engineering Sciences Data, Structures series, section 21. IHS, London
7. ESDU 76016 (2013) Generation of smooth continuous stress-strain curves for metallic materials. Engineering Sciences Data, Structures series, section 2. IHS, London
8. ESDU 77014 (2012) Flat panels in shear. Post-buckling analysis. Engineering Sciences Data, Structures series, section 9. IHS, London
9. ESDU 83044 (2014) Plasticity correction factors for plate buckling. Engineering Sciences Data, Structures series, section 2. IHS, London
10. ESDU 98016 (2008) Elastic buckling of flat isotropic stiffened panels and struts in compression. Engineering Sciences Data, Structures series, section 6. IHS, London
11. ESDU STRUCT 02.03.02 (1983) Flat panels in shear. Buckling of long panels with transverse stiffeners. Engineering Sciences Data, Structures series, section 9. IHS, London
12. Farrar DJ (1949) The design of compression surfaces for minimum weight. J R Aeronaut Soc 53(10):1041–1052
13. Gerard G, Becker H (1957) Handbook of structural stability. Part I—buckling of flat plates. NACA tech. Note 3781
14. Kuhn P (1956) Stresses in aircraft and shell structures. McGraw-Hill, New York
15. Ramberg W, Osgood, WR (1943) Description of stress-strain curves by three parameters. NACA tech. Note 902
16. Rees DWA (2009) Mechanics of optimal structural design: minimum weight structures. Wiley, New York
17. Rothwell A (1978) A design space representation of stiffened shear webs. Aeronaut J 82:359–363
18. Timoshenko SP, Gere JM (1961) Theory of elastic stability. McGraw-Hill, New York (reprinted by Dover Publications, 2009)
19. Vinson JR (2005) Plate and panel structures of isotropic, composite and piezoelectric materials, including sandwich construction. Springer, Dordrecht

Chapter 8
Composite Laminates

Abstract Lamination theory for the stress in the individual layers of a composite laminate is reviewed, and appropriate failure criteria are introduced. Optimization of lay-up, based on lamination theory, involves the orientation of the different layers and the number of plies per layer to best tailor the material to the specific application. Different methods of optimization are introduced, including use of a simplified netting analysis for an initial estimate of the required lay-up. Starting from this, an iterative procedure using full lamination theory is described, making step-by-step adjustments to the number of plies to reach a minimum. The discrete ply thickness is a significant complication in formal optimization methods. With integer variables for the number of plies, the 'branch-and-bound' method progressively eliminates combinations of variables that cannot lead to a solution in the search for an optimum. Alternatively, a genetic algorithm—a semi-random process—retains a family of designs and repeatedly combines the best of them to arrive at an optimum. A spreadsheet program is presented for the analysis and optimization of a composite laminate under in-plane load, with discrete ply thickness and internal stresses due to change in temperature, based on either the Tsai–Hill or the Tsai–Wu failure criterion.

Composite materials, consisting of thin fibres embedded in a softer matrix material, play an ever-increasing role in many branches of engineering design. In fact, we could hardly think of aerospace and many other advanced structures these days without some use of composites. Originally used for their ability to be formed into complex curved shapes, now they are employed much more for their high strength and stiffness and their low density. In more demanding applications, composites are almost invariably used as laminates, made up of individual layers of unidirectional or woven material at different angles. These may be manufactured by laying up a sequence of plies to form the laminate or by processes such as tow placement, filament winding or a wide variety of other means. Composite materials offer the designer greater freedom to satisfy conflicting requirements by allowing the material itself as well as the structure to be designed to suit its purpose. By appropriate choice of lay-up, the laminate can be 'tailored' to match as closely as possible the specific loading and other requirements on it.

© Springer International Publishing AG 2017

A. Rothwell, *Optimization Methods in Structural Design*, Solid Mechanics and Its Applications 242, DOI 10.1007/978-3-319-55197-5_8

Numerical optimization methods play an essential role in the design of composite structures. Even in quite simple situations, it can soon go beyond the natural intuition of the designer to decide what changes should be made to a design to achieve the desired result. Changes in the size and shape of a structure, together with the many different lay-ups possible in a composite laminate, make design on a 'trial-and-error' basis difficult as well as time consuming. Practical restrictions on lay-up may be imposed to avoid both delamination and unwanted deformation and, of course, for reasons of manufacture. The designer is offered the opportunity to tailor both the material and the structure to best suit the application, but at the same time is faced with the challenge of developing a design procedure to make this possible.

Here, we restrict the problem to that of the design and optimization of a single laminate. Lamination theory, reviewed in the following section, enables the stiffness properties of a laminate and the stress in each layer to be calculated, taking account of the interaction between layers such that a consistent strain distribution through the thickness is obtained. For a more comprehensive account of lamination theory than is possible here, the reader is referred to the classic text of Jones [1], Daniel and Ishai [2] and many others. Based on the layer stresses calculated by lamination theory, various failure criteria are available to predict the strength of a laminate layer by layer. The design of a laminate involves the orientation of each layer, the number of plies within each layer, the stacking sequence of the different layers, as well as material selection for the laminate or for each layer contained within it. However, for a laminate made of a single material under only in-plane load, we can define the lay-up more simply by the different angles at which plies are laid up and the number of plies in each of these directions. The stacking sequence then becomes irrelevant, except that the lay-up should remain symmetric about the middle plane if bending deformation of the laminate is to be avoided. For a balanced laminate, that is to say one with no coupling between direct and shear strains, we require equal numbers of plies at the same positive and negative ply angles. A balanced, symmetric laminate is a common requirement in laminate design, helping to avoid unwanted deformations both in manufacture and under load.

The steps in the analysis of a composite laminate are illustrated in Fig. 8.1. Starting with material data and the elastic properties of the individual layers, these are assembled at appropriate orientations to form an initially chosen laminate. The stiffness coefficients of the assembled laminate are then derived, with which the strain components in the laminate under specified applied load can be calculated. The stresses in each layer are obtained from these strain components. With ply strength data obtained by test, a failure criterion is used to detect failure in any layer under the combination of stresses in that layer. At this stage, the initial analysis of the laminate is complete. To proceed with optimization, the value of the failure criterion in each layer serves as constraint, together with any strain limitation and practical constraints on lay-up, also taking into account the discrete ply thickness.

Fig. 8.1 Steps in the analysis of a composite laminate

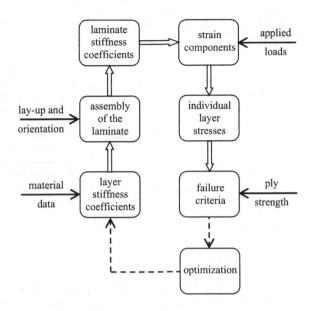

8.1   Lamination Theory

The elastic properties of a single layer of a composite laminate are defined by:

$$\varepsilon_1 = \frac{\sigma_1}{E_1} - v_{21}\frac{\sigma_2}{E_2},$$

$$\varepsilon_2 = \frac{\sigma_2}{E_2} - v_{12}\frac{\sigma_1}{E_1}, \tag{8.1}$$

$$\gamma_{12} = \frac{\tau_{12}}{G_{12}}.$$

As shown in Fig. 8.2, axes 1, 2 refer to the principal axes of the layer, that is along and perpendicular to the fibre direction in a unidirectional material or in the directions of the weave for a conventional woven material.[1] With respect to these axes, the layer is termed 'orthotropic', that is, there is no coupling between the shear strain γ_{12} and direct stresses σ_1 and σ_2 or between the direct strains ε_1 and ε_2 and the shear stress τ_{12}. The elastic constants E_1, E_2 and G_{12} in the formulae above are usually referred to as 'engineering constants'. Note in particular the subscripts on the two Poisson's ratio terms v_{12} and v_{21}. By the reciprocal theorem, these are related to the elastic moduli E_1 and E_2 by:

$$v_{12}E_2 = v_{21}E_1,$$

---

[1]While Figs. 8.2 and 8.3 show a layer composed of unidirectional plies, unless otherwise stated the formulae in this chapter apply to both unidirectional and conventional woven materials.

Fig. 8.2 Stress components referred to the fibre direction

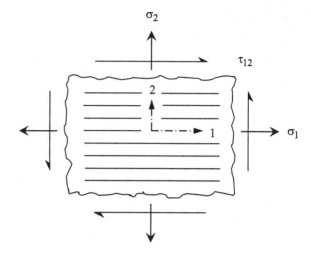

and there are, therefore, only four independent elastic constants for an orthotropic material. For unidirectional plies, E_1 is generally much greater than E_2 and the shear modulus G_{12}, and ν_{12} is much greater than ν_{21}. For this reason, ν_{12} is usually referred to as the major Poisson's ratio.

When assembled into a laminate and subject to in-plane load, the stresses differ from layer to layer due to their different stiffness properties, but the strain is assumed constant through the thickness, that is, if there is perfect bonding between the individual layers and of course if there is no bending of the laminate. For this reason, it is more convenient to express stress in terms of strain and by solution of Eq. (8.1):

$$
\begin{aligned}
\sigma_1 &= \frac{E_1}{\mu}\varepsilon_1 + \frac{\nu_{21}E_1}{\mu}\varepsilon_2, \\
\sigma_2 &= \frac{E_2}{\mu}\varepsilon_2 + \frac{\nu_{12}E_2}{\mu}\varepsilon_1, \\
\tau_{12} &= G_{12}\gamma_{12},
\end{aligned}
\tag{8.2}
$$

where

$$
\mu = 1 - \nu_{12}\nu_{21}
$$

is introduced simply as a convenient constant. Equation (8.2) can readily be expressed in matrix notation:

$$
\begin{Bmatrix} \sigma_1 \\ \sigma_2 \\ \tau_{12} \end{Bmatrix} =
\begin{bmatrix}
E_1/\mu & \nu_{21}E_1/\mu & 0 \\
\nu_{12}E_2/\mu & E_2/\mu & 0 \\
0 & 0 & G_{12}
\end{bmatrix}
\begin{Bmatrix} \varepsilon_1 \\ \varepsilon_2 \\ \gamma_{12} \end{Bmatrix},
\tag{8.3}
$$

or more briefly in terms of a stiffness matrix $\mathbf{Q}$:

$$\sigma_{12} = \mathbf{Q}\,\varepsilon_{12}. \tag{8.4}$$

Matrix notation is introduced not only because computers can readily work in it, but also because we shall shortly be faced with multiplying matrices by each other and other matrix manipulations, as will be seen in the following section.

8.1.1 Transformed Stiffness Matrix

The different layers making up the laminate will be laid up at different orientations to achieve the desired properties. This means that the layer stiffness properties in Sect. 8.1 have to be transformed into a common axis system, the reference axes x, y of the laminate, as shown in Fig. 8.3. Stresses and strains with no subscript will be taken to refer to laminate axes and those with subscripts 1, 2 to layer axes. The standard transformation formulae are

$$\begin{aligned} \varepsilon_{12} &= \mathbf{T}\,\varepsilon, \\ \sigma &= \mathbf{T}^{\mathsf{T}}\sigma_{12}, \end{aligned} \tag{8.5}$$

where the transformation matrix $\mathbf{T}$ is

$$\mathbf{T} = \begin{bmatrix} \cos^2\theta & \sin^2\theta & \sin\theta\cos\theta \\ \sin^2\theta & \cos^2\theta & -\sin\theta\cos\theta \\ -2\sin\theta\cos\theta & 2\sin\theta\cos\theta & (\cos^2\theta - \sin^2\theta) \end{bmatrix} \tag{8.6}$$

and $\mathbf{T}^{\mathsf{T}}$ is its transpose (interchange of rows and columns). Expanding the second of Eq. (8.5), we obtain for the stress components in laminate axes:

$$\begin{aligned} \sigma_x &= \sigma_1\cos^2\theta + \sigma_2\sin^2\theta - 2\tau_{12}\sin\theta\cos\theta, \\ \sigma_y &= \sigma_1\sin^2\theta + \sigma_2\cos^2\theta + 2\tau_{12}\sin\theta\cos\theta, \\ \tau_{xy} &= \sigma_1\sin\theta\cos\theta - \sigma_2\sin\theta\cos\theta + \tau_{12}(\cos^2\theta - \sin^2\theta). \end{aligned} \tag{8.7}$$

The above formulae are the standard ones from the theory of elasticity (see Timoshenko and Goodier [3]) and are unrelated to the material properties, whether isotropic, orthotropic, or more generally anisotropic. It is left to the reader to verify these relations by resolving stress resultants or displacements due to strain components into the new directions. It is important to note that the angle θ in Fig. 8.3 is the angle from the laminate x-axis to the fibre direction 1 measured positive in the anticlockwise direction. Transformation of stress or strain back from laminate axes to layer axes is achieved simply by replacing θ by $-\theta$ in the $\mathbf{T}$ matrix.

Fig. 8.3 Rotation of axes

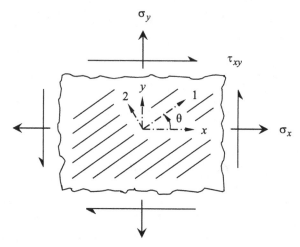

With the above transformation formulae and matrix $\mathbf{Q}$ in Eq. (8.4), we can express stress in terms of strain, both in laminate axes, by:

$$\boldsymbol{\sigma} = \mathbf{T}^{\mathrm{T}}\boldsymbol{\sigma}_{12} = \mathbf{T}^{\mathrm{T}}\mathbf{Q}\,\boldsymbol{\varepsilon}_{12} = \mathbf{T}^{\mathrm{T}}\mathbf{Q}\,\mathbf{T}\,\boldsymbol{\varepsilon},$$

or

$$\boldsymbol{\sigma} = \bar{\mathbf{Q}}\,\boldsymbol{\varepsilon}, \tag{8.8}$$

where

$$\bar{\mathbf{Q}} = \mathbf{T}^{\mathrm{T}}\mathbf{Q}\,\mathbf{T}$$

is the 'transformed stiffness matrix' for the layer. Note that both $\mathbf{Q}$ and $\bar{\mathbf{Q}}$ are symmetric matrices. While in the computer the above matrix form for $\bar{\mathbf{Q}}$ is the most appropriate, it is also useful to write down the explicit form of the terms of the $\bar{\mathbf{Q}}$ matrix, as follows:

$$\bar{Q}_{11} = \frac{1}{\mu}\left[E_1\cos^4\theta + E_2\sin^4\theta + (2\nu_{12}E_2 + 4\mu G_{12})\sin^2\theta\cos^2\theta\right],$$

$$\bar{Q}_{12} = \bar{Q}_{21} = \frac{1}{\mu}\left[(E_1 + E_2 - 4\mu G_{12})\sin^2\theta\cos^2\theta + \nu_{12}E_2\left(\cos^4\theta + \sin^4\theta\right)\right],$$

$$\bar{Q}_{13} = \bar{Q}_{31} = \frac{1}{\mu}\left[(E_1 - \nu_{12}E_2 - 2\mu G_{12})\sin\theta\cos^3\theta - (E_2 - \nu_{12}E_2 - 2\mu G_{12})\sin^3\theta\cos\theta\right],$$

$$\bar{Q}_{22} = \frac{1}{\mu}\left[E_1\sin^4\theta + E_2\cos^4\theta + (2\nu_{12}E_2 + 4\mu G_{12})\sin^2\theta\cos^2\theta\right],$$

$$\bar{Q}_{23} = \bar{Q}_{32} = \frac{1}{\mu}\left[-(E_2 - \nu_{12}E_2 - 2\mu G_{12})\sin\theta\cos^3\theta + (E_1 - \nu_{12}E_2 - 2\mu G_{12})\sin^3\theta\cos\theta\right],$$

$$\bar{Q}_{33} = \frac{1}{\mu}\left[(E_1 + E_2 - 2\nu_{12}E_2)\sin^2\theta\cos^2\theta + \mu G_{12}\left(\cos^2\theta - \sin^2\theta\right)^2\right],$$

where $\bar{Q}_{ij}$ is the term in row i, column j of matrix $\bar{\mathbf{Q}}$. Note that as a consequence of the rotation of axes, all nine terms of matrix $\bar{\mathbf{Q}}$ are now present.

8.1.2 Laminate Stiffness Coefficients

With the stiffness $\bar{\mathbf{Q}}$ of the individual layers now transformed into a common laminate axis system, the stress in the layers can be summed to give a resultant load per unit width $\mathbf{N}$ on the laminate:

$$\mathbf{N} = \left\{ \begin{array}{c} N_x \\ N_y \\ N_{xy} \end{array} \right\} = \sum_{k=1}^{k=L} t_k\, \boldsymbol{\sigma},$$

where the summation is over the L layers making up the laminate and t_k is the thickness of each layer. With the layer stress:

$$\boldsymbol{\sigma} = \bar{\mathbf{Q}}\, \boldsymbol{\varepsilon}$$

from Eq. (8.8) and the strain $\boldsymbol{\varepsilon}$ common to all layers if there is no bending of the laminate, we have

$$\mathbf{N} = \sum_{k=1}^{k=L} t_k\, \boldsymbol{\sigma} = \sum_{k=1}^{k=L} t_k\, \bar{\mathbf{Q}}\, \boldsymbol{\varepsilon}. \tag{8.9}$$

This can be written more briefly:

$$\mathbf{N} = \mathbf{A}\, \boldsymbol{\varepsilon}, \tag{8.10}$$

where $\mathbf{A}$ is the symmetric, in-plane stiffness matrix of the laminate, with

$$A_{ij} = \sum_{k=1}^{k=L} t_k\, \bar{Q}_{ijk}. \tag{8.11}$$

In expanded form, Eq. (8.10) gives us the load–strain relations:

$$\left\{ \begin{array}{c} N_x \\ N_y \\ N_{xy} \end{array} \right\} = \left[\begin{array}{ccc} A_{11} & A_{12} & A_{13} \\ A_{21} & A_{22} & A_{23} \\ A_{31} & A_{32} & A_{33} \end{array} \right] \left\{ \begin{array}{c} \varepsilon_x \\ \varepsilon_y \\ \gamma_{xy} \end{array} \right\} \tag{8.12}$$

The $\mathbf{A}$-matrix is sufficient to calculate the resulting strain in the laminate under given applied load and from this the stresses in the individual layers, again provided that no bending takes place. Under in-plane load, this is so if the laminate is symmetric about its middle plane (for every layer in the upper half, there is an identical layer in the lower half, at the corresponding position in the stacking sequence). Further, if the laminate is balanced (an equal number of identical $+\theta$ and $-\theta$ layers in the laminate), the terms A_{12}, A_{21}, A_{13}, and A_{31} are all zero, and the laminate (not only the individual layers) also has orthotropic properties. This implies that for a balanced laminate, forces N_x and N_y produce no shear strain, and shear force N_{xy} produces no direct strains. The **A**-matrix is actually a sub-matrix of a larger, the so-called **ABD** matrix, where the **D** matrix defines the bending stiffness of the laminate and the **B** matrix the coupling between in-plane and bending deformation for an unsymmetric laminate. Both are derived from the $\bar{\mathbf{Q}}$ matrix for each layer.

Strains in the laminate, under in-plane load, are obtained by inverting Eq. (8.10):

$$\boldsymbol{\varepsilon} = \mathbf{A}^{-1}\mathbf{N}, \tag{8.13}$$

where $\mathbf{A}^{-1}$ is the inverse of the stiffness matrix **A**. Strains can also be calculated by means of the elastic constants of the laminate. For a balanced laminate, Eq. (8.12) can be reduced to:

$$N_x = A_{11}\varepsilon_x + A_{12}\varepsilon_y = \bar{\sigma}_x t,$$
$$N_y = A_{21}\varepsilon_x + A_{22}\varepsilon_y = \bar{\sigma}_y t,$$
$$N_{xy} = A_{33}\gamma_{xy} = \bar{\tau}_{xy} t,$$

where $\bar{\sigma}_x$, $\bar{\sigma}_y$, $\bar{\tau}_{xy}$ are *average* stresses through the thickness of the laminate (not the stress in any particular layer) and t is its thickness. Solving the above equations for ε_x, ε_y and γ_{xy} in terms of the average stresses:

$$\varepsilon_x = \frac{\left(A_{22}\bar{\sigma}_x - A_{12}\bar{\sigma}_y\right)t}{A_{11}A_{22} - A_{12}A_{21}} = \frac{\bar{\sigma}_x}{E_x} - v_{yx}\frac{\bar{\sigma}_y}{E_y},$$

$$\varepsilon_y = \frac{\left(A_{11}\bar{\sigma}_y - A_{21}\bar{\sigma}_x\right)t}{A_{11}A_{22} - A_{12}A_{21}} = \frac{\bar{\sigma}_y}{E_y} - v_{xy}\frac{\bar{\sigma}_x}{E_x},$$

$$\gamma_{xy} = \frac{t}{A_{33}}\bar{\tau}_{xy} = \frac{\bar{\tau}_{xy}}{G_{xy}},$$

from which the following formulae for the 'engineering' elastic constants of the laminate are deduced:

$$E_x = \frac{1}{t}\cdot\frac{A_0}{A_{22}}, \quad E_y = \frac{1}{t}\cdot\frac{A_0}{A_{11}}, \quad G_{xy} = \frac{A_{33}}{t},$$
$$v_{xy} = \frac{A_{12}}{A_{22}}, \quad v_{yx} = \frac{A_{12}}{A_{11}}. \tag{8.14}$$

For convenience, we write

$$A_0 = A_{11}A_{22} - A_{12}^2.$$

The above elastic constants apply *only* to a balanced laminate.

The stress in the individual layers, for a symmetric laminate under in-plane load, is obtained from the strain in the laminate given by Eq. (8.13):

$$\varepsilon = \mathbf{A}^{-1}\mathbf{N}.$$

Alternatively, for a balanced laminate, the strains may more easily be calculated by means of the elastic constants in Eq. (8.14). Either way, the strains in laminate axes have first to be changed back into the different layer axes by the first of Eq. (8.5):

$$\varepsilon_{12} = \mathbf{T}\varepsilon$$

(since we are now rotating from laminate axes to layer axes, θ has to be replaced by $-\theta$ in the T matrix). The corresponding layer stresses are then calculated with the layer properties by Eq. (8.2).

Example 8.1 A balanced, symmetric laminate composed of unidirectional plies of carbon fibre composite with the properties given in Table 8.1[2] has 12 plies at $0°$, 2 pairs of plies at $\pm 45°$ and 4 plies at $90°$, all of thickness 0.125 mm. Calculate the terms of the **A**-matrix and the elastic constants of the laminate.

With 20 plies, the thickness of the laminate is

$$t = 20 \times 0.125 = 2.5\,\text{mm}.$$

Calculate first the minor Poisson's ratio:

$$v_{21} = v_{12} \cdot \frac{E_2}{E_1} = 0.3 \times \frac{10,000}{150,000} = 0.02$$

and the constant μ in Eq. (8.2):

$$\mu = 1 - v_{12}v_{21} = 0.994.$$

Next, we evaluate each term of the **Q**-matrix in Eq. (8.3) by substituting the appropriate elastic constants from Table 8.1:

---

[2]The properties of composite materials vary widely. The data in Table 8.1 is intended only for use in the examples and exercises in this chapter, and should not be taken as design data.

Table 8.1 Material data for a unidirectional carbon fibre composite (for use in examples and exercises in this chapter)

Elastic modulus in fibre direction	$E_1 = 150{,}000$ N/mm^2
Elastic modulus in transverse direction	$E_2 = 10{,}000$ N/mm^2
Shear modulus	$G_{12} = 6000$ N/mm^2
Major Poisson's ratio	$v_{12} = 0.30$
Tensile strength in fibre direction	$X_t = 2000$ N/mm^2
Compressive strength in fibre direction	$X_c = 1200$ N/mm^2
Tensile strength in transverse direction	$Y_t = 80$ N/mm^2
Compressive strength in transverse direction	$Y_c = 200$ N/mm^2
Shear strength	$S = 160$ N/mm^2
Ply thickness	$t_{ply} = 0.125$ mm

$$\mathbf{Q} = \begin{bmatrix} 150{,}905 & 3018 & 0 \\ 3018 & 10{,}060 & 0 \\ 0 & 0 & 6000 \end{bmatrix} \text{N/mm}^2.$$

Since all plies have the same material properties, the above matrix applies to all plies in the laminate, regardless of their orientation. Note that $Q_{21} = Q_{12}$, as required.

We define the x-axis of the laminate as the $0°$ direction, in which case no transformation of axes is necessary for the $0°$ plies. Therefore:

$$\bar{\mathbf{Q}}_0 = \mathbf{Q}_0.$$

For the $+45°$ plies, we require the transformation matrix $\mathbf{T}$ in Eq. (8.6) to transform the ply stiffness into laminate axes. Substituting $\theta = 45°$ in the matrix:

$$\mathbf{T}_{+45} = \begin{bmatrix} 0.5 & 0.5 & 0.5 \\ 0.5 & 0.5 & -0.5 \\ -1.0 & 1.0 & 0 \end{bmatrix}.$$

The formula for the transformed stiffness matrix $\bar{\mathbf{Q}}$ in Eq. (8.8) is

$$\bar{\mathbf{Q}} = \mathbf{T}^T \mathbf{Q}\, \mathbf{T}.$$

The matrix multiplications in the above formula may be performed by hand or more easily by the matrix multiplication function in Excel. We obtain

$$\bar{\mathbf{Q}}_{+45} = \begin{bmatrix} 47{,}750 & 35{,}750 & 35{,}211 \\ 35{,}750 & 47{,}750 & 35{,}211 \\ 35{,}211 & 35{,}211 & 38{,}732 \end{bmatrix} \text{N/mm}^2.$$

Note that the $\bar{\mathbf{Q}}$-matrix no longer has the zero terms in the last row and column. Alternatively, the explicit formulae for $\bar{\mathbf{Q}}$ can be used. For example, for the $\bar{Q}_{11}$ term:

$$\bar{Q}_{11} = \frac{1}{\mu}\left[E_1\cos^4\theta + E_2\sin^4\theta + (2\nu_{12}E_2 + 4\mu G_{12})\sin^2\theta\cos^2\theta\right]$$

$$= \frac{1}{0.994}\left[\frac{150,000}{4} + \frac{10,000}{4} + \frac{2\times0.3\times10,000 + 4\times0.994\times6000}{4}\right]$$

$$= 47,750\,\text{N/mm}^2$$

Similarly, for the $-45°$ plies, we obtain

$$\bar{\mathbf{Q}}_{-45} = \begin{bmatrix} 47,750 & 35,750 & -35,211 \\ 35,750 & 47,750 & -35,211 \\ -35,211 & -35,211 & 38,732 \end{bmatrix}\text{N/mm}^2.$$

This is identical to the matrix for the $+45°$ plies except for the sign of the off-diagonal terms in the last row and column.

For the $90°$ plies, the transformed stiffness matrix is readily obtained by interchanging E_1 and E_2 in the $\mathbf{Q}$-matrix to give:

$$\bar{\mathbf{Q}}_{90} = \begin{bmatrix} 10,060 & 3018 & 0 \\ 3018 & 150,905 & 0 \\ 0 & 0 & 6000 \end{bmatrix}\text{N/mm}^2.$$

The terms of the $\mathbf{A}$-matrix in Eq. (8.10) are now obtained by summing the corresponding terms of the $\bar{\mathbf{Q}}$-matrices above for each of the four ply directions, with the appropriate number of plies and ply thickness, as in Eq. (8.11). For example:

$$A_{11} = 12\times0.125\times150,905 + 4\times0.125\times47,750 + 4\times0.125\times10,060$$
$$= 255,263\,\text{N/mm}.$$

For the complete $\mathbf{A}$-matrix, we find

$$\mathbf{A} = \begin{bmatrix} 255,263 & 23,911 & 0 \\ 23,911 & 114,418 & 0 \\ 0 & 0 & 31,366 \end{bmatrix}\text{N/mm}.$$

Note the zero terms in the last row and column, confirming a balanced laminate.

Finally, the elastic constants can be calculated by Eq. (8.14). With:

$$A_0 = A_{11}A_{22} - A_{12}^2 = 255,263\times114,418 - (23,911)^2 = 28,635\times10^6$$

and $t = 2.5$ mm:

$$E_x = \frac{1}{t} \cdot \frac{A_0}{A_{22}} = \frac{1}{2.5} \cdot \frac{28,635 \times 10^6}{114,418} = 100,106 \, \text{N/mm}^2.$$

Similarly:

$$E_y = 44,871 \, \text{N/mm}^2, \quad v_{xy} = 0.209,$$
$$G = 12,546 \, \text{N/mm}^2, \quad v_{yx} = 0.094.$$

We see that there is a significant modulus E_y in spite of the relatively few plies at 90°. This is due to interaction between plies in the different directions.

The example is continued in Example 8.2, where the tensile strength of the laminate is calculated.   ∎

8.1.3 Failure Criteria

Failure modes in a composite laminate are highly complex, since they involve the fibres, the matrix, and the interface between the two. In addition, the stacking sequence can also influence the failure mode by interaction between the different layers. The actual failure mode depends critically on the type of loading. Within an individual layer, if it is of unidirectional material with a conventional, relatively soft matrix, it is the fibres that provide the main tensile strength in the fibre direction and are mainly responsible for the compressive strength in that direction. The strength in other directions—in the transverse direction or in shear—is largely dominated by the properties of the matrix, resulting in much lower strength values. For a woven material, the strength in the transverse direction is, of course, also provided by the fibres. Due to the different stress components in each layer, arising from their different stiffness properties, each layer has to be checked for failure. However, failure of a particular layer, referred to as 'first ply failure', does not necessarily result in failure of the laminate as a whole, but only in loss of stiffness, the extent of which depends on the mode of failure in that layer. Even so, it may be preferred to avoid initial failure in this way, and here, we shall be concerned principally with first ply failure.

It is well known that ply properties are generally much more variable than those of metallic materials, due to fibre misalignment, broken fibres, voids and other defects. Furthermore, properties are greatly affected by environmental conditions—temperature and water absorption. In particular, this applies to the matrix material. To allow for such uncertainties, reduced strain limits on the laminate as a whole are commonly imposed, as alternative to or as well as a more detailed layer stress analysis. Further, it should be pointed out that lamination theory is concerned only with stresses in the plane of the laminate. Various measures have to be taken to reduce the risk of failure due to stresses through the thickness and the possibility of delamination, as will be discussed later in Sect. 8.1.5.

Just as for metallic materials, reliable ply data has to be obtained from tests on suitable specimens of the material. The specific strength data usually required are the tensile and compressive strengths X_t and X_c in the fibre direction (or in the direction of the weave for a woven material), the tensile and compressive strengths Y_t and Y_c in the transverse direction, and the shear strength S. These values cannot, of course, relate directly to combinations of stress in a given layer. A simple maximum stress criterion—stating that failure occurs when any of the individual stress components in a layer reach its maximum value—generally giving unduly optimistic results. Further, a maximum stress criterion leads to discontinuities in the failure envelope, as one stress component takes over from another, and this is not reflected in actual measurements. To improve on this, various interactive failure criteria have been developed. Here, we shall be concerned with the two most generally accepted—the Tsai–Hill criterion and the Tsai–Wu criterion.

The Tsai–Hill criterion originates from the von Mises criterion for metallic materials, discussed in Chap. 7, modified by Hill [4] for anisotropic materials and applied by Azzi and Tsai [5] to an orthotropic composite. A quadratic function of stress is assumed:

$$F(\sigma_2 - \sigma_3)^2 + G(\sigma_1 - \sigma_3)^2 + H(\sigma_1 - \sigma_2)^2 + 2L\tau_{23}^2 + 2M\tau_{13}^2 + 2N\tau_{12}^2 = 1,$$

to predict failure of a layer. This formula includes those stresses and combinations of stresses that cause *distortion* of the material. The similarity to the von Mises criterion is apparent. The stresses σ_1, σ_2 and σ_3 are in layer axes (with σ_1 in the fibre direction), so are not in general principal stresses. Therefore, the three shear stresses are added. As in the von Mises criterion, this is initially for a three-dimensional state of stress. By substituting:

$$\sigma_1 = X, \quad \sigma_2 = \sigma_3 = Y, \quad \tau_{23} = \tau_{13} = \tau_{12} = S$$

for each stress component applied individually, we obtain six equations for the six unknown constants F, G, H, L, M and N, and after some algebra, the following formula for failure of a layer is obtained:

$$\frac{\sigma_1^2}{X^2} - \frac{\sigma_1 \sigma_2}{X^2} + \frac{\sigma_2^2}{Y^2} + \frac{\tau_{12}^2}{S^2} = 1. \tag{8.15}$$

Note that it is assumed above that the properties of the layer in the transverse direction and through the thickness are the same, that is, $\sigma_2 = \sigma_3$ and $\tau_{23} = \tau_{13} = \tau_{12}$. For this reason, the Tsai–Hill criterion applies in principle to unidirectional plies and not to woven materials. Clearly, the above formula satisfies the basic condition:

$$\sigma_1 = X, \quad \sigma_2 = Y, \quad \tau_{12} = S$$

under the individual stress components, while if applied to an isotropic material ($X = Y$ and $S = X/\sqrt{3}$), the Tsai–Hill criterion in Eq. (8.15) reduces to the von Mises criterion.

However, the quadratic form of the Tsai–Hill criterion means that it is unable to distinguish between tension and compression. Usual practice to allow for differences in tensile and compressive strength is to use X_t for X when σ_1 is positive (tensile) and X_c when σ_1 is negative (compressive) and to use Y_t for Y when σ_2 is positive and Y_c when σ_2 is negative. The signs of σ_1 and σ_2 in the numerator of the second term must, of course, be respected. Due again to the quadratic form of the Tsai–Hill criterion, to calculate the failure load, each applied load should be divided by the *square root* of the value of the Tsai–Hill criterion.

The Tsai–Wu criterion [6], on the other hand, can account for differences in tensile and compressive properties. For an orthotropic layer of composite material, a general second degree expression is adopted:

$$a_{11}\sigma_1^2 + 2a_{12}\sigma_1\sigma_2 + a_{22}\sigma_2^2 + b_{12}\tau_{12}^2 + b_1\sigma_1 + b_2\sigma_2 = 1$$

Note that terms in $\sigma_1\tau_{12}$, $\sigma_2\tau_{12}$ and τ_{12} have been removed from the above formula, because the sign of τ_{12} does not affect the strength of the layer. The linear terms in σ_1 and σ_2 enable the distinction between tension and compression to be made. Substitution of

$$\sigma_1 = X_t \text{ or } -X_c,$$
$$\sigma_2 = Y_t \text{ or } -Y_c,$$
$$\tau_{12} = S$$

in turn under the individual stress components in the above expression gives five of the coefficients:

$$a_{11} = \frac{1}{X_t X_c}, \quad b_1 = \frac{1}{X_t} - \frac{1}{X_c},$$
$$a_{22} = \frac{1}{Y_t Y_c}, \quad b_2 = \frac{1}{Y_t} - \frac{1}{Y_c},$$
$$b_{12} = \frac{1}{S^2}.$$

The remaining term a_{12} has to be found from biaxial test data. However, it is commonly assumed:

$$a_{12} = -\frac{1}{2}\sqrt{a_{11}a_{22}},$$

in which case the Tsai–Wu criterion again reduces to the von Mises criterion when applied to an isotropic material.

As might be expected, neither the Tsai–Hill nor the Tsai–Wu criteria match the experimental data perfectly. Both give generally acceptable predictions when the loading is predominantly tensile, especially in view of uncertainties in the basic material data and the likely presence of defects. When the loading is predominantly compressive, there are larger differences. Use of one or other criterion remains largely a matter of experience.

Example 8.2 Calculate the maximum value of the Tsai–Hill criterion for the laminate in Example 8.1 under a tensile load of 2000 N/mm applied in the 0° direction.

With thickness $t = 2.5$ mm, the average stress in the laminate is

$$\sigma_x = \frac{2000}{2.5} = 800 \, \text{N/mm}^2.$$

For a balanced, symmetric laminate, we make use of the elastic constants E_x, E_y, ν_{xy} calculated in Example 8.1 for the corresponding strains:

$$\varepsilon_x = \frac{800}{100106} = 0.007992,$$
$$\varepsilon_y = -0.209 \times 0.007992 = -0.001670,$$
$$\gamma_{xy} = 0.$$

Alternatively, we could have used Eq. (8.13), requiring inversion of the **A**-matrix, for the same result.

The above strains are also the strains in the 0° plies. For the 90° plies, we have only to interchange ε_x and ε_y. The stresses in these plies are then readily calculated by Eq. (8.2) with the ply data in Table 8.1. We obtain for the stresses in the 0° plies:

$$\sigma_1 = 1201.0 \, \text{N/mm}^2,$$
$$\sigma_2 = 7.3 \, \text{N/mm}^2,$$
$$\tau_{12} = 0,$$

and for the 90° plies:

$$\sigma_1 = -227.9 \, \text{N/mm}^2,$$
$$\sigma_2 = 75.4 \, \text{N/mm}^2,$$
$$\tau_{12} = 0.$$

Note that stress σ_1 in the 90° plies is in compression.

For the $\pm 45°$ plies, we require the transformation matrix already calculated in Example 8.1. The strains in ply axes are given by the first of Eq. (8.5):

$$\varepsilon_{12} = T\varepsilon.$$

Performing the matrix multiplication with the laminate strains above, we obtain

$$\sigma_1 = 486.5 \, \text{N/mm}^2,$$
$$\sigma_2 = 41.3 \, \text{N/mm}^2,$$
$$\tau_{12} = -58.0 \, \text{N/mm}^2.$$

Note that the shear stress in the $\pm 45°$ plies is not zero. The above stresses are for the $\pm 45°$ plies, and for the $-45°$ plies, the sign of τ_{12} is changed.

With the material data in Table 8.1, we calculate now the value of the Tsai–Hill criterion in the different ply directions.

$$0° \text{ plies:} \left(\frac{1201.0}{2000}\right)^2 - \frac{1201.0 \times 7.3}{(2000)^2} + \left(\frac{7.3}{80}\right)^2 = 0.367,$$

$$90° \text{ plies:} \left(\frac{-227.9}{1200}\right)^2 - \frac{(-227.9 \times 75.4)}{(1200)^2} + \left(\frac{75.4}{80}\right)^2 = 0.936,$$

$$\pm 45° \text{ plies:} \left(\frac{486.5}{2000}\right)^2 - \frac{486.5 \times 41.3}{(2000)^2} + \left(\frac{41.3}{80}\right)^2 + \left(\frac{-58.0}{160}\right)^2 = 0.452.$$

It is seen that the greatest value of the Tsai–Hill criterion occurs in the 90° plies, even though no load is applied in this direction. This is again due to the complicated interaction between plies in the different directions. If we increase the applied load by a factor of $\sqrt{\frac{1}{0.936}} = 1.034$, the Tsai–Hill criterion is increased to unity, since all stresses are increased in proportion. The maximum load on the laminate is therefore 2068 N. However, the 12 plies in the 0° direction are alone more than sufficient to carry the applied load. It is unlikely, therefore, that first ply failure in the 90° plies will lead to failure of the laminate as a whole. ∎

8.1.4 Change in Temperature

Change in temperature causes internal stresses in a laminate due to differences in the thermal expansion or contraction of layers with different orientations and perhaps different properties. Commonly, this is due to curing of the laminate at an elevated temperature during manufacture, leaving residual stresses in the laminate. Stresses due to change in temperature can conveniently be calculated in a two-step process. First, it is supposed that thermal expansion (or contraction) is fully restrained. The stress induced in each layer in this way, and the resulting forces on the laminate, is calculated. Equal and opposite forces are then applied to the laminate to counteract those forces, causing stresses in the layers in addition to the previously calculated restraint stresses. The sum of these stresses is the internal stress in a laminate free of external load. Finally, the internal stresses can be added to stresses caused by external loading on the laminate.

The strains ε_1 and ε_2 in an individual layer, if free to expand, due to a temperature increase ΔT are

$$\begin{aligned}
\varepsilon_1 &= \alpha_1 \Delta T, \\
\varepsilon_2 &= \alpha_2 \Delta T,
\end{aligned} \tag{8.16}$$

where α_1 and α_2 are the coefficients of expansion of the layer in the longitudinal and transverse directions. If now the layer is fully restrained, from Eq. (8.2) the stresses needed to counteract the strains ε_1 and ε_2 above are

$$\begin{aligned}
\sigma_1 &= -\left(\frac{E_1}{\mu} \varepsilon_1 + \frac{\nu_{21} E_1}{\mu} \varepsilon_2 \right), \\
\sigma_2 &= -\left(\frac{E_2}{\mu} \varepsilon_2 + \frac{\nu_{12} E_2}{\mu} \varepsilon_1 \right).
\end{aligned} \tag{8.17}$$

Transforming these stresses into laminate axes by the second of Eq. (8.5), we find

$$\begin{aligned}
\sigma_x &= \sigma_1 \cos^2 \theta + \sigma_2 \sin^2 \theta, \\
\sigma_y &= \sigma_1 \sin^2 \theta + \sigma_2 \cos^2 \theta,
\end{aligned}$$

and the resulting forces on the laminate to restrain expansion, from Eq. (8.9), are

$$\begin{aligned}
R_x &= \sum_{k=1}^{k=L} \sigma_x t_k, \\
R_y &= \sum_{k=1}^{k=L} \sigma_y t_k.
\end{aligned}$$

Note that if it is a balanced laminate, there is no resulting shear force on the laminate due to change in temperature. Further, it is assumed that the laminate is symmetric, so that change in temperature does not cause bending of the laminate. To eliminate the restraining forces R_x and R_y, equal and opposite forces:

$$\begin{aligned}
N_{x_t} &= -R_x, \\
N_{y_t} &= -R_y
\end{aligned}$$

have to be applied to the laminate. The forces N_{x_t} and N_{y_t} can be added to any externally applied loads, and the resulting stress in each layer is calculated as in Sect. 8.1.2. Finally, the restraint stresses in Eq. (8.17) have to be added to the stresses obtained above for the total stress in each layer.

As already stated, internal stresses are created in a laminate if it is cured at elevated temperature during manufacture, in which case it is usually taken to be in a stress-free condition at or near the curing temperature. Cooling to normal temperatures induces residual stresses, depending amongst other things on the curing cycle. These residual stresses can be calculated in the same way as other changes in temperature and have to be taken into account in any subsequent strength analysis. In the long term, residual stresses may be reduced by gradual deformation of the matrix material. Moisture absorption by the laminate can cause internal stresses in a similar way to change in temperature. The thermal strains in Eq. (8.16) have then to be replaced by a 'hygrothermal strain', calculated for each layer according to both moisture content and temperature.

8.1.5 Practical Restrictions on Lay-up

This section cannot attempt to summarize all the good design practice built up over many years, but summarizes some aspects of particular relevance to optimization of the lay-up of a composite laminate. Lamination theory is essentially concerned with in-plane stress and strain. However, failure is commonly caused by through-the-thickness stresses, both interlaminar shear stress and peel stress, resulting in delamination. While much research has been devoted to the analysis of these stresses, some well-accepted, practical design 'rules' have emerged to reduce the likelihood of such failures. These are reviewed here and for convenience also summarized in Table 8.2. For more detailed information, the reader is referred to Niu [7] and Kassapoglou [8].

Interlaminar shear stress is the result of transfer of load from one layer to an adjacent one, for example with change in load direction around a hole in the laminate, or some other discontinuity. The magnitude of the load to be transferred is directly related to the number of plies in the layer, which in practice is generally limited to a maximum of four plies per layer. To ensure that transfer of load is to an adjacent layer with adequate properties in the new direction as the local load direction changes, usual practice is that there should not be more than 45° difference in fibre direction between adjacent layers. Further, to avoid undue weakness in any direction in the laminate, the '10% rule' requires that there should be at least 10% of the plies in each of four main directions (e.g. 0°, +45°, −45° and 90°). Delamination can also be caused by unwanted deformations in the laminate, that is, incompatible deformation of two adjacent layers caused, for example, by out-of-plane bending of the laminate, even if subject only to in-plane loads. For this (and other) reasons, it is usually preferred that the lay-up of the laminate should be both symmetric and balanced.

Table 8.2 Design 'rules' for composite laminates

10% rule Ensure at least 10% of each of 4 different ply orientations in the laminate	To avoid undue weakness in any direction
Maximum number of plies Ensure not more than 4 plies of the same orientation in each layer	To limit the magnitude of the load to be transferred to an adjacent layer
Angle between layers Ensure not more than 45° difference in ply orientation between adjacent layers	To ensure load transfer is to an adjacent layer with adequate properties in the necessary directions

While we have been concerned here in principle with those 'rules' that directly affect the lay-up of a laminate, there are many other sources of through-the-thickness stresses that can cause delamination. Ply drop-off—local reduction in the number of plies—also causes interlaminar shear stress, as well as tensile 'peel' stresses. To reduce the effects of both, ply drop-off should best be one ply at a time and should take place gradually. Ply drop-off can better be internally within the laminate rather than at the surface, to help restrain peeling. A free edge also leads to transfer of load between layers, from a $+\theta$ to a $-\theta$ layer, or due to differences in Poisson's ratio, and is best avoided where possible. Voids and other defects can be a source of local delamination. Delamination can also be the direct result of impact on the surface of a laminate. 'Barely visible impact damage', even at relatively low impact levels, can propagate through the laminate, eventually causing failure. Strain limits are commonly imposed to reduce the likelihood of this kind of failure.

8.2 Laminate Optimization

Different methods of optimization of a composite laminate are described, ranging from a simple but practical iterative procedure to more formal optimization methods. The steps necessary to analyse the layer stresses by lamination theory in the earlier parts of this chapter then have to be repeated many times during optimization. The failure criteria described in Sect. 8.1.3 serve as constraints for optimization, while other more general requirements may refer to a maximum strain limitation and other restrictions on lay-up discussed in Sect. 8.1.5. A particular problem in the optimization of composite laminates is the discrete ply thickness. This demands some substantial adaptation of conventional optimization methods, or alternatively, we can turn to what might well be considered unconventional methods like the 'Evolutionary' optimization algorithm now implemented in Solver. However, before going on to study numerical optimization of a composite laminate in more detail, we shall consider first a simplified method of laminate design which, although based on an approximate analysis, can lead to a useful first estimate of the required lay-up. This is by 'netting analysis', in which the ply properties are simplified to the extent that direct hand calculations are readily made, at the same time giving some insight into the lay-up to be obtained by a formal optimization.

8.2.1 Netting Analysis

Netting analysis was introduced by Cox [9] for the analysis of fibrous materials such as paper and can equally be applied to composite laminates. The strength of conventional, unidirectional plies is much greater in the fibre direction than in other directions, due to the relatively weak matrix. In netting analysis, the basic assumption is made that individual plies provide strength *only* in the direction of the fibres. We refer here specifically to unidirectional plies, as netting analysis has limited application for woven materials. Consider first a single layer of unidirectional material, at angle θ to the principal loading directions, as in Fig. 8.4. The stress $s = \sigma_1$ in the fibre direction (using symbol s for stress to indicate netting analysis) can be transformed into stress components in the loading directions simply by neglecting the stresses σ_2 and τ_{12} in Eq. (8.7). If the layer has a thickness t, the corresponding components of load (per unit width) are

$$N_x = s\,t\,\cos^2\theta,$$
$$N_y = s\,t\,\sin^2\theta,$$
$$N_{xy} = s\,t\,\sin\theta\,\cos\theta.$$

If we now have three such layers of unidirectional plies at angles θ_1, θ_2, θ_3 with thicknesses t_1, t_2, t_3 and stresses s_1, s_2, s_3, respectively, the resulting load components become

$$N_x = s_1\,t_1\,\cos^2\theta_1 + s_2\,t_2\,\cos^2\theta_2 + s_3\,t_3\,\cos^2\theta_3,$$
$$N_y = s_1\,t_1\,\sin^2\theta_1 + s_2\,t_2\,\sin^2\theta_2 + s_3 t_3\,\sin^2\theta_3, \tag{8.18}$$
$$N_{xy} = s_1\,t_1\,\cos\theta_1\,\sin\theta_1 + s_2\,t_2\,\cos\theta_2\,\sin\theta_2 + s_3\,t_3\,\cos\theta_3\,\sin\theta_3.$$

The three ply stresses s_1, s_2, s_3 can be solved for any combination of loads N_x, N_y, and N_{xy} if the layer thicknesses and ply angles are already given. This means that a three-fibre system (three distinct fibre directions) is sufficient provided, of course, that the individual plies are strong enough for the resulting stress in their respective fibre directions. In fact, we now have a solution based entirely on equilibrium (not surprisingly with much in common with some of the simple truss structures earlier in this book). In this way, we have bypassed virtually the whole of lamination theory, with no recourse to strains in the laminate to calculate the stresses in the plies. At the same time, we have ignored transverse and shear stresses in the plies which, as we have seen, can play a significant role in the failure criterion.

For a laminate to carry a given set of in-plane loads, with chosen fibre angles $\theta_1, \theta_2, \theta_3$ and ply strengths X_t and X_c (in tension and compression) in the fibre direction, we can substitute $s_1, s_2, s_3 = X_t$ or X_c as appropriate in Eq. (8.18) to calculate the required layer thicknesses t_1, t_2, t_3. By analogy with a statically determinate truss, this might be regarded as a 'fully stressed design'.

Example 8.3 A $0°$, $\pm 45°$ laminate has to carry loads $N_x = 2400$ N/mm, $N_y = 400$ N/mm, $N_{xy} = 800$ N/mm. The strength of the unidirectional ply material in the fibre direction is $X_t = 2000$ N/mm^2 in tension and $X_c = -1200$ N/mm^2 in compression. Use netting analysis to calculate the required layer thicknesses.

By substituting $\theta_1 = 0°$, $\theta_2 = +45°$, $\theta_3 = -45°$ in Eq. (8.18), these reduce to:

$$s_1 t_1 = N_x - N_y,$$
$$s_2 t_2 = N_y + N_{xy},$$
$$s_3 t_3 = N_y - N_{xy}.$$

Putting now $s_1 = X_t$, $s_2 = X_t$ and $s_3 = X_c$ (the choice of X_t or X_c in each case is made to avoid negative thicknesses), we obtain

$$t_1 = \frac{N_x - N_y}{X_t} = \frac{2400 - 400}{2000} = 1.0 \, \text{mm},$$
$$t_2 = \frac{N_y + N_{xy}}{X_t} = \frac{400 + 800}{2000} = 0.6 \, \text{mm},$$
$$t_3 = \frac{N_y - N_{xy}}{X_c} = \frac{400 - 800}{-1200} = 0.33 \, \text{mm},$$

with a total thickness of 1.93 mm.

Clearly, the results obtained by netting analysis are entirely dependent on the choice of fibre directions, which have to be chosen appropriately. Optimum fibre angles, at least within the assumption of netting analysis, can be deduced by transforming the loading into components N_1 and N_2 in their principal directions. (Note that loads per unit length N_x, N_y, N_{xy} on a supposed unit thickness can be treated simply as stress components.) Using the standard formula for principal stresses, we obtain

$$N_{1,2} = \frac{2400 + 400}{2} \pm \sqrt{\left(\frac{2400 - 400}{2}\right)^2 + (800)^2},$$

$$N_1 = 2681 \, \text{N/mm}, \quad N_2 = 119 \, \text{N/mm}$$

Fig. 8.4 Netting analysis—
stress s in the fibre direction

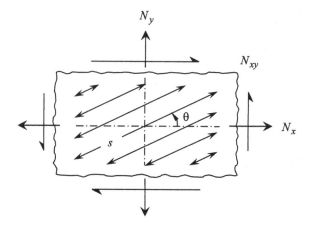

at angles:

$$\theta = \frac{1}{2}\tan^{-1}\left(\frac{2 \times 800}{2400 - 400}\right),$$

$$\theta = 19.3° \quad \text{and} \quad -70.7°.$$

Both N_1 and N_2 are in tension. The required thicknesses if the plies are placed with their fibres along the principal directions are then:

$$t_1 = \frac{N_1}{X_t} = \frac{2681}{2000} = 1.34\,\text{mm},$$

$$t_2 = \frac{N_2}{X_t} = \frac{119}{2000} = 0.06\,\text{mm}.$$

This is now the optimum laminate for the given applied loading, with a thickness of 1.40 mm. This is a reduction of 0.53 mm compared with the previous netting analysis design. However, since there are now only two fibre directions in the laminate, it is unable (within the assumption of netting analysis) to carry any loads other than those applied along the same principal directions. It has to be concluded, therefore, that such a laminate would be highly impractical. ∎

As we have seen, netting analysis is limited to at most three fibre directions, but in practice, we are unlikely to be satisfied with such a laminate. The '10% rule' from Sect. 8.1.5 requires a laminate to have at least four different fibre directions. However, with four fibre directions, we can no longer make use of the simple equilibrium conditions in Eq. (8.18), and a true netting analysis solution is then not possible. Even so, we can still take advantage of the concept of netting analysis for a more practical solution in the following way. If we confine our attention to a 0°, ±45°, 90° lay-up, we make the assumption that fibres in the 0° direction are *entirely* responsible for the load N_x and fibres in the 90° direction for load N_y. The shear load N_{xy} can be resolved into principal components of the same magnitude in the $\pm 45°$ directions, and it is the fibres in those directions that take the shear load. The method is best illustrated in the following example.

Example 8.4 Calculate the required layer thicknesses for a $0°$, $\pm 45°$, $90°$ laminate to carry the same loads as in the previous example, again by netting analysis, using the same material data but taking into account now a ply thickness $t_{ply} = 0.125$ mm. The chosen laminate should be balanced and symmetric.

The required thickness of $0°$ plies is

$$t_0 = \frac{N_x}{X_t} = \frac{2400}{2000} = 1.20 \, \text{mm} \quad \rightarrow \quad 10 \, \text{plies at } 0°$$

after rounding up to a whole number of plies. Similarly, the thickness of $90°$ plies is

$$t_{90} = \frac{N_y}{X_t} = \frac{400}{2000} = 0.20 \, \text{mm} \quad \rightarrow \quad 2 \, \text{plies at } 90°.$$

The applied shear load is resolved into tensile and compressive loads of 800 N/mm in the directions of the $\pm 45°$ plies. For a balanced laminate, we require an equal number of plies in the $\pm 45°$ directions, and the required ply thickness has therefore to be based on the lower compressive ply strength:

$$t_{+45} = t_{-45} = \frac{N_{xy}}{X_c} = \frac{800}{1200} = 0.67 \, \text{mm} \quad \rightarrow \quad 6 \, \text{plies}$$

in each of the $\pm 45°$ directions. With an even number of plies in each direction, no change is necessary for a symmetric laminate. The final laminate has a total of 24 plies, with a thickness of 3.00 mm. This can be interpreted as the following lay-up[3]:

$$[\pm 45/0_3/\pm 45/90/\pm 45/0_2]_s,$$

conforming to the practical restrictions on lay-up in Sect. 8.1.5. The thickness of the laminate is significantly greater than in the previous example as a consequence of the preferred four fibre directions and the discrete ply thickness. ∎

It has to be emphasized that netting analysis neglects the interaction between the different ply directions, in particular the transverse and shear stresses induced in each layer. It was already seen in Sect. 8.1.3 that these play an important role in the failure criterion (whether this is Tsai–Hill, Tsai–Wu or some other). The laminate thickness calculated by netting analysis can, therefore, prove in some cases to be a substantial underestimate of the actually required thickness. As will be found in Example 8.5 in the following section, this is also the case in the present example. The most common use of netting analysis is then as an initial estimate for further

---

[3]See 'Principal Notation' at the beginning of this book.

design and optimization using full lamination theory, such as by the iterative redesign procedure in the following section or by the formal numerical optimization in Sects. 8.2.3 and 8.2.4. However, we shall see that an optimized laminate commonly has no more than three fibre directions, justifying to some degree the concept of netting analysis.

8.2.2 Iterative Redesign

The procedure described in the previous section, based on netting analysis, can give a useful initial design of laminate, but the analysis is of course a severe simplification of standard lamination theory. An initial netting analysis design can be refined by an iterative redesign procedure, employing full lamination theory for an accurate analysis. The method described here follows closely that proposed by Morton and Webber [10] which, as well as being for in-plane loading as here, deals extensively with laminate selection under bending. This method may be reminiscent of the simple truss design in Chap. 1, but with some essential differences. As in the last example, we will consider only a $0°$, $\pm45°$, $90°$ balanced, symmetric laminate under in-plane load. Again, it is assumed that the laminate is composed entirely of unidirectional plies, although a similar procedure can be adopted for woven materials.

In this iterative redesign procedure, the laminate is analysed, and the value of a chosen failure criterion is calculated for each layer. In the example that follows, the Tsai–Wu criterion is used. The layer with the largest value of the failure criterion under any loading case is the most critical layer (assuming that at least one layer has a value greater than unity). The Tsai–Wu criterion has a single value for each layer, but in an attempt to identify a so-called principal cause of failure, the stress ratios σ_1/X, σ_2/Y and τ_{12}/S are calculated for this most critical layer (using tensile or compressive values $X = X_t$ or X_c and $Y = Y_t$ or Y_c as appropriate). Failure indicated by each of these stress ratios is referred to as longitudinal, transverse and shear failure, respectively. The first of these might be regarded as fibre failure, the other two as matrix failure. The largest of the three stress ratios is identified, and this is treated as the dominant failure mode. Either the layer which is found to be the most critical or another layer is thickened according to a certain set of rules based on the dominant failure mode, as will be discussed shortly. The amount of thickening can be based on the appropriate stress ratio above, but it may be more convenient simply to increase the thickness by one or more whole plies. The new laminate is then analysed again, and the process repeated.

It will be observed that the procedure as described here is essentially one of progressively *increasing* the strength of a laminate until it reaches the required strength. For this reason, it is desirable that the initial design of laminate should be understrength, not overstrength. In the latter case (if the failure criterion is less than unity in all layers), the laminate could be reduced in thickness by removing plies from the layer with the *smallest* value of the failure criterion. However, for sufficient freedom to converge to a satisfactory laminate, it is recommended to choose an initial design in which most or all layers have a failure criterion greater than unity. In this way, a better selection of the number of plies in each layer is likely to be made.

The redesign rules for which layer to thicken are an essential part of this procedure. Unlike a truss structure, for which each member is increased or reduced in area according to the stress in that same member, for a laminate a different approach is necessary. Although the analysis is now based on full lamination theory, the redesign rules borrow from netting analysis the concept that individual plies provide strength only in their own fibre direction. If, therefore, we have longitudinal failure in the 0° plies (σ_1/X is largest), then it is the thickness of those layers that should be increased. If, however, we have transverse failure in the 0° plies (σ_2/Y is largest), then it is the thickness of the 90° layers that should be increased, since this relieves the transverse stress in the 0° layers. Shear failure in the 0° plies (τ_{12}/S is largest) means that it is the ±45° layers that should be increased. Failure in the 90° plies is treated similarly. The ±45° plies provide the shear strength. If we have failure in the ±45° plies, then either the ±45° or the 0°/90° layers should be increased, according to the failure mode in the ±45° plies. These rules are summarized in Table 8.3.

Example 8.5 Repeat Example 8.4 by the iterative redesign procedure, with the same material and ply thickness, and under the same loading.

An initial lay-up is chosen in Table 8.4 with the same number of plies as the netting analysis solution in Example 8.4 in each of the 0°, ±45°, 90° directions. This lay-up has a maximum Tsai–Wu value of 1.68 (but less than unity in the 0° layer). The number of plies is progressively increased in the appropriate layer, according to the rules in Table 8.3. The spreadsheet program 'Composite Laminate' described later in Sect. 8.3.1 is used to analyse the laminate, giving Tsai–Wu values for each layer. Values of the stress ratios σ_1/X, σ_2/Y and τ_{12}/S are calculated from

Table 8.3 Redesign rules

Layer type (°)	Dominant failure mode	Layer type to be increased (°)
0	Longitudinal	0
	Transverse	90
	Shear	±45
±45	Longitudinal	±45
	Transverse	±45
	Shear	0/90[a]
90	Longitudinal	90
	Transverse	0
	Shear	±45

[a]Layer type with the smaller number of plies

Table 8.4 Iterative redesign procedure in Example 8.5

$\theta°$	Number of plies	Largest Tsai–Wu value	Dominant failure mode	Action
0	10	1.68 in ±45° plies	Transverse $\sigma_2/Y_t = 1.20$	Increase ±45° plies to 8[b]
±45[a]	6			
90	2			
0	10	1.41 in 90° plies	Transverse $\sigma_2/Y_t = 1.05$	Increase 0° plies to 11[c]
±45	8			
90	2			
0	11	1.26 in 90° plies	Transverse $\sigma_2/Y_t = 0.98$	Increase 0° plies to 12
±45	8			
90	2			
0	12	1.15 in ±45° plies	Transverse $\sigma_2/Y_t = 0.92$	Increase ±45° plies to 10
±45	8			
90	2			
0	12	1.08 in 90° plies	Transverse $\sigma_2/Y_t = 0.86$	Increase 0° plies to 13
±45	10			
90	2			
0	13	0.98 in 90° plies	Transverse $\sigma_2/Y_t = 0.81$	No further action
±45	10			
90	2			

[a]The number of ±45° plies is the number in each of the ±45° and −45° directions
[b]An even number of ±45° plies are required for a symmetric laminate
[c]An odd number of 0° plies are permissible for a symmetric laminate, if located on the middle plane and provided that the number of 90° plies is even

the stresses found by the spreadsheet for the layer with the largest Tsai–Wu value, to identify the dominant failure mode. The action to be taken at each following step is noted in Table 8.4. The number of plies is increased at each step by the minimum number to preserve a balanced, symmetric laminate (a single 0° or 90° ply or two +45° plies together with two −45° plies). It will be seen that in this example, the dominant failure mode is transverse at every step, as is commonly the case, illustrating the fact that while failure has occurred in a particular layer, it is frequently a different layer that has to be increased. The procedure converges to a laminate thickness of 4.375 mm after five iterations, with a Tsai–Wu value of 0.98.

At the end of this procedure, it will be seen that there has been no increase in the number of 90° plies (even though these have several times shown the largest Tsai–Wu value). It might be, therefore, that the number of 90° plies could be reduced. If doing so leads to a Tsai–Wu value again greater than unity in any layer, then the procedure already followed has to be continued. In the present example, reducing the number of 90° plies from two plies to one would require an increase in the number of 0° plies from 13 to 14 to maintain a symmetric laminate, with no reduction in thickness (an odd number of both 0° and 90° plies are not allowed).

The final laminate, at the end of Table 8.4, can be interpreted as the following lay-up:

$$[\pm45_2/0_3/\pm45/90/\pm45/0_2/\pm45/0/\bar{0}]_s.$$

The thickness of the laminate is considerably greater than that found in Example 8.4, which has a thickness of 3.00 mm. However, the laminate in Example 8.4 has a Tsai–Wu value of 1.68, meaning that it is around 30% under strength. This arises from neglect of transverse and shear stresses in the netting analysis solution and accounts for greater thickness in the present example.   ∎

8.2.3   Numerical Optimization

The iterative redesign procedure of the previous section will in most cases provide a satisfactory laminate, in other words one satisfying the specified failure criterion in all layers without undue use of material. However, it does not necessarily yield an optimum laminate, that is, one with the true minimum number of plies. Furthermore, it does not optimize ply angles. Turning now to formal numerical optimization, a particular problem that arises is the discrete ply thickness, requiring special methods of optimization which can deal with discrete variables. With discrete ply thickness, there may be a number of alternative laminates, all of the same thickness but with different ply angles and numbers of plies in each layer, so that there is no single, unique optimum. This can occur when none of those laminates has a thickness close to what would be the optimum if the discrete ply thickness were ignored, leaving sufficient margin for alternatives to exist. It is also worth noting that alternative laminates having the same thickness and satisfying or more than satisfying the failure criterion are nevertheless unlikely to have the same margin of safety. Furthermore, due to the complex interactions between different ply directions in lamination theory and the many possible combinations of ply angles and numbers of plies, local minima may exist with altogether different lay-up and not necessarily the same number of plies. If this is suspected, optimization should be repeated from different starting points until it is clear that the true optimum has been found. The 'Multistart' option in the GRG method in Solver, described further in the Appendix, may offer an outcome here, although the time taken to reach an optimum is greatly increased. Some of these aspects will be explored further in the examples that follow later in this section.

Solver provides an option for optimization with integer variables (in this case a whole numbers of plies) by the 'branch-and-bound' method. The original algorithm was developed by Land and Doig [11]. Useful descriptions of the method can be found in Gürdal et al. [12], Reklaitis et al. [13] and various other texts. This is a logical adaptation of a conventional optimization routine, by which it is possible to extract from it an integer solution. In effect, it searches all combinations of integer variables, eliminating as quickly as possible those which cannot lead to the required optimum. For simplicity, it will be assumed that all variables are integer variables, although the method works in the same way with mixed continuous and integer variables. Assuming that we are searching for a minimum value (here, it is the minimum number of plies) rather than a maximum, the method works as follows. First, an optimization is performed with all-continuous variables. Since integer constraints on the variables can only worsen the result, the minimum found in this optimization must be a *lower bound* to the solution of the integer variable problem. One of the variables is then selected, and with this variable, two new sub-problems are defined. In one of these, the selected variable is constrained to be less than or equal to the first integer value below its current value, and in the other, the same variable is constrained to be greater than or equal to the first integer value above its current value. This is what is meant by 'branching'. Each of these sub-problems is then optimized with the added constraint. Branching is continued by selecting further variables until a feasible, all-integer solution is obtained. This then becomes an *upper bound* to the true integer optimum. However, it does not rule out other all-integer solutions with a minimum between the newly found upper bound and the earlier lower bound. Branching is continued by branching from the so-called nodes at the ends of existing branches. From this point onwards, any branch that leads to a minimum value greater than the upper bound can be terminated. This is what is meant by 'bounding'. A branch is also terminated if no feasible solution to the sub-problem can be found. If a better all-integer solution is found, the previous upper bound can be updated. Branching is continued until all branches have been terminated—or 'fathomed', as it is called. The required optimum is the best integer solution at one of the fathomed nodes. Solver applies the branch-and-bound method to both the GRG nonlinear and the simplex LP methods, but it will be clear that optimization of the sub-problems has to be performed many times and computing times will be much longer than for a similar continuous variable problem.

An alternative method implemented in Solver is the 'evolutionary' or 'genetic' algorithm. This is a semi-random procedure intended in particular for problems with non-smooth or discontinuous functions. Since a genetic algorithm can work directly with integer variables, it is of course suited to problems such as that of a composite laminate. Furthermore, since the method explores the whole extent of the design space, it is more likely to detect local optima. The genetic algorithm is described in the following section, but first we shall examine use of the GRG method in Solver with the option for integer variables for laminate optimization in the examples that follow.

Example 8.6 Use the spreadsheet program 'Laminate Optimization' to optimize the $0°$, $\pm45°$, $90°$ laminate in Example 8.5, with the same ply material and under the same loading.

The Tsai–Wu criterion is chosen on the spreadsheet, and constraints are entered in the Solver dialogue box to limit the Tsai–Wu values in all plies to less than or equal to unity, to limit numbers of plies n_i to positive or zero values and for discrete ply thickness by setting n_i to integer. Further constraints are added to set ply angles $\theta_1 = 0°$, $\theta_2 = 45°$ and $\theta_3 = 90°$; alternatively, the ply angles could simply be removed from the set of design variables. All options in Solver are left at their default values. The initial number of plies is chosen to be the same as found in Example 8.5. Results of the optimization are shown in the first row of Table 8.5, with an optimum laminate thickness of 4.00 mm (note that n_i refers to the number of ply *pairs*, so that the laminate in the first row of Table 8.5 consists of 32 plies; the seven 0° plies implies seven at +0°, seven at −0°, making fourteen 0° plies in total, and the nine plies at ±45° being nine plies in each of these directions). The lay-up is reduced to only three fibre directions, as normally occurs when only a single loading case is specified, bringing to mind the earlier netting analysis with three distinct fibre directions being sufficient for equilibrium. However, this lay-up with an odd number of ±45° plies cannot be symmetric (note that the program ensures a balanced laminate, but not necessarily a symmetric one). Therefore, an additional constraint is added to set the number of ±45° to an even number above or below its current value, and the optimization is repeated. The result of this second optimization, with ten ±45° plies, is shown in the second row of Table 8.5. The thickness is increased to 4.25 mm, with no further change in lay-up. However, by examining the Tsai–Wu values in each of the ply directions, it will be noticed that the maximum Tsai–Wu value occurs in the 90° ply direction, in which the number of plies has been reduced to zero! For this reason, the optimization is repeated again, as in the third row of Table 8.5, by changing the constraint $\theta_3 = 90°$ to $\theta_3 = 45°$ (or 0°) as a means of removing the 90° ply direction. The thickness is reduced again to 4.00 mm, but with a significantly different lay-up to that found in the previous optimizations (first two rows of Table 8.5) and with a reduced maximum value of the Tsai–Wu criterion. It is also worth noting that the Tsai–Wu value of the 0° plies remains relatively low.

There is a specific explanation for this behaviour. The 90° ply direction is retained in the solution, even though there are no actual plies in this direction. Values of strain are calculated for the 90° direction, giving rise to 'apparent'

Table 8.5 Results using the GRG nonlinear method in Example 8.6

Ply angle θ_i (°)	Number of ply pairs n_i	Laminate thickness (mm)	Tsai–Wu value
0	7	4.00	0.204
±45	9		0.972
90	0		0.978
0	7	4.25	0.171
±45	10		0.874
90	0		0.960
0	4	4.00	0.171
±45	12		0.874

stresses in this ply direction. The Tsai–Wu criterion for the 'missing' 90° plies is still calculated and is incorrectly included in the optimization, leading to what may be a 'false' optimum. At the same time, this may serve as a warning that excessive strains can occur in directions in which there are no fibres, justifying the 'rule' in Sect. 8.1.5 that in practice there should be a minimum of 10% of plies in each of the four directions.

The laminate in the third row of Table 8.5 can be interpreted as the following lay-up:

$$[\pm 45_3/0_3/\pm 45_3/0]_s,$$

needing in this case no further modification for a symmetric laminate. The present laminate has three plies less than the final laminate in Example 8.5. However, in both cases, it is likely that a number of 90° plies would have to be added for a more practical laminate, which may well reduce the difference between the two. Finally, if the laminate is now optimized with the 'integer' constraint for discrete ply thickness removed (as before retaining only the 0° and ±45° ply directions), a minimum laminate thickness of 3.813 mm is obtained—showing in this case a relatively small penalty for discrete ply thickness.   ∎

Example 8.7 Use the spreadsheet program 'Laminate Optimization' with the GRG Nonlinear method in Solver to optimize a laminate under the following alternative loading:

$$\begin{aligned} \text{case 1}: \quad & N_x = 2400, \quad N_y = 800 \quad N_{xy} = 0\,\text{N/mm}, \\ \text{case 2}: \quad & N_x = 0, \quad N_y = 1600, \quad N_{xy} = 400\,\text{N/mm}. \end{aligned}$$

with no restriction on ply angles. The properties of the ply material and the ply thickness are the same as in the previous examples.

Note the conflicting nature of the two load cases—each on its own would require a very different type of laminate. When more one than loading case is defined, the program finds the minimum thickness of laminate that will satisfy *all* loading cases when applied individually. From different starting points, three alternative local optima are found, as shown in Table 8.6, all with a laminate thickness of 3.50 mm but with different Tsai–Wu values. The first two of these have reduced to three sets of plies, while the third has retained four distinct ply directions, with a different distribution of plies in each case. Due to the discrete thickness, there will be a range of ply angles in the vicinity of those in Table 8.6 for the same number of plies. ∎

8.2.4 Genetic Algorithm

A genetic—or evolutionary—algorithm is a semi-random procedure developed to deal with discontinuous or otherwise irregular problems for which gradient-based

Table 8.6 Results using the GRG Nonlinear method in Example 8.7

Ply angle θ_i (°)	Number of ply pairs n_i	Laminate thickness (mm)	Maximum Tsai–Wu value
0	6	3.50	0.984
±66.1	8		
0	7	3.50	0.946
±75.6	7		
0	7	3.50	0.947
±74.7	6		
90.0	1		

methods are inappropriate. Commonly, these are problems with multiple local minima requiring a wider search over the whole design space. The name derives directly from the fact that the method imitates the natural evolutionary process. Unlike a conventional optimization routine, a genetic algorithm retains a sufficiently large 'population' of designs covering more effectively the complete design space and increasing the likelihood of detecting different local optima. Furthermore, a genetic algorithm can work in discrete variables from the outset and is therefore appropriate for the optimization of composite laminates with whole numbers of plies. Starting from a randomly chosen set of designs, by exchanging properties—or 'genes'—between different members of the population, new designs are created, some better and some worse. For this, the selection of pairs of designs between which genes are to be exchanged is biased towards the already 'good' designs, thus emulating Darwin's principle of 'survival of the fittest'. Genetic algorithms were originally developed by Holland [14]. The description of their application to composite laminates below follows that of Gürdal et al. [12].

The number of different designs to be retained in a genetic algorithm varies with the number of design variables. Design variables representing each of those designs are commonly converted first into binary form. Since we are using discrete variables, this is possible with only a limited number of digits, depending on the range of values specified for the design variables. For a composite laminate under in-plane load, the design variables are simply the number of plies at each ply angle. If the ply angles are also to be treated as variables, they can be represented in discrete form as well, in other words by a chosen set of discrete ply angles. As before, under in-plane load, the stacking sequence is irrelevant provided that the lay-up remains symmetric, and the number of variables is then significantly reduced. Design variables representing a single design are assembled sequentially into a single string regarded as a 'genetic chain', as indicated in Fig. 8.5.

Initial designs are produced by a random number generator, within the appropriate range of each design variable. These are then assessed for their 'fitness', meaning in fact how 'good' the design is. Conventionally, we speak of maximizing fitness rather minimizing some other objective function. For a composite laminate, fitness relates to the total number of plies n_{lam}, in the laminate—the smaller the number of plies, the greater the fitness. Fitness f is then defined as:

Fig. 8.5 Genetic chains—
alternative designs

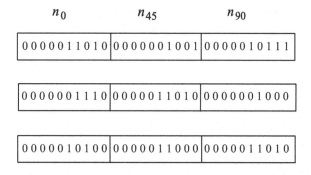

Fig. 8.6 Augmented fitness
function

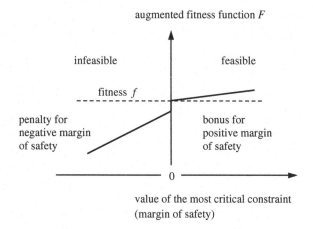

$$f = c - n_{\text{lam}}$$

where the constant c is chosen to ensure that f remains positive. However, not all randomly generated designs will be feasible, so a penalty function is introduced to penalize infeasible designs. A simple form of penalty function can be represented as shown in Fig. 8.6. The augmented fitness F is the sum of the original fitness f and the penalty function. On the infeasible side of the diagram, a penalty is applied for a negative margin of safety, in other words a negative constraint value. This refers to the most critical constraint for that design. In addition, the small step in the penalty function is to ensure that *all* infeasible designs carry some small penalty. The 'bonus' on the feasible side of the diagram is to reward designs of equal fitness but with a higher margin of safety. As with all penalty functions, some degree of tuning of the parameters involved is usually necessary to achieve the best results. Parameters should be chosen so that an infeasible design can never have an augmented fitness value greater than the optimum design, but should not be so severe that a design cannot return from the infeasible region, since some of these may well be close to the optimum.

Selection of pairs of designs between which genes are to be exchanged is on the basis of their augmented fitness. The most fit 'parent' designs are selected the most often, to increase the chance of producing even fitter 'child' designs. This process can be visualized in terms of the 'roulette wheel' in Fig. 8.7. The size of the different sectors of the wheel is in proportion to the augmented fitness F of each design, thus favouring the most fit. Some parent designs may not be selected at all and others selected more times. Interchange of genes within the selected pair is illustrated in Fig. 8.8. A random cut-off point is located, and 'crossover' between the two is performed. One or both of the parent designs is replaced by one or both of the child designs, and the process continued until a complete new generation has been obtained. The usual procedure also includes occasional 'mutation' of designs, that is to say, randomly changing one digit in the string. Less fit designs may well have characteristics that would otherwise be lost in the selection process, and mutation increases the chance of recovering these. Throughout the procedure, the number of designs in the original population is retained, but each generation of new designs will include some improved ones. The process is continued until no improvement can be found in the best of the designs.

It will be apparent that the procedure typically requires a very large number of constraint evaluations to find optimum or near-optimum designs, but by choosing a sufficiently large population, it is able to locate all or some of the local optima, from which the true optimum can be selected. When several designs of equal fitness

Fig. 8.7 'Roulette wheel'

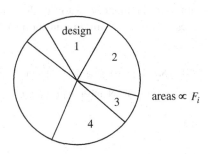

Fig. 8.8 Interchange of genes

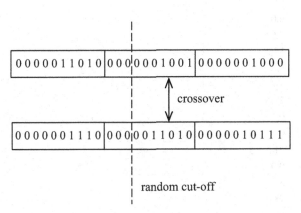

emerge, the design with the best margin of safety, or alternatively one that might be considered the most practical, can be chosen. A version of the genetic algorithm, with various enhancements, is now implemented in Solver. Use of this method for laminate optimization in the 'Laminate Optimization' program will be described further in the example that follows.

Example 8.8 Use the Evolutionary method in Solver to repeat the optimization in Example 8.7.

Design variables are the ply angles $\theta_1, \theta_2, \theta_3$, which can take any values, and the numbers of ply pairs n_1, n_2, n_3, constrained to integer values. In the evolutionary method, upper and lower bounds have to be set on the variables, to limit the extent of the design space. Ply angles are constrained to $0 \leq \theta_i \leq 90°$ and further constraints added to define bounds $0 \leq n_i \leq 10$ for this problem. All options in Solver are left at their default values. The evolutionary method in Solver is chosen simply by selecting 'Evolutionary' in the Solver dialogue box.

When the procedure is completed, Solver selects the best from its population of designs. By setting the Random Seed in Options to zero, a different initial population of designs is generated each time an optimization is performed. A selection of results of repeated optimizations is shown in Table 8.7. In one solution, a laminate thickness of 3.75 mm is obtained and in all others a thickness of 3.50 mm, in agreement with Example 8.7. The number of 0° (or near 0°) ply pairs varies from five to seven (10–14 plies), with a wide range of other ply angles. The results in Table 8.7 illustrate the large number of alternative lay-ups that can exist for an optimum laminate, due to the discrete ply thickness and in many cases due to the presence of local minima in the problem itself.   ∎

Table 8.7 Results using the evolutionary method in solver in Example 8.8

Ply angle θ_i (°)	Number of ply pairs n_i	Laminate thickness (mm)	Maximum Tsai–Wu value
0.0	6	3.50	0.990
65.4	4		
71.2	4		
0.0	7	3.50	0.948
72.4	5		
87.2	2		
0.0	2	3.50	0.998
1.7	4		
70.4	8		
0.1	3	3.50	0.997
0.7	4		
71.5	7		
0.0	5	3.75	0.980
38.2	4		
78.5	6		

8.3 Spreadsheet Program

Lamination theory and different methods of applying this to laminate optimization have been discussed in the previous sections. Here, a spreadsheet is presented to perform the numerical optimization of a composite laminate under any in-plane loading, by either the GRG Nonlinear method or the Evolutionary method in Solver.

8.3.1 'Composite Laminate'

The spreadsheet uses Solver to optimize the composite laminate in Fig. 8.9, consisting of individual layers of unidirectional plies. The laminate is defined as balanced and symmetric, with three pairs of fibre directions. These may be distributed between the layers in any required manner, only provided that the laminate remains symmetric to avoid unwanted bending deformation. Fibre directions may be fixed or variable. A discrete ply thickness may be specified. The spreadsheet is shown in Fig. 8.10.

8.3.1.1 Modelling

Analysis of the laminate is by standard lamination theory, as in Sect. 8.1, based on the elastic properties of the ply material and ply strength data entered by the user. For a given number of plies in each of the three fibre directions, the spreadsheet

Fig. 8.9 Fibre angles in a composite laminate

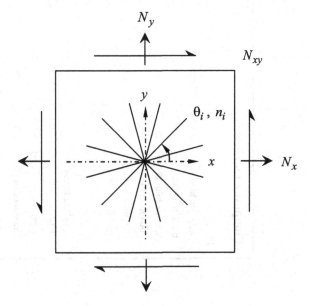

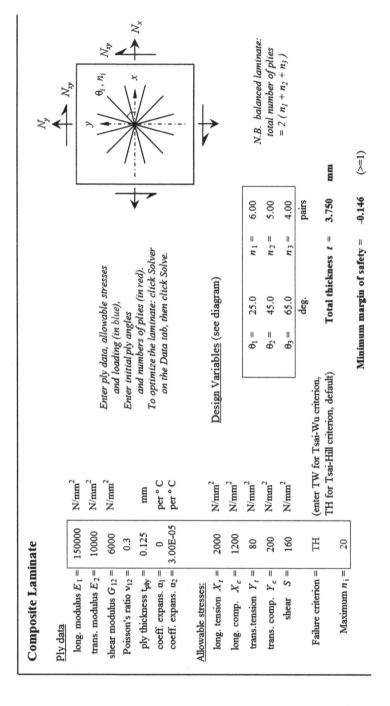

Composite Laminate

Ply data

long. modulus E_1 =	150000	N/mm^2	
trans. modulus E_2 =	10000	N/mm^2	
shear modulus G_{12} =	6000	N/mm^2	
Poisson's ratio ν_{12} =	0.3		
ply thickness t_{ply} =	0.125	mm	
coeff. expans. α_1 =	0	per ° C	
coeff. expans. α_2 =	3.00E-05	per ° C	

Allowable stresses:

long. tension X_t =	2000	N/mm^2
long. comp. X_c =	1200	N/mm^2
trans.tension Y_t =	80	N/mm^2
trans. comp. Y_c =	200	N/mm^2
shear S =	160	N/mm^2

Failure criterion =   TH   (enter TW for Tsai-Wu criterion,
                            TH for Tsai-Hill criterion, default)

Maximum n_i =   20

*Enter ply data, allowable stresses
and loading (in blue),
Enter initial ply angles
and numbers of plies (in red).
To optimize the laminate: click Solver
on the Data tab, then click Solve.*

Design Variables (see diagram)

θ_1 =	25.0	n_1 =	6.00
θ_2 =	45.0	n_2 =	5.00
θ_3 =	65.0	n_3 =	4.00
	deg.		pairs

Total thickness t =   **3.750**   **mm**

Minimum margin of safety =   -0.146   (>=1)

*N.B. balanced laminate:
total number of plies
= 2 (n_1 + n_2 + n_3)*

Fig. 8.10 Spreadsheet 'Composite Laminate'

calculates the terms of the **A**-matrix in Visual Basic function Stiffness and the elastic constants. Up to four loading cases may be defined, for each of which the spreadsheet calculates the strain and the stress in the fibre direction, in transverse direction and in shear for each set of plies. A temperature difference ΔT from the internal stress-free state, curing temperature or otherwise, may also be specified (positive ΔT indicates increase in temperature *above* the stress-free state). Restraint forces due to difference in temperature are calculated in Visual Basic function Restraint, and the resulting stresses are included in the stresses calculated under the externally applied loads. With given allowable stresses, the spreadsheet calculates either the Tsai–Hill or the Tsai–Wu failure criterion for each set of plies, in each loading case. The maximum value of the failure criterion in any layer and under any loading case is displayed.

8.3.1.2 Optimization

Design variables are the fibre angles $\pm\theta_1, \pm\theta_2, \pm\theta_3$ and corresponding numbers of plies n_1, n_2, n_3 at each angle. Note that these refer to the number of *pairs* of plies, so that the total number of plies is $2 \times (n_1 + n_2 + n_3)$. For example, if $\theta_1 = 0°$ and $n_1 = 1$, there are *two* $0°$ plies. Constraints are the values of the Tsai–Hill or Tsai–Wu criterion for each set of plies, in each loading case. Additional constraints are to restrict n_1, n_2, n_3 to integer values for a discrete ply thickness, to limit fibre angles to between $0°$ and $90°$ ($-90°$ for the $-\theta$ plies) and to set a maximum number of plies at any ply angle. For angle plies, with a discrete ply thickness, an even number of pairs of plies are necessary to maintain a symmetric laminate. The total thickness of the laminate is minimized.

Optimization may be performed either by the GRG Nonlinear method or by the Evolutionary method. More local minima commonly exist in this problem, both due to the nature of the problem itself and as a result of a discrete ply thickness. For the same reasons, more laminates may be found with the same minimum thickness but different lay-ups. Even though the computing time is much longer, it may be preferred therefore to use first the Evolutionary method and then to refine the result by the GRG Nonlinear method. It is known that local minima with thickness greater than the optimum commonly occur at or near $\theta = 0°$ and $\theta = \pm90°$. It is advisable, therefore, to select mid-range values of θ for initial design variables (e.g. $\theta = 25°, 45°, 65°$), in which case ply angles $\theta = 0°$ or $\theta = 90°$ will normally be found when these angles do in fact correspond to the true optimum.

As stated above, the numbers of plies are limited to integer values for a discrete ply thickness. If this is not required, then these three constraints should be deleted from the list of constraints in the Solver dialog box. Similarly, if fixed fibre angles are required, these may be added to the list of constraints, or fibre angles may be removed from the set of design variables. If allowable strains are to be specified as constraints, rather than stresses expressed in the failure criteria, the laminate strain components already calculated in the spreadsheet can readily be substituted for failure criteria in the list of constraints.

Table 8.8 Data entry for spreadsheet program 'Composite Laminate'

Parameters	
Applied loads per unit length N_x, N_y, N_{xy} in load cases 1 up to 4	Enter values in cells C28:F30 (tension positive, cells may be zero or left blank as necessary)
Ply properties: Longitudinal modulus E_1 Transverse modulus E_2 Shear modulus G_{12} Major Poisson's ratio v_{12}	Enter values in cells C5:C8
Ply thickness t_{ply}	Enter the value in cell C9
Temperature difference ΔT	Enter the value in cell E32 (positive *above* the stress-free state)
Longitudinal coeff. of expansion α_1 Transverse coeff. of expansion α_2	Enter values in cells C10:C11 (may be positive or negative)
Allowable stresses: Longitudinal tension X_t Longitudinal compression X_c Transverse tension Y_t Transverse compression Y_c Shear S	Enter values in cells C19:C23 all as positive numbers
Failure criterion	Enter TW for Tsai–Wu criterion or TH for Tsai–Hill criterion (default) in cell C20
Maximum number of *pairs* of plies n_{max} in each fibre direction	Enter the value in cell C22
Design variables	
Ply angles θ_1, θ_2, θ_3	Enter initial values in cells C16:C18 ($0 \leq \theta \leq 90°$)
Number of *pairs* of plies n_1, n_2, n_3 at each ply angle	Enter initial values in cells J16: J18 ($0 \leq n \leq n_{max}$)

Parameters and design variables to be entered are listed in Table 8.8. These include the elastic properties of the ply material, allowable stresses and other data. After optimization, initial values of the design variables are replaced by their optimum values and the total thickness of the laminate is given. The minimum margin of safety for first ply failure with the Tsai–Hill criterion is also given. With the Tsai–Wu criterion, the margin of safety, if required, has to be found by suitable factoring of the set of applied loads (with no further optimization) until the maximum value of the Tsai–Wu criterion is 1.00.

8.4 Summary

Lamination theory for analysis of the stress in the different layers of a composite laminate is reviewed. The stiffness properties of each layer, defined in layer axes in terms of its elastic constants, have to be transformed into global, laminate axes before the individual layers can be assembled into the complete laminate.

A standard transformation matrix is used for this, transforming either stress or strain into one or other axis system. The resulting laminate stiffness matrix is inverted to give the strains in the laminate under specified in-plane loads. These strains are then transformed back into layer axes for the strains in each layer. Using again the layer stiffness properties, the stress components in each layer are then calculated. Different failure criteria predict the combination of stresses to initiate failure of a layer, in terms of measured data for the strength of the ply material under the individual stress components. Internal stresses are developed in a laminate with change in temperature, due to the different coefficients of expansion of the layers in the longitudinal and transverse directions. In particular, residual stresses occur with cooling of a laminate after curing and should be taken into account. Some design 'rules' for lay-up are presented, intended to reduce the likelihood of failure due to interlaminar stresses. Such stresses arise due to transfer of load between layers and are not accounted for by standard lamination theory.

Ply properties are generally much more variable than those for metallic materials, and substantial reductions in properties are usually made to allow for this. In addition, the strength and other properties of conventional composite materials are highly sensitive to both temperature and humidity, and further reductions in properties have to be made for this. Alternatively, a maximum allowable strain is often introduced, substantially less than the maximum strain of the material. Strain limits are also imposed to limit delamination growth for composite laminates subject to impact conditions.

Different methods for the design of a composite laminate are presented, ranging from a simple method for an initial design to optimization with Solver in the spreadsheet 'Composite Laminate'. Netting analysis assumes that only the fibres provide the strength of a laminate, and if there are no more than three fibre directions, this leads directly to a set of equilibrium equations from which the required number of plies can be calculated. However, since netting analysis neglects transverse and shear stresses in the plies, in many cases a principal cause of failure, this commonly leads to an underestimate of the required number of plies. A simple, iterative redesign scheme using full lamination theory can be employed to progressively improve an initial design of laminate, by identifying the most critical plies and adding the plies that are the most effective in reducing the stress in those plies. In other words, plies are not necessarily added in the ply direction found to be critical, but in another direction. Plies may also have to be removed. This process does not, however, guarantee an optimum laminate, even if it generally produces a usable one.

While the above methods can readily deal with whole numbers of plies, discrete ply thickness presents a significant obstacle in a formal optimization procedure. The various examples illustrate the often substantial increase in thickness resulting from discrete ply thickness, especially when a balanced, symmetric laminate is required. In the GRG Nonlinear method in Solver, this problem is resolved by the branch-and-bound method. This progressively sets the variables to integer values, performing a new optimization at each step. Upper and lower bounds are established, to eliminate as quickly as possible solutions that cannot lead to the required result. The method can accurately locate a local optimum, but this may not be the true global optimum if more local optima exist. From different starting points, different solutions may be obtained. An alternative way of dealing with discrete ply thickness is by use of a genetic algorithm (the Evolutionary method in Solver), which can work with integer variables from the outset. This method retains a population of designs, to make a more complete search of the whole design space, and can therefore better detect local minima and alternative designs that may arise with discrete ply thickness. However, being a semi-random process, computing time can rise dramatically with the number of design variables. While the method inevitably produces an improved design, generally one close to the global optimum, it may not succeed in finding an exact solution to the problem.

Exercises

Use the material data in Table 8.1 *in the exercises below.*

8.1 Calculate the A_{11} and A_{33} terms of the A-matrix for a symmetric laminate consisting of 4 plies at $+60°$, 4 plies at $-60°$ and 8 plies at $0°$, with a ply thickness of 0.125 mm. Use the spreadsheet 'Composite Laminate' to verify the values obtained. With the full A-matrix in the spreadsheet, calculate the elastic constants of the laminate.

Follow the procedure in Example 8.1.

8.2 If the laminate in Exercise 8.1 carries a tensile load of 1000 N/mm in the $0°$ direction, calculate the stresses in the $0°$ and $\pm 60°$ layers. Use the spreadsheet to verify the values obtained.

Follow the procedure in Example 8.2.

8.3 Calculate the values of the Tsai–Hill criterion for each layer of the laminate in Exercise 8.2, under the given tensile load. If the load is increased, at what load does first ply failure occur according to the Tsai–Hill criterion?

Identify the stress component making the largest contribution to the Tsai–Hill criterion in the most critical layer.

8.4 Compare the failure envelopes of the Tsai–Hill and Tsai–Wu criteria for different combinations of stress σ_1 and σ_2 in a single layer of a composite laminate.

Use the appropriate tensile or compressive strength data in Table 8.1. *For a series of values of σ_1, calculate the corresponding pair of values of σ_2 to just satisfy each criterion. (This may more easily be done by Goal Seek in Excel.) To compare the two criteria, make a plot of the different combinations of σ_1 and σ_2 at failure on the same polar plot with axes σ_1 and σ_2. Notice the discontinuity in the Tsai–Hill curve as σ_1 or σ_2 becomes negative.*

8.5 Derive the first row of the $\mathbf{T}^T$ (transpose) matrix in the second of Eq. (8.5) to transform stresses in layer axes into stresses in laminate axes.

Sketch a right-angled triangle with sides normal to stresses σ_1 and σ_2 (at angle θ to the laminate axes) and a third side normal to stress σ_x. Calculate the resulting forces on all three sides of the triangle under stresses σ_1, σ_2 and shear stress τ_{12}. Assume some arbitrary size and thickness of the triangle. Calculate the horizontal components of the resulting forces on all three sides of the triangle. By equilibrium of the three horizontal force components, deduce the three required terms of the $\mathbf{T}^T$ matrix.

8.6 Use the method in Example 8.4 for the initial design of a $0°$, $\pm45°$, $90°$ laminate under the following alternative loading cases:

$$N_x = 2000\,\text{N/mm}, \quad N_{xy} = 1000\,\text{N/mm},$$
$$N_y = 500\,\text{N/mm}, \quad N_y = 1500\,\text{N/mm},$$
$$N_{xy} = 1000\,\text{N/mm}, \quad N_{xy} = 500\,\text{N/mm}.$$

The laminate should be balanced and symmetric, with a ply thickness of 0.125 mm. *Select the larger number of plies in each fibre direction from the two loading cases. From the solution choose an actual lay-up, observing the practical restrictions in Table 8.2.*

8.7 Use the iterative redesign procedure in Sect. 8.2.2 to improve the design in Exercise 8.6 above.

Use the program 'Composite Laminate' to calculate Tsai–Hill values at each step.

8.8 Use the program 'Composite Laminate' with the Tsai–Hill criterion to optimize the $0°$, $\pm45°$, $90°$ balanced, symmetric laminate in Exercise 8.6, under the same loading and with the same ply thickness.

Use the GRG Nonlinear method in Solver. Add constraints for the fixed ply angles. Repeat the optimization from a few different starting points.

8.9 Repeat Exercise 8.8 using the Evolutionary method in Solver.

Repeat the optimization a few times with 'Population Size' (in Options/ Evolutionary) set to 10 and 'Seed' set to zero to generate different starting points.

8.10 Calculate the minimum thickness of the laminate in Exercises 8.8 and 8.9 above if there is no restriction on ply angles and the discrete ply thickness is ignored.

Use the GRG Nonlinear method in Solver. Delete the integer constraints on numbers of plies in Solver and constraints on fixed ply angles (if previously added). Select 'Multistart' (in Options/GRG Nonlinear) to assist in locating a true optimum. Set 'Population Size' to 10 and 'Random Seed' to zero.

References

1. Jones RM (1999) Mechanics of composite materials. Taylor & Francis
2. Daniel IM, Ishai O (2006) Engineering mechanics of composite materials. Oxford University Press
3. Timoshenko SP, Goodier JN (1970) Theory of elasticity. McGraw-Hill
4. Hill R (1948) A theory of the yielding and plastic flow of anisotropic metals. Proc R Soc Ser A 193:281–297
5. Azzi VD, Tsai SW (1965) Anisotropic strength of composites. Exp Mech 5(9):283–288
6. Tsai SW, Wu EM (1971) A general theory of strength for anisotropic materials. J Compos Mater 5:58–80
7. Niu, MCY (1992) Composite airframe structures. Hong Kong Conmilit Press
8. Kassapoglou C (2013) Design and analysis of composite structures with applications to aerospace structures. John Wiley and Sons, Chichester, UK
9. Cox HL (1952) The elasticity and strength of paper and other fibrous materials. Brit J Appl Phys 3:72–79
10. Morton SK, Webber JPH (1991) Heuristic methods in the design of composite laminated plates. Compos Struct 19:207–265
11. Land AH, Doig AG (1960) An automatic method for solving discrete programming problems. Econometrica 28:497–520
12. Gürdal Z, Haftka RT, Hajela P (1999) Design and optimization of laminated composite materials. John Wiley & Sons, New York
13. Reklaitis GV, Ravindran A, Ragsdell KM (1983) Engineering optimization. John Wiley and Sons, New York
14. Holland JH (1975) Adaptation in natural and artificial systems. University of Michigan Press

Chapter 9
Optimization With Finite Element Analysis

Abstract Particular methods are called for when structural optimization is coupled with finite element analysis. A sensitivity analysis can be employed to evaluate the constraint gradient data needed in a conventional, gradient-based optimization. This is to avoid the repeated finite element analysis required if constraint gradients were to be obtained by finite difference. By setting up a matrix of derivatives of constraints with respect to the displacements of the finite element model, the inverted stiffness matrix obtained in the normal course of finite element analysis is reused in a sensitivity analysis. In one method, the individual columns of this matrix of derivatives are treated as a set of 'dummy loads' with which the constraint gradients can be calculated. The other method is the so-called direct method. The computation can be further reduced by an active constraint strategy, in which constraint gradients are evaluated only for those constraints that are active or near-active at any stage. The number of variables involved in the optimization may be reduced by design variable linking. This is by defining some variables as slave variables that are then related to the remaining master variables for optimization, while for accuracy all variables are retained in the finite element analysis.

The analysis of complex structures, any other than the relatively simple ones chosen in this book as examples to illustrate different optimization methods, is these days almost inevitably by finite element analysis. This involves a detailed model and a substantial amount of computer processing time. An efficient optimization of a structure when the analysis is by the finite element method requires therefore some particular methods. We have been considering structures where the aim is to minimize weight subject to constraints on strength, stiffness, dimensional limitations, and perhaps other conditions. The weight of a finite element model, whether or not it fully represents the real structure, is generally straightforward to calculate. This presents no problem. The problem lies in the calculation of constraints, or more specifically in the calculation of constraint *gradients*.

In a gradient-based optimization, from an initial point in the design space, we determine a suitable search direction and conduct a line search along that direction to locate the next point. To find the search direction requires calculation of the

© Springer International Publishing AG 2017   283
A. Rothwell, *Optimization Methods in Structural Design*, Solid Mechanics
and Its Applications 242, DOI 10.1007/978-3-319-55197-5_9

constraint gradients. With n design variables, recalculation of constraint gradients for the next search direction, if done by finite difference, would require at least n additional finite element analyses for each constraint. This would be too expensive for any but the smallest finite element models (such as the beam under lateral load in Chap. 6, which does necessarily use finite difference). To get around this, we perform a 'sensitivity analysis' for the gradient data, from which the new search direction is then calculated. A sensitivity analysis provides the rate of change of each constraint with each design variable in an efficient way. By efficient is meant here without repeated finite element analysis of the structure.

Methods have been developed whereby use is made of a single finite element solution for a sensitivity analysis, by which constraint gradients at each new point in the design space are recalculated. Two variations of the method exist: the 'direct method' and a method which makes use of so-called dummy loads. These are described in the following section and illustrated in a simple example. It will be appreciated that this example does not have the complexity of a typical finite element problem but is simple enough to enable the calculation to be performed by hand, without undue matrix manipulation. It is not the intention in this short chapter to review the finite element method itself. It is assumed that those reading this chapter will already have some familiarity with it. Nevertheless, only a quite rudimentary knowledge of the finite element method will be sufficient to follow what is presented here.

9.1 Sensitivity Analysis

A 'design sensitivity analysis' can be performed to evaluate the gradients of constraint functions with respect to the design variables. When the constraints are implicit functions of the design variables, as in a finite element analysis, special methods have been developed (see [1, 2]). A constraint g_j is considered to be a function of both the design variables $\mathbf{x}$ and the displacements $\mathbf{u}$ (including rotations where appropriate) of the finite element analysis. Design variables may appear explicitly in a constraint, for example if a maximum value of the diameter to thickness ratio of a tube is specified. Other quantities appearing in constraints, such as the stress in a bar or the bending moment in a beam, are derived from the nodal displacements $\mathbf{u}$ obtained from the finite element analysis. Therefore, we may write:

$$g_j = g_j(\mathbf{x}, \mathbf{u}).$$

The constraint gradient is defined by dg_j/dx_i. Differentiating the above formula:

$$\frac{dg_j}{dx_i} = \frac{\partial g_j}{\partial x_i} + \left(\frac{\partial g_j}{du_1} \cdot \frac{\partial u_1}{\partial x_i} + \frac{\partial g_j}{du_2} \cdot \frac{\partial u_2}{\partial x_i} + \cdots \right), \tag{9.1}$$

following the usual rules for partial differentiation. Notice that the left-hand side is the *total* differential of g_j with respect to x_i. The first term on the right-hand side represents the explicit dependence of the constraint on the design variables and is in general readily evaluated. In many cases, this first term will be zero. For example, a stress constraint $\sigma_{max} \leq \sigma_{all}$ does not refer directly to any design variable, so $\partial g_j / \partial x_i = 0$ for that constraint. On the other hand, the stress in an element is obtained from the displacements in a finite element analysis, so at least some of the terms in brackets on the right-hand side remain present.

We have therefore to evaluate $\partial g_j / \partial u_1$ and $\partial u_1 / \partial x_i$ in the above equation and all following terms within the brackets. We express the first part of each pair of terms in Eq. (9.1) in a column matrix:

$$\mathbf{h}_j = \left\{ \frac{\partial g_j}{\partial u_1}, \frac{\partial g_j}{\partial u_2}, \dots \right\}.$$

The individual terms of $\mathbf{h}_j$ are obtained either directly or from the element stiffness matrices, as will be seen in the following example. When $\partial g_j / \partial x_i = 0$, by Eq. (9.1) we can write:

$$\frac{dg_j}{dx_i} = \mathbf{h}_j^T \cdot \frac{\partial \mathbf{u}}{dx_i}, \tag{9.2}$$

where superscript T denotes the transpose of $\mathbf{h}_j$. The term $\partial \mathbf{u} / \partial x_i$ in the above equation represents the second part of each pair of terms within brackets in Eq. (9.1):

$$\frac{\partial \mathbf{u}}{\partial x_i} = \left\{ \frac{\partial u_1}{\partial x_i}, \frac{\partial u_2}{\partial x_i}, \dots \right\}.$$

For this, we refer to the general finite element formulation:

$$\mathbf{K}\mathbf{u} = \mathbf{f}, \tag{9.3}$$

where $\mathbf{f}$ is the set of applied forces (and moments) at the nodal points. To solve for the displacements $\mathbf{u}$, we have as usual to invert the stiffness matrix $\mathbf{K}$:

$$\mathbf{u} = \mathbf{K}^{-1} \mathbf{f}.$$

If we now differentiate Eq. (9.3) by the product rule, we obtain:

$$\mathbf{K}\frac{\partial \mathbf{u}}{\partial x_i} + \frac{\partial \mathbf{K}}{\partial x_i}\mathbf{u} = \frac{\partial \mathbf{f}}{\partial x_i}.$$

The right-hand side of the above equation is zero provided the applied forces are constant, in which case the above equation reduces to:

$$\mathbf{K}\frac{\partial \mathbf{u}}{\partial x_i} = -\frac{\partial \mathbf{K}}{\partial x_i}\mathbf{u}.$$

Premultiplying by the inverted stiffness matrix we have:

$$\frac{\partial \mathbf{u}}{\partial x_i} = -\mathbf{K}^{-1}\frac{\partial \mathbf{K}}{\partial x_i}\mathbf{u}.$$

Substituting for $\partial \mathbf{u}/\partial x_i$ in Eq. (9.2), we obtain:

$$\frac{dg_j}{dx_i} = \mathbf{h}_j^T \frac{\partial \mathbf{u}}{\partial x_i} = -\mathbf{h}_j^T \mathbf{K}^{-1}\frac{\partial \mathbf{K}}{\partial x_i}\mathbf{u}. \qquad (9.4)$$

This is now the required sensitivity of constraint g_j to the variable x_i. The explicit term $\partial g_j/\partial x_i$ in Eq. (9.1) may be added if necessary. As already stated, Eq. (9.4) may be computed either by a 'direct' method or by the 'dummy load method', as shown below. In both methods, we require both the inverted stiffness matrix $\mathbf{K}^{-1}$ and the displacements $\mathbf{u}$ under the applied load, calculated during the normal finite element analysis. These have to be recalculated at each step in the optimization process. The matrix $\partial \mathbf{K}/\partial x_i$ is generated directly from the stiffness matrix of each element. While each full matrix $\partial \mathbf{K}/\partial x_i$ is the same size as the assembled stiffness matrix, it contains only terms relevant to that element, all other terms being zero.

In the direct method, we perform the series of matrix multiplications in Eq. (9.4) to evaluate the required dg_j/dx_i for each pair (i,j). If performed in order, each multiplication produces no more than a column matrix. Alternatively, in the dummy load method, we rewrite Eq. (9.4) as:

$$\frac{dg_j}{dx_i} = -\left\{\mathbf{K}^{-1}\mathbf{h}_j\right\}^T\frac{\partial \mathbf{K}}{\partial x_i}\mathbf{u}. \qquad (9.5)$$

Again the first term in Eq. (9.1) should be added if necessary. This alternative form of the equation is only possible because $\mathbf{K}$ is a symmetric matrix. We now treat $\mathbf{h}_j$ as a set of dummy loads so that the term within braces above can be computed at the *same time* as the actual loads, using the same already inverted stiffness matrix. It then remains to complete the matrix multiplications in Eq. (9.5) for dg_j/dx_i. The dummy load method is generally more efficient when the number of constraints is smaller than the number of design variables.

Example 9.1 The two-bar truss in Fig. 9.1 has a constraint on the vertical deflection $\delta \le \delta_0$ under a load P at the tip. Calculate the sensitivity of this constraint to the cross-sectional area A_1 of the lower bar.

Fig. 9.1 Two-bar truss

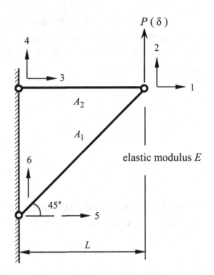

The stiffness matrix for the truss, after elimination of degrees of freedom at the supports, is:

$$\mathbf{K} = \frac{E}{L} \begin{bmatrix} \frac{A_1}{2\sqrt{2}} + A_2 & \frac{A_1}{2\sqrt{2}} \\ \frac{A_1}{2\sqrt{2}} & \frac{A_1}{2\sqrt{2}} \end{bmatrix},$$

using the standard finite element stiffness matrix for a bar pinned at each end.

Differentiating the above matrix with respect to the chosen variable A_1 and simplifying, we have:

$$\frac{\partial \mathbf{K}}{\partial A_1} = \frac{E}{L} \begin{bmatrix} 0.3536 & 0.3536 \\ 0.3536 & 0.3536 \end{bmatrix}.$$

The vertical displacement $\delta = u_2$, so the constraint becomes:

$$g = \delta_0 - u_2 \geq 0$$

(suffix j is omitted on the assumption; there are no more constraints) giving for this simple constraint:

$$\frac{\partial g}{\partial u_1} = 0, \quad \frac{\partial g}{\partial u_2} = -1.$$

The column matrix $\mathbf{h}$ is then:

$$\mathbf{h} = \{0, -1\}.$$

To compute Eq. (9.4), we now require the inverse of the stiffness matrix $\mathbf{K}$ of the truss and the displacements $\mathbf{u}$ under load P. These are, of course, part of the normal finite element solution of the truss. We shall assume that the two bars have the same cross-sectional area:

$$A_1 = A_2 = A.$$

The inverse of $\mathbf{K}$ is then:

$$\mathbf{K}^{-1} = \frac{L}{EA} \begin{bmatrix} 1 & -1 \\ -1 & 3.828 \end{bmatrix}.$$

(In this example, matrix inversion is readily performed by hand, or by means of the matrix inverse function in Excel if preferred.) Displacements $\mathbf{u}$ are:

$$\mathbf{u} = \mathbf{K}^{-1}\,\mathbf{f},$$

where the force vector $\mathbf{f}$ (under a single vertical load P) is:

$$\mathbf{f} = P \begin{bmatrix} 0 \\ 1 \end{bmatrix}.$$

This gives for $\mathbf{u}$:

$$\mathbf{u} = \frac{PL}{EA} \begin{bmatrix} 1 & -1 \\ -1 & 3.828 \end{bmatrix} \cdot \begin{bmatrix} 0 \\ 1 \end{bmatrix} = \frac{PL}{EA} \begin{bmatrix} -1 \\ 3.828 \end{bmatrix}.$$

We are now ready to evaluate Eq. (9.4).

By the so-called 'direct method', writing the equation out in full we have:

$$\frac{dg}{dA_1} = -\begin{bmatrix} 0 & -1 \end{bmatrix} \cdot \frac{L}{EA} \begin{bmatrix} 1 & -1 \\ -1 & 3.828 \end{bmatrix} \cdot \frac{E}{L} \begin{bmatrix} 0.3536 & 0.3536 \\ 0.3536 & 0.3536 \end{bmatrix} \cdot \frac{PL}{EA} \begin{bmatrix} -1 \\ 3.828 \end{bmatrix}$$

$$= 2.828\, \frac{PL}{EA^2}.$$

This is the required sensitivity of the given constraint on deflection δ to the cross-sectional area A_1 of the lower bar. Note that there is no explicit dependence of the constraint on A_1 in this example and that load P remains constant. The positive value of dg/dA_1 implies that with increase in area the deflection is reduced, and the value of the constraint is therefore increased.

In the dummy load method, $\mathbf{h}$ is the set of dummy loads (in this example a single column matrix). We calculate $\mathbf{K}^{-1}\mathbf{h}$ at the same time as the solution of the structure under the actual loads. In the present example, this amounts to calculating:

$$\begin{bmatrix} 1 & -1 \\ -1 & 3.828 \end{bmatrix} \cdot \begin{bmatrix} 0 \\ -1 \end{bmatrix}.$$

Taking the transpose of the result and completing the calculation of Eq. (9.5) gives, of course, the same result as before. However, the difference in the two methods is largely lost in this simple example.

The analytical formula for the vertical deflection at the tip of the truss is:

$$\delta = \frac{PL}{E}\left(\frac{2\sqrt{2}}{A_1} + \frac{1}{A_2}\right).$$

Differentiating with respect to A_1:

$$\frac{d\delta}{dA_1} = -2\sqrt{2}\,\frac{PL}{EA_1^2}.$$

With $dg/dA_1 = -d\delta/dA_1$ (constraint reduces with increase in deflection) and $A_1 = A$, this agrees with the result above. ■

It is generally not necessary to perform a sensitivity analysis for all constraints. In many problems, we may know that some constraints cannot be active at or near the optimum, while other constraints may be inactive in the whole design space. For a more economical computation, it is common to follow an 'active constraint' strategy, as already introduced in Chap. 5. This means that we include in the optimization, and therefore in a sensitivity analysis as well, only those constraints actually active at any stage, and those considered to be 'near-active'. By appropriate selection of constraints, the size of the optimization problem and the computation time may be significantly reduced. However, this also implies that at each step in the optimization constraints have to be checked to determine which constraints must remain in the active set, which can leave it, and which have to be added to it.

A sensitivity analysis implies that a linear approximation to the actual constraints is made at each step in the optimization, that is if they are in fact non-linear. The degree of linearity of constraints can often be improved by the use of inverse variables, again as previously discussed in Chap. 5. This means replacing simple variables such as cross-sectional area A or thickness t by $\frac{1}{A}$ or $\frac{1}{t}$. The effect of this is that move limits chosen for the optimization can be increased and that the calculated constraint gradients remain longer valid. This can again lead to a smaller number of finite element analyses being required to reach an optimum.

9.2   Reduction in Design Variables

As well as the reduction in the number of constraints in the previous section, it is also possible to reduce the number of design variables to be included in an optimization. For an accurate finite element analysis, a fine mesh is required, with a sufficiently large number of elements, and in principle design variables have to be assigned to each of them. To reduce the number of variables for optimization, we define some variables as 'slave' variables, and the remainder as 'masters'. The slave variables are related to the masters by interpolation between adjacent elements. Only the master variables are included as variables in the actual optimization.

To illustrate this, we consider part of a tapered beam, as shown in Fig. 9.2. If the individual elements are uniform, with bending stiffness EI, the finite element model might appear as in the upper part of the figure. Suppose for optimization, we select the bending stiffness EI_2 and EI_3 as slave variables. By linear interpolation, these are expressed in terms of the master variables EI_1 and EI_4 by:

$$EI_2 = \frac{2}{3}EI_1 + \frac{1}{3}EI_4,$$

$$EI_3 = \frac{1}{3}EI_1 + \frac{2}{3}EI_4$$

(assuming all elements to be of equal length). The optimization model then appears as in the lower part of Fig. 9.2. Of course, the true optimum may not be an exact linear taper, but for a large problem, this enables us to reduce the optimization problem to an acceptable level without impairing the accuracy of the finite element analysis.

The relation between the full set of variables and the master variables can conveniently be expressed in the form of a 'linking matrix'. For the beam in Fig. 9.3, with seven elements, we define the bending stiffness of three of these, every alternate element, as slave variables. These are shown shaded in the figure. In

Fig. 9.2 Tapered beam

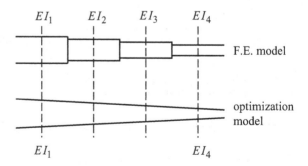

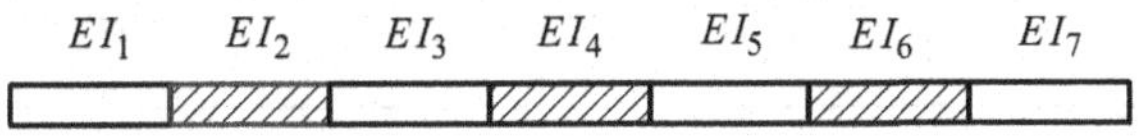

$$EI_1 \qquad EI_2 \qquad EI_3 \qquad EI_4 \qquad EI_5 \qquad EI_6 \qquad EI_7$$

Fig. 9.3 Design variable linking

this case, again for a linear taper, the slave variables are simply the average of the two adjacent variables:

$$EI_2 = \frac{EI_1 + EI_3}{2},$$

and similarly for variables EI_4 and EI_6. In matrix form, this becomes:

$$
\begin{bmatrix} EI_1 \\ EI_2 \\ EI_3 \\ EI_4 \\ EI_5 \\ EI_6 \\ EI_7 \end{bmatrix}
=
\begin{bmatrix}
1 & 0 & 0 & 0 \\
1/2 & 1/2 & 0 & 0 \\
0 & 1 & 0 & 0 \\
0 & 1/2 & 1/2 & 0 \\
0 & 0 & 1 & 0 \\
0 & 0 & 1/2 & 1/2 \\
0 & 0 & 0 & 1
\end{bmatrix}
\cdot
\begin{bmatrix} EI_1 \\ EI_3 \\ EI_5 \\ EI_7 \end{bmatrix}.
$$

A similar linking matrix is used in the spreadsheet program in the next section.

9.3 Spreadsheet Program

The spreadsheet modifies the earlier spreadsheet 'Beam under Lateral load' in Chap. 6 to illustrate reduction in the number of design variables by selection of master and slave variables, with the aim of making the minimum of change to the original spreadsheet.

9.3.1 'Design Variable Linking'

The linking matrix is placed on a new page 5 of the spreadsheet. This is a rectangular matrix of dimension $(n \times m)$, where n is the total number of design variables and m is the number of master variables. For the present example, an appropriate linking matrix has already been entered in the spreadsheet, as shown in Fig. 9.4, but of course this normally has to be set up by the user for the particular problem. The $n = 24$ design variables are the second moments of area of each of the 24 elements in the original spreadsheet. The $m = 14$ master variables selected in the example are listed in cells P16:P29 and again in cells V14:AI14 (these are given for reference only). Suitable initial values of the master variables are entered in cells

Reduction in Design Variables

Master Variables

element	I
1	1.00E+04
2	1.00E+04
4	1.00E+04
6	1.00E+04
8	1.00E+04
10	1.00E+04
12	1.00E+04
14	1.00E+04
16	1.00E+04
18	1.00E+04
19	1.00E+04
21	1.00E+04
23	1.00E+04
24	1.00E+04
	mm^4

Linking matrix

element \ master variable	1	2	4	6	8	10	12	14	16	18
1	1.00	0.00	0.00	0.00	0.00	0.00	0.00	0.00	0.00	0.00
2	0.00	1.00	0.00	0.00	0.00	0.00	0.00	0.00	0.00	0.00
3	0.00	0.50	0.50	0.00	0.00	0.00	0.00	0.00	0.00	0.00
4	0.00	0.00	1.00	0.00	0.00	0.00	0.00	0.00	0.00	0.00
5	0.00	0.00	0.50	0.50	0.00	0.00	0.00	0.00	0.00	0.00
6	0.00	0.00	0.00	1.00	0.00	0.00	0.00	0.00	0.00	0.00
7	0.00	0.00	0.00	0.50	0.50	0.00	0.00	0.00	0.00	0.00
8	0.00	0.00	0.00	0.00	1.00	0.00	0.00	0.00	0.00	0.00
9	0.00	0.00	0.00	0.00	0.50	0.50	0.00	0.00	0.00	0.00
10	0.00	0.00	0.00	0.00	0.00	1.00	0.00	0.00	0.00	0.00
11	0.00	0.00	0.00	0.00	0.00	0.50	0.50	0.00	0.00	0.00
12	0.00	0.00	0.00	0.00	0.00	0.00	1.00	0.00	0.00	0.00
13	0.00	0.00	0.00	0.00	0.00	0.00	0.50	0.50	0.00	0.00
14	0.00	0.00	0.00	0.00	0.00	0.00	0.00	1.00	0.00	0.00
15	0.00	0.00	0.00	0.00	0.00	0.00	0.00	0.50	0.50	0.00
16	0.00	0.00	0.00	0.00	0.00	0.00	0.00	0.00	1.00	0.00
17	0.00	0.00	0.00	0.00	0.00	0.00	0.00	0.00	0.50	0.50

Fig. 9.4 Spreadsheet 'Design Variable Linking'

Table 9.1 Data entry for spreadsheet program 'Design Variable Linking'

Parameters	
Parameters are unchanged from spreadsheet program 'Beam under Lateral load' (see Table 6.3 in Chap. 6)	
Variables	
Master variables: second moments of area I	Enter initial values in cells Q16:Q29 See note in Table 6.3 Make no entries in cells H16:H39
Construct the linking matrix in cells V16:AI39 Note: If a different number of master variables are chosen, new cell ranges for the linking matrix and master variables have to be selected The matrix multiplication in cells H16:H39 must then be deleted and re-entered with the new cell references New cell references to the master variables have to be entered in Solver The total number of design variables must remain 24	

Q16:Q29. Since all elements are of the same length, the 10 slave variables (all remaining variables) take values equal to a simple average of the second moments of area of their adjacent elements, as given in the linking matrix. Use of slave variables is avoided at the point of loading and close to the supported ends of the beam. The full set of variables is calculated with the linking matrix in cells H16: H39. *Note that these are no longer the cells in which initial values of design variables are entered.* In the present example, initial values of *master* variables are entered in cells Q16:Q29, as stated above. This is also the cell range for variables to be entered in the Solver dialogue box, to replace those present in the original spreadsheet. When master variables have been selected and the linking matrix set up, analysis and optimization proceeds in precisely the same manner as before. Note that only one change has been necessary in the original spreadsheet, that is instead of design variables being entered directly in cells H16:H39, they are now generated by the linking matrix. The required data input for the spreadsheet is listed in Table 9.1.

Performing the optimization with the same load, constraints and other data as in the original spreadsheet, we obtain the results for the optimum second moments of area given in Table 9.2. These are compared in the table with the previous results with no design variable reduction. While there are some differences in stiffness distribution, we find only a small difference in the mass of the beam, at 2.88 kg with design variable reduction compared with 2.85 kg with no reduction (these masses are before the recalculation performed in the original spreadsheet). With 10 of the 24 variables now being slaves, it is seen that we still have a reasonable approach to the previous optimum. Also we might be reminded that the number of elements in the example is small compared with that of a more representative finite element problem. With more elements, the use of slave variables will give a better representation of the optimum stiffness distribution.

Table 9.2 Second moments of area after optimization with and without reduction in the number of design variables

Element	Second moment of area I (mm^4)	
	No reduction in design variables	With reduction in design variables
1	10,383	11,386
2	8357	9269
3	6447	7332
4	4671	5395
5	3050	3766
6	1625	2137
7	473	1497
8	1230	858
9	2582	2276
10	4147	3693
11	5877	5492
12	7747	7290
13	9738	9351
14	11,837	11,412
15	14,034	13,681
16	16,320	15,951
17	18,689	18,396
18	21,136	20,841
19	21,136	20,841
20	16,575	16,489
21	12,310	12,137
22	8388	8477
23	4885	4817
24	1939	1912
	$W = 2.853$ kg	$W = 2.881$ kg

9.4 Summary

Due to the extensive computation involved in a typical finite element analysis, some special methods are necessary to optimize a structure when the analysis is by the finite element method. This is to avoid the repeated re-analysis of the structure if constraint gradients were to be obtained by finite difference. By a sensitivity analysis, a complete set of gradient data, or derivatives of each constraint with respect to each of the design variables, can be computed in a single finite element analysis at each step in the optimization. Coupled with an efficient gradient-based optimization routine, an adequate optimization of the structure can usually take place within relatively few iterations, that is to say, relatively few finite element analyses.

In addition to this, economies in computing can be made by following an 'active constraint' strategy, in which only those constraints active at any stage and those which are considered near-active are included in the optimization and associated sensitivity analysis. At each step, constraints have then to be re-evaluated to determine which constraints must remain in the active set, those that can leave it, and those that have to be added to it. Further economies can be made by defining some design variables as slave variables and the remainder as master variables. The slave variables are linked to the master variables by linear or other form of interpolation. This can be expressed in a linking matrix. Only the master variables are included in the actual optimization, while constraints and other quantities are computed by finite element analysis with the full set of variables at each stage.

Since Solver generates gradient data by finite difference and does not permit gradient data to be input directly, it is not possible to incorporate a sensitivity analysis into an optimization by Solver. Similarly, we cannot implement an active constraint strategy within Solver. However, use of design variable linking has been demonstrated by modification of a previous spreadsheet program for a beam by finite element analysis, using Solver for the optimization. The methods described in this chapter, with many refinements, are widely used in large, commercial finite element packages for structural analysis and optimization.

Exercises

9.1 Derive the stiffness matrix for the truss in Example 9.1.
The stiffness matrix for a bar with pinned ends is given in any textbook on finite element methods. Assemble first the complete stiffness matrix, then delete rows and columns corresponding to displacements at the two supports.

9.2 Calculate the sensitivity of the constraint in Example 9.1 to the cross-sectional area A_2 of the upper bar. Verify the result by the analytical formula in Example 9.1.
Follow the method in Example 9.1.

9.3 Consider why, in a large problem, it would be more economical to use the dummy load method when the number of constraints is less than the number of design variables.
Examine the number of individual calculations in successive matrix multiplications when the number of constraints is less than the number of design variables, and when the number of constraints is greater than the number of design variables.

9.4 Repeat the optimization of the beam in the spreadsheet 'Design Variable Linking', selecting a different set of master and slave variables.
Modify the linking matrix in the spreadsheet. If the number of master variables remains the same, use the same cell ranges in the spreadsheet for the linking matrix and master variables. If a different number of master variables are chosen see Table 9.1. The total number of variables should remain 24 if further changes to the original spreadsheet are to be avoided.

References

1. Adelman H, Haftka RT (1986) Sensitivity analysis of discrete structural systems. AIAA J 24 (5):823–832
2. Arora JS, Haug EJ (1979) Methods of design sensitivity analysis in structural optimization. AIAA J 17(9):970–974

Appendix

Abstract The use of Solver and the different optimization methods is described, together with setting up the objective function, variables and constraints in the Solver Parameters box. Various options can be chosen, including the maximum time and number of iterations to reach a solution, proceeding step-by-step through an optimization, automatic scaling of constraints and use of the Multistart option when more local maxima or minima are suspected. The spreadsheet program 'seven-bar truss' in Chap. 1 is used as an example of setting up the Solver Parameters box. Some further notes are included on matrix calculation in Excel, and on custom functions.

A.1 Use of Solver in Microsoft Excel

The Solver tool in Microsoft Excel provides three general purpose optimization methods linked to formulae entered by the user in an Excel spreadsheet. In Excel 2016, the Solver command is found in the Analysis group on the 'Data' tab. Solver is an 'Add-in' and if not already available has to be loaded before use. To do this:

- on 'File' tab, click 'Options'
- click 'Add-ins'
- in the 'Manage' box, select 'Excel Add-ins' and click 'Go'
- under 'Add-ins available', check 'Solver Add-in' and click 'OK'.

(Note that these instructions can differ between different versions of Excel.)
The three options for optimization offered by Solver are as follows:

- GRG Nonlinear (generalized reduced gradient) method,
- Simplex LP (linear programming) method,
- Evolutionary (genetic algorithm) method.

For general use, the 'GRG Nonlinear' option is often the preferred choice, and the one mainly used in this book. This is a constraint-following algorithm in which the objective function and constraints may be linear or nonlinear functions of the optimization variables. The problem may also be unconstrained. The use of the

© Springer International Publishing AG 2017   297
A. Rothwell, *Optimization Methods in Structural Design*, Solid Mechanics
and Its Applications 242, DOI 10.1007/978-3-319-55197-5

GRG Nonlinear method requires the objective function and all constraints to be smooth functions of the optimization variables. There is an option to restrict chosen variables to integer values. The generalized reduced gradient method is described in Sect. 5.1.2. The 'Simplex LP' method requires the objective function and all constraints to be linear functions of the design variables. When these conditions are met, this is an efficient method for large problems. The 'Evolutionary' method is a random-based method suitable for problems where the objective function or constraints are not smooth functions of the design variables and is a useful alternative to the GRG Nonlinear method when more local maxima or minima exist or when the variables take discrete values. However, being random based, the Evolutionary method can be expected to take considerably more computing time than the GRG Nonlinear method. The Evolutionary method or genetic algorithm is described in Sect. 8.2.4.

Before running Solver, a spreadsheet must be prepared in which the objective function and constraints are calculated for any values of the optimization variables. The spreadsheet for a 'seven-bar truss' in Chap. 1 (shown again in Fig. A.1) is used as an example in all figures in this Appendix. Click 'Solver' on the Data tab to start Solver (see Fig. A.2). The objective function, constraints and variables are referred to by their cell references in the 'Solver Parameters' dialog box that appears (see Fig. A.3). If named variables are used, these may replace cell references.

The formula for the objective function refers either directly or indirectly through other formulae to the chosen design variables and any other parameters. The objective function is entered by its cell reference in the box 'Set Objective'. A maximum or a minimum of the objective function is chosen by clicking the appropriate button under this box. In the example, we require the minimum of the objective function, the total volume of the members of the truss, which is calculated in cell I20.

Variables are defined by their cell references, which in the example are the dimensions D and H in cells F6 and F7. Variables may be entered in any order, with cell references separated by commas, in the box labelled 'By Changing Variable Cells'. Variables may also be entered by a range of adjacent cells. In the standard version of Solver, the number of variables is limited to 200. Cells containing the optimization variables must contain only numerical values, i.e. may not contain formulae referring to other cells.

Constraints are added, changed or deleted with the buttons in the Solver dialog box, again using the appropriate cell references. The number of constraints in the standard version of Solver is limited to 100, not including numerical upper and lower bounds on variables. To add a constraint, click 'Add' to the side of the 'Constraints' box. In the example, constraints are $D \leq L$ (F6 $\leq$ C7), $D \geq 0$ (F6 $\geq$ 0) and $H \geq L/100$ (F7 $\geq$ C7/100). The cell reference to each constraint is entered in the 'Add Constraint' box that appears (see Fig. A.4). Note that this may contain nothing other than a cell reference to the quantity that has to be constrained. In the drop-down menu to the right, we select 'less than', 'equals to' or 'greater than', according to the type of constraint. The required value of the constraint is entered under 'Constraint'. This may be a numerical value, another variable, or a formula in

Seven-bar Truss

Parameters

$P =$	10000	N
$L =$	1.000	m
$\sigma_0 =$	400	MPa
$\rho =$	2780	kg/m³

Variables

$D =$	0.200	m
$H =$	0.500	m

Enter Parameters (in blue) and Variables (in red).
To optimize the truss: click Solver on the Data tab, then click Solve.

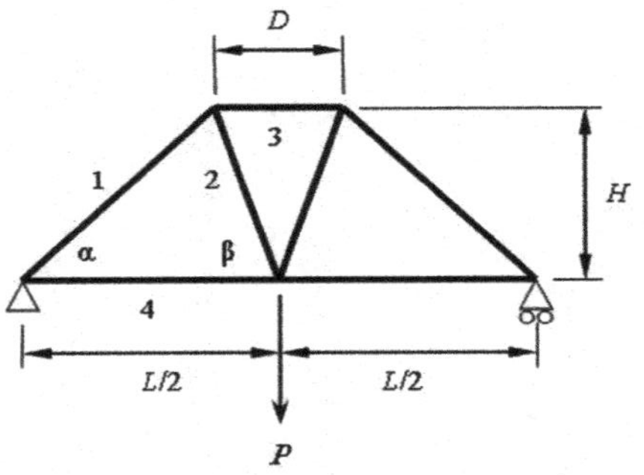

Analysis

bar	l_i	F_i	A_i	V_i
1	0.6403	-6403	16.01	10250
2	0.5099	5099	12.75	6500
3	0.2000	-5000	12.50	2500
4	0.5000	4000	10.00	5000
	m	N	mm²	mm³

Optimization

$V =$	46000	mm³
$\alpha =$	51.34	deg
$\beta =$	78.69	deg
$n =$	1.840	
$P/W =$	7974	

Fig. A.1 Spreadsheet to illustrate the use of Solver (Figs. A.1, A.2, A.3, A.4, A.5, A.6, A.7, A.8 and A.9 are used with permission from Microsoft and from Frontline Solvers)

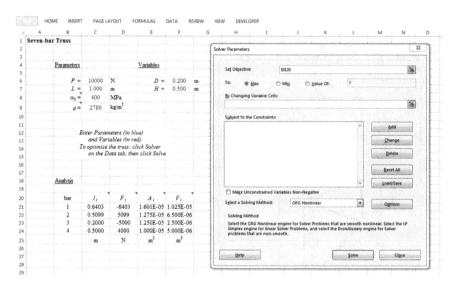

Fig. A.2 Solver optimization tool

terms of the variables and other parameters. By clicking 'OK', the constraint is added to the list of constraints in the Solver Parameters box. Alternatively, by clicking 'Add', another constraint may be added. By selecting 'int' in the drop-down menu, specified variables are restricted to integer values. By checking 'Make Unconstrained Variables Non-Negative', below the Constraints box, a zero lower bound is applied to those variables not otherwise constrained, useful when negative values are not allowed or would lead to an invalid calculation. Use 'Change' or 'Delete' to modify or remove a constraint already entered. 'Reset All' can be used to remove all entries. Finally, we have to select an optimization method in the drop-down menu at the bottom of the dialogue box. The choices are as already described above.

With 'Load/Save', the Solver Parameters box may be saved and reloaded. To save the current Solver Parameters, select an empty range on the sheet with the requested number of cells and click Save. To reload the Solver Parameters, select a previously selected range of cells, click Load and then click Replace or Merge.

The 'Options' button allows certain control parameters to be entered, for example the maximum time and number of iterations to reach a solution (see Fig. A.5). The default values (where given) are generally satisfactory for most problems. Check 'Show Iteration Results' to proceed step-by-step through the solution. Click 'Continue' to proceed to the next iteration or 'Stop' to end the process. Use 'Automatic Scaling' to reduce all constraints to a similar magnitude prior to optimization. This is for a more efficient optimization when there are excessive differences in the values of constraints. On the GRG Nonlinear tab in Options, 'Convergence' specifies the relative change to be allowed in the last five iterations before Solver reaches a solution. Under 'Derivatives', either forward or

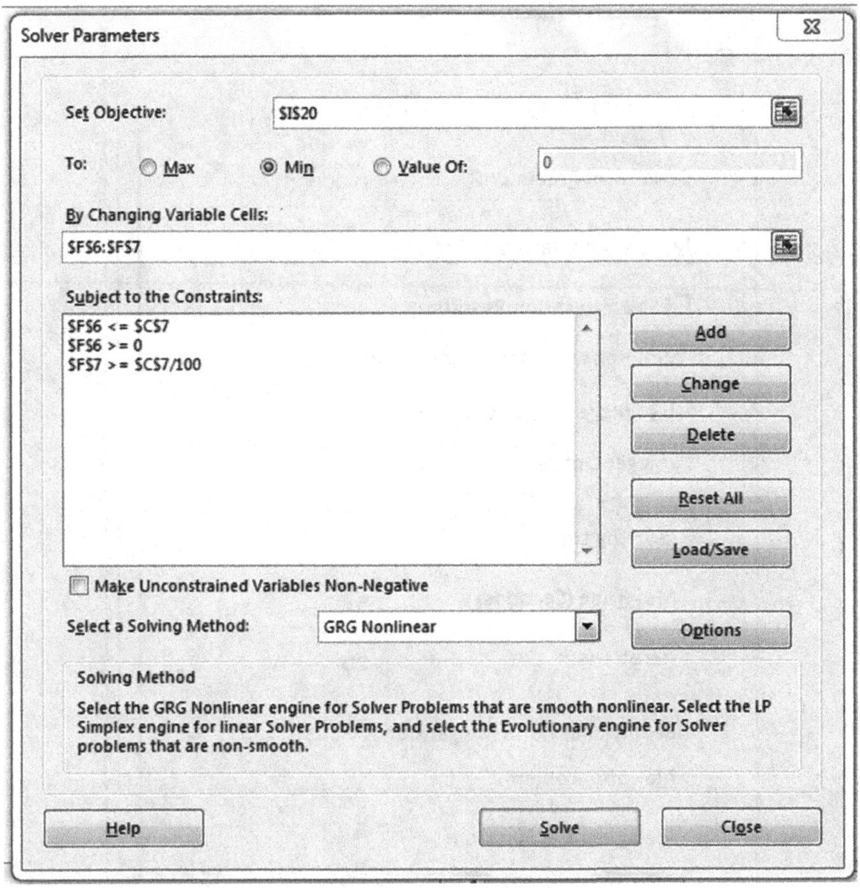

Fig. A.3 'Solver parameters' box

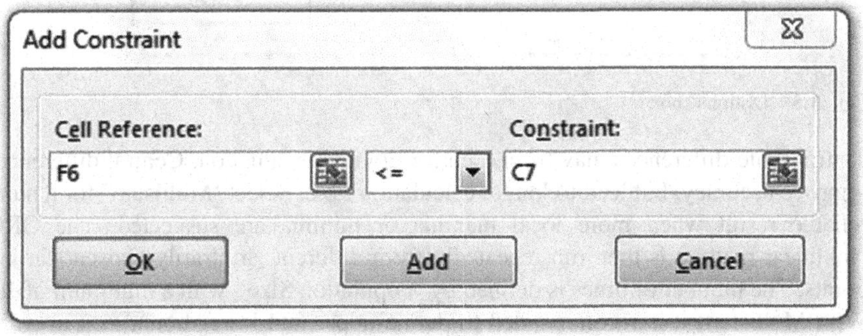

Fig. A.4 'Add constraint' box

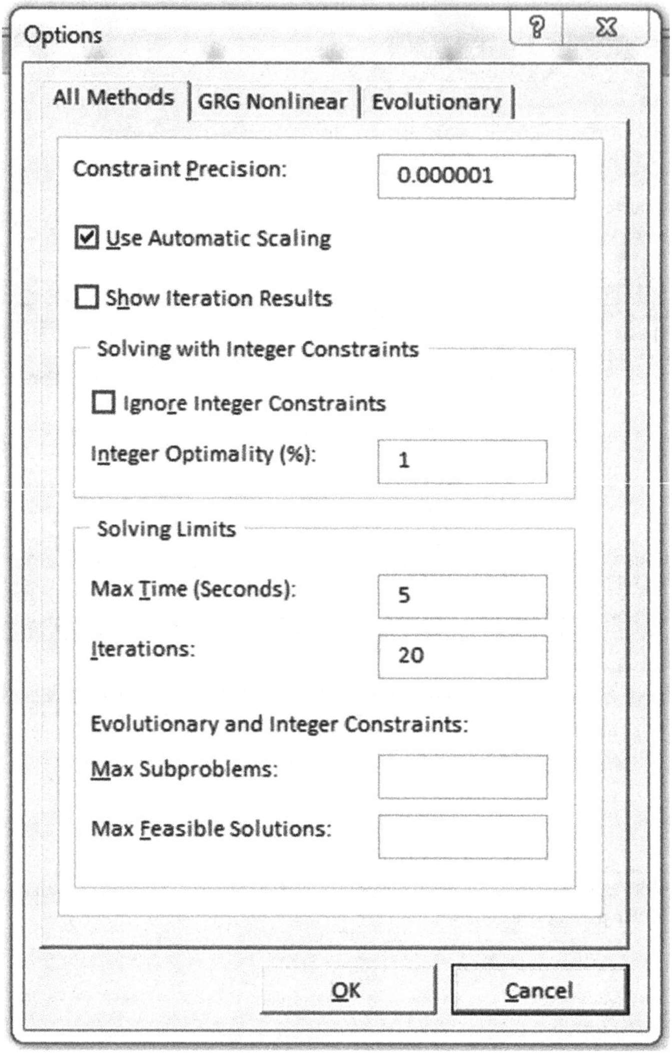

Fig. A.5 'Options' box

central finite differences may be chosen for gradient evaluation. Central differences improve accuracy, but lead to longer calculation time. Select 'Multistart' for a more reliable result when more local maxima or minima are suspected. The GRG Nonlinear method is then run repeatedly from different, arbitrarily chosen starting points. The number of times is defined by 'Population Size', with a minimum of 10. With Multistart, it is recommended to define upper and lower bounds on all variables for a better result. Similar options can be set on the Evolutionary tab in Options. Again with the Evolutionary method, it is recommended that upper and lower bounds on variables should be defined to limit the extent of the search space.

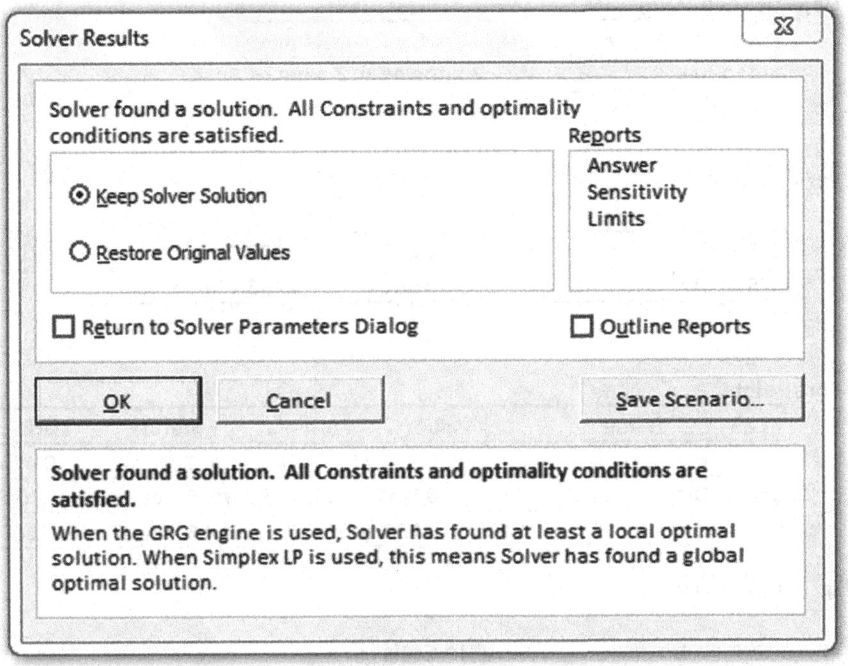

Fig. A.6 'Solver results' box

Click 'Solve' to start the optimization. When complete, initial values of the variables and the objective function are replaced by their optimized values in the spreadsheet, and all other values in the spreadsheet are adjusted accordingly. A valid solution is confirmed by the message 'Solver found a solution' in the 'Solver Results' box (see Fig. A.6). The message 'Solver could not find a feasible solution' implies that no valid solution exists. Commonly, this is due to constraints that exclude any feasible region in the design space, but may also be due to an unsuitable starting point. The message 'Solver has converged to the current solution' may sometimes appear. This means that the maximum time or number of iterations has been reached. A simple method to continue optimization is to click 'Keep Solver Solution' and then click Solve again to restart from that point. Alternatively, the maximum time or number of iterations may be increased in Options. Press 'Esc' if it is necessary to stop Solver while it is still running.

When a solution has been found, any of the three reports may be selected in the Solver Results box. In the 'Answer Report' (Fig. A.7), initial and final values of variables and constraints are tabulated, together the value of each constraint and whether or not the constraint is active at the optimum. In the 'Sensitivity Report' (Fig. A.8), values of the Lagrange multiplier are given for each constraint. In the 'Limits Report' (Fig. A.9), upper and lower limits on the values of the variables and objective function in the solution are given. After selecting any required reports, the

Objective Cell (Min)

Cell	Name	Original Value	Final Value
$I$20	V =	4.600E-05	4.330E-05

Variable Cells

Cell	Name	Original Value	Final Value	Integer
$F$6	D =	0.200	0.500	Contin
$F$7	H =	0.500	0.433	Contin

Constraints

Cell	Name	Cell Value	Formula	Status	Slack
$F$7	H =	0.433	$F$7>=$C$7,	Not Binding	0.423
$F$6	D =	0.500	$F$6<=$C$7	Not Binding	0.5
$F$6	D =	0.500	$F$6>=0	Not Binding	0.500

Fig. A.7 Answer report

Fig. A.8 Sensitivity report

Variable Cells

Cell	Name	Final Value	Reduced Gradient
$F$6	D =	0.5	0
$F$7	H =	0.4330122	0

Constraints

Cell	Name	Final Value	Lagrange Multiplier
$F$7	H =	0.4330122	0

solution may be kept or the original values restored by clicking the appropriate button in Solver Results.

With the GRG Nonlinear method in particular, the optimum found by Solver may be dependent on the starting point chosen. This is the case when more local maxima or minima exist, when Solver typically finds the one closest to the starting point, also with the use of discrete variables when more solutions of equal value may exist. Unless the Multistart option is chosen, it is generally advisable to repeat an optimization from different starting points to ensure that a consistent result is obtained.

Many of the spreadsheets supplied with this book contain macros. Depending on the chosen security settings, a message may appear in the Message Bar: 'Macros

	Objective			
Cell	Name	Value		
$I$20	V =	4.330E-05		

	Variable		Lower	Objective	Upper	Objective
Cell	Name	Value	Limit	Result	Limit	Result
$F$6	D =	0.500	0.000	0.000	1.000	0.000
$F$7	H =	0.433	0.010	0.001	#N/A	#N/A

Fig. A.9 Limits report

have been disabled'. Click 'Enable Content' to continue. Note that earlier versions of Solver may differ from the version in Excel 2016 referred to in this Appendix. Further help with Solver can be found on internet at: http://www.solver.com/excel-solver-help.

A.2 Matrix Calculation and Custom Functions

Excel offers a number of standard functions for matrix calculation. These are MMULT(...), MINVERSE(...), MDETERM(...) and TRANSPOSE(...) for matrix multiplication, matrix inversion, for the determinant of a matrix and to interchange rows and columns, respectively. These are used in certain spreadsheets in this book. Some brief notes are included here for those unfamiliar with matrix calculations in Excel.

A matrix is defined in Excel as an array, by selecting an appropriate range of cells. Suppose, for example, that to multiply a matrix **A** by matrix **B**, the range (A1: C4) is selected for a (4×3) matrix **A** and range (E1:J3) for a (3×6) matrix **B**. The required numerical data is entered in the two ranges. Note that the number of columns in **A** is equal to the number of rows in **B**, as required by the rules for matrix multiplication. The result of the multiplication will be a (4×6) matrix, so to perform the multiplication, we first select an appropriate range of cells, say this is the range (A6:F9), in which the result will appear. With cells (A6:F9) still selected, we create the formula:

$$\{= \text{MMULT}(A1:C4, E1:J3)\},$$

entering it by ctrl+shift+enter. Note the braces that appear to indicate an array formula. Note also the sequence of the cell references in the above function to evaluate the matrix product $[\mathbf{A}][\mathbf{B}]$. The other matrix functions above are used in a similar manner. For matrix inversion and for the determinant of a matrix, we must have a square matrix (the same number of rows and columns).

It is important to note that, while the numerical data in an array may be changed as required, no change may be made to an array formula once it has been entered. If necessary, the array formula should be deleted and a new array formula entered. Failure to do this will result in an error message.

Finally, it will be seen that many of the spreadsheets in this book make use of custom functions created in VBA to perform specific calculations. To view the code in these functions or to make changes:

– click 'Visual Basic' on the 'Developer' tab to open the Visual Basic Editor
– open the named VBAProject, then go to 'Modules' and the relevant module.

To return to the spreadsheet, close the Visual Basic Editor. If the Developer tab is not available on the spreadsheet:

– on the 'File' tab, click 'Options'
– click 'Customize Ribbon'
– under 'Main Tabs', check 'Developer' and click 'OK'.

Recommended Further Reading

Optimization theory and applications:

Haftka, R.T. and Gurdal, Z. Elements of structural optimization. Kluwer Academic Publishers, 1992.

Arora, J.S. Introduction to optimum design. Academic Press, 2012.

Vanderplaats, G.N. Numerical optimization techniques for engineering design. McGraw-Hill, 1984.

Rees, D.W.A. Mechanics of optimal structural design: minimum weight structures. Wiley, 2009.

Structural analysis and stability:

Megson, T.H.G. Aircraft structures for engineering students. Arnold, 1999.

Hoff, N.J. The analysis of structures. Wiley, 1956.

Timoshenko, S.P. and Gere, J.M. Theory of elastic stability. McGraw-Hill, 1961 (reprinted by Dover Publications, 2009).

Composite materials and structures:

Jones, R.M. Mechanics of composite materials. Taylor and Francis, 1999.

Daniel, I.M. and Ishai, O. Engineering mechanics of composite materials. Oxford University Press, 2006.

Kassapoglou, C. Design and analysis of composite structures with applications to aerospace structures. Wiley, 2013.

© Springer International Publishing AG 2017   307
A. Rothwell, *Optimization Methods in Structural Design*, Solid Mechanics
and Its Applications 242, DOI 10.1007/978-3-319-55197-5

Solutions to Selected Exercises

Chapter 1

1.2 $n \approx 1.38$
1.3 $L_{max} = 277.7\,\text{m}$
1.4 $V_{min} = 1.657 \frac{PL}{\sigma_0}$
1.5 $L_{max} = 3224\,\text{m}$
1.6 $L_{max} = 136.5\,\text{m}$
1.9 $x = 2, y = 2, f_{min} = 6$

Chapter 2

2.1 $\frac{P}{L^2} = 0.103\,\text{N/mm}^2$
2.2 $\eta = 1.112$

Chapter 3

3.1 $x_1 = \frac{1}{2}, x_2 = 1, x_3 = \frac{3}{2}, f(\mathbf{x})_{min} = \frac{7}{2}, \lambda = 1$
3.2 $f(x, y)_{min} = 4, x = 2, y = 1, \lambda_1 = 0.333, \lambda_3 = 1.333$
3.3 $A_{min} = 5.536\,V_0^{2/3}, R = (V_0/2\pi)^{1/3}, L = (4V_0/\pi)^{1/3}, \lambda = (16\pi/V_0)^{1/3}$
3.4 € 360, $x = z = 2\text{m}, y = 1\,\text{m}, \lambda = 60, €/\text{m}^3, €420$
3.5 $R = (PL^2/\pi^3 Et)^{1/3}, t = (P/2\pi KE)^{1/2}, \lambda_1 = 2L^2/3\pi^2 ER^2, \lambda_2 = R/3KEt$
3.7 € 360, € 417.7

Chapter 4

4.6 $x = \frac{1}{\sqrt{2}} = 0.7071, f_{min} = 0.5858$

Chapter 5

5.1 $\lambda_1 = \frac{8}{9}, \quad \lambda_2 = \frac{4}{9}$
5.2 $x_1 = \frac{1}{5}, x_2 = \frac{2}{5}, f_{min} = \frac{1}{5}$
5.3 $G_r = (0.1666, \quad 0.9445, \quad 0.3889)$

$x_1 = 1.338, x_2 = x_3 = 0.662, x_4 = x_5 = 0, f(\mathbf{x}) = 2.6667$

© Springer International Publishing AG 2017
A. Rothwell, *Optimization Methods in Structural Design*, Solid Mechanics
and Its Applications 242, DOI 10.1007/978-3-319-55197-5

5.9 $b = 88.76$ mm (outside dimension), $t = 2.20$ mm, $A = 761.0$ mm^2

Chapter 6

6.1 $n_h = 3.214$

6.2 $C = 0.07958$, $n_g = 3.691$ (solid section)
     $C = 0.1326$, $n_g = 2.890$ (hollow section)

6.3 $V_{min} = 0.2221\, n_g \left(\frac{PL}{\sigma_0}\right)^{2/3} L$

6.4 $M_y = 0.2141\, a^3 \sigma_y$

6.5 $M_1 = M_2 = 75 \times 10^3$ Nmm, $M_3 = 225 \times 10^3$ Nmm, $V = 92.66 \times 10^3$ mm^3
     $M_1 = M_2 = M_3 = 131.3 \times 10^3$, $V = 106.0 \times 10^3$ mm^3 (uniform beam)

6.8 (a) $W = 2.683$ kg, $W = 2.353$ kg (recalculated for taper)
     (b) $W = 1.112$ kg, $W = 0.941$ kg
     (c) $W = 1.327$ kg, $W = 1.148$ kg

6.9 $W = 0.936$ kg, $W = 0.802$ kg (recalculated for taper)

6.10 $W = 0.872$ kg, $W = 0.800$ kg (recalculated for taper)

Chapter 7

7.1 $\sigma_F = 58.47$ N/mm^2, $\sigma_L = 179.7$ N/mm^2, $\eta = 0.416$

7.2 $\tau' = 40.64$ N/mm^2, 63.54 N/mm^2, 99.24 N/mm^2
     $n = 0.644$, $\alpha = 1.90$

7.3 $\eta = 0.546$ (constrained)
     $\eta = 0.774$ (unconstrained)
     $\eta_{max} = 0.810$

7.9 $\eta = 0.541$, $\eta_{max} = 0.932$

7.10 $M = 68.20$ kg/m, $M = 67.28$ kg/m (unconstrained)

Chapter 8

8.1 $A_{11} = 171,627$ N/mm, $A_{33} = 36,547$ N/mm
     $E_x = 81,192$ N/mm^2, $E_y = 47\,877$ N/mm^2, $G_{xy} = 18,275$ N/mm^2
     $\nu_{xy} = 0.302$, $\nu_{yx} = 0.178$

8.2 $0°$ layers: $\sigma_1 = 923.7$ N/mm^2, $\sigma_2 = 0$, $\tau_{12} = 0$
     $\pm 60°$ layers: $\sigma_1 = 34.2$ N/mm^2, $\sigma_2 = 42.2$ N/mm^2, $\tau_{12} = 41.7$ N/mm^2

8.3 Tsai–Hill criterion: $0°$ layers: 0.2133, $\pm 60°$ layers: 0.3469
     First-ply failure at 1698 N

8.6 8 plies at $0°$, 8 plies at $\pm 45°$, 6 plies at $90°$, $t = 3.75$ mm

8.7 7 plies at $0°$, 12 plies at $\pm 45°$, 6 plies at $90°$, $t = 4.625$ mm

8.8 12 plies at $0°$, 10 plies at $\pm 45°$, 4 plies at $90°$, $t = 4.5$ mm

8.9 12 plies at $0°$, 10 plies at $\pm 45°$, 4 plies at $90°$, $t = 4.5$ mm

8.10 3.16 plies at $0°$, 9.56 plies at $\pm 30.7°$, 7.88 plies at $90°$, $t = 3.77$ mm

Chapter 9

9.2 $\dfrac{dg}{dA_2} = 1 \times \dfrac{PL}{EA^2}$

Index

A

Achieved efficiency, 35, 77, 213
Active constraint, 59, 70
Active constraint strategy, 109, 289
Angle of twist, 65
Angle-section column, 35
Augmented Lagrangian penalty function, 135

B

Beams
  buckling, 150
  collapse mode, 164
  cross-sectional shape, 148
  finite element model, 173
  form factor, 163
  geometrically similar sections, 153
  limit load, 163
  minimum weight, 149
  shape efficiency, 153
  spanwise stiffness distribution, 154
  statically determinate, 155
  statically indeterminate, 158
  strength to weight ratio, 157
  thin-walled, 150
  yield moment, 161
Box beam, 2, 56, 183, 201
Branch-and-bound method, 267
Broyden, Fletcher, Goldfarb, Shanno (BFGS)
  formula, 92
Buckling in compression, 201
Buckling in shear, 208
Buckling under combined stress, 204

C

Circular tube, 30, 132, 140
Coefficient n, 13, 42
Collapse mode, 164
Comparison of layouts, 16
Composite laminates

A-matrix, 247
  balanced laminate, 248
  delamination, 258
  discrete ply thickness, 267
  engineering constants, 243
  failure criteria, 252
  interlaminar shear stress, 258
  internal stresses, 256
  iterative redesign, 264
  laminate stiffness coefficients, 247
  layer stiffness matrix, 245
  netting analysis, 260
  optimization, 259
  orthotropic laminate, 243
  practical restrictions on lay-up, 258
  temperature change, 256
  transformation matrix, 245
  transformed stiffness matrix, 246
Composite materials, 241
Conjugate gradient methods, 92
Conservation of energy, 41
Constrained optimization, 107
Constraint, 31
Constraint-following methods, 108, 120
Constraint gradient, 283
Constraint linearization, 109, 113, 129
Constraint normalization, 134
Constraint selection, 40
Continuous variable, 21
Conventional design process, 3
Convex constraint, 59
Criterion for maximum stiffness, 40
Crossover, 273

D

Davidon, Fletcher, Powell (DFP) formula, 92
Dependent and independent variables, 120
Design space, 31, 58, 212
Design variables, 31, 57

© Springer International Publishing AG 2017
A. Rothwell, *Optimization Methods in Structural Design*, Solid Mechanics
and Its Applications 242, DOI 10.1007/978-3-319-55197-5

Diagonal tension, 218
Discrete stiffeners, 189
Discrete variable, 21, 142
Dummy load method, 286

E
Eccentrically loaded column, 74
Effective modulus, 205
Efficiency, 33, 210, 214, 222
Elimination of design variables, 61
Equality constraint, 58
Equivalent shear stress, 216
Equivalent stress, 169
Equivalent thickness, 189, 211, 215
Euler's formula, 30
Evolutionary algorithm, 268, 270
Evolutionary method, 297
Excel add-ins, 297
Exploratory move, 84, 102
Exterior penalty function, 133

F
Factor of safety, 5
Feasibility, 15
Feasible directions method, 126
Feasible region, 32, 58
Feasible search direction, 110
Finite difference, 85, 89
Finite element analysis, 283
Finite element model of a beam, 173
First-order methods, 85
Fitness, 271
Fletcher-Reeves method, 90
Flexural buckling, 30, 202, 206, 222
Form factor, 163
Fully stressed design, 3, 5, 29
Fuselage section, 230

G
Generalized reduced gradient method, 108, 120
General optimization problem, 57
Genetic algorithm, 108, 268, 270
Geometrically–similar sections, 175
Geometric similarity, 153, 214
Golden section method, 95
Gradient, 87
Gradient-based methods, 85
Gradient projection method, 109
GRG Nonlinear method, 297

H
Hessian matrix, 64, 92
Hooke and Jeeves method, 84

I
Imperfection sensitivity, 76, 213
Inactive constraint, 59
Index m, 77, 187
Inequality constraint, 58
Interior penalty function, 130
Interpolation formula, 223
Intersection optimum, 59
Inverse variable, 129, 289
I-section beam, 149
I-section column, 35
Iterative design process, 3

K
Kuhn–Tucker conditions, 73, 136

L
Lagrange multiplier, 61, 67, 68, 87, 111
Lagrangian function, 62, 87, 111
Lamination theory, 242
Lateral buckling, 151
Layout, 13, 20
Limit load, 163
Linearization of constraints, 109, 113, 129
Linear programming, 108
Line search, 84, 94
Linking matrix, 290
Loading intensity, 211
Local buckling, 31, 150, 202, 222

M
Master variable, 290
Material breaking length, 14
Material limitation, 38
Mathematical optimum, 59
Matrix calculation, 305
Maximum stiffness, 40
Michell structure, 17
Move limits, 113, 129

N
Nonlinear constraint, 118
Normalization of constraints, 134

O
Objective function, 57
Optimality criteria, 29, 32, 44, 60

P
Partial structures, 6, 17
Pattern move, 84, 102
Penalty function (genetic algorithm), 272
Penalty function methods, 108, 129, 140
Polynomial interpolation, 95, 97

Post-buckled design, 218
Projection matrix, 112, 116

Q
Quadratic interpolation, 98
Quasi-Newton method, 92

R
Ramberg–Osgood formula, 76, 170, 187
Rate of twist, 198
Rectangular box beam, 227
Rectangular section beam, 148
Rectangular torsion box, 65
Reduced gradient, 120
Reduced modulus, 187, 222
Reduction in design variables, 290
Region elimination, 95
Restoration move, 113
Robust feasible directions method, 128

S
Saddle-point, 64
Search direction, 85, 112
Secant formula, 74
Secant modulus, 170, 187
Sensitivity analysis, 284
Sensitivity of the optimum, 68
Sequential quadratic programming, 126
Sequential unconstrained optimization, 130
Seven-bar truss, 21, 48, 298
Shape optimization, 20
Shear flow, 192
Shear stress distribution, 192
Shear web, 208
Shell structures
    bending stress, 185
    discrete stiffeners, 189
    product second moment, 186
    rate of twist, 198
    second moment of area, 186, 206
    shear stress distribution, 192
    warping displacement, 198
Side-constraint, 58
Simplex LP method, 297
Simultaneous modes, 32
Sizing, 21
Slave variable, 290
Solver, 21, 108, 297
Solver options button, 300
Solver parameters dialog box, 298
Solver results box, 303
Spanwise stiffness distribution, 154
Specific strain energy, 44
Specific weight, 13

Spreadsheet programs
    beam under lateral load, 158, 173
    circular and square tubes, 44
    circular fuselage section, 230
    composite laminate, 275
    design variable linking, 291
    eccentrically loaded column, 74
    Hooke and Jeeves method, 100
    I-section beam, 169
    penalty function method, 140
    rectangular box beam, 227
    seven-bar truss, 21
    stiffened panel, 221
    truss with tubular members, 48
Square–cube law, 36
Square section tube, 34
Statically indeterminate truss, 3
Steepest descent direction, 109
Steepest descent method, 85
Stiffened panel, 201, 210
Stiffener criterion (shear web), 214
Strain energy, 44
Strength to weight ratio, 12, 157, 216
Structural index, 13, 34, 156, 212, 216
Structure made of different materials, 7
Structure under alternative loads, 9
Substitute stiffener, 190
Substitution of variables, 129
Surplus variables, 122

T
Tangent modulus, 76, 188, 207
Tapered beam, 290
Thin-walled beam, 150
Three-bar truss, 7
Topology, 19
Torsional stiffness, 56, 66, 198
Torsion box, 65
Truss structures
    iterative design process, 3
    long truss structure, 18
    made of different materials, 7
    maximum span, 15, 19
    optimum layout, 42
    seven-bar truss, 21
    statically indeterminate, 3
    three-bar truss, 7
    under alternative loads, 9
    with tubular members, 48
Tsai-Hill criterion, 253
Tsai-Wu criterion, 254
Twisting moment, 195

U
Unconstrained optimization, 84
Update formulae, 92
Usable search direction, 111

V
Visual Basic, 306
von Mises criterion, 169

W
Weighted second moment, 188
Weight lines, 32, 59

X
X–section column, 35

Y
Yielding, 76, 187
Yield moment, 161

Z
Zero-order methods, 85

Printed in the United States
By Bookmasters